CEREAL HUSBANDRY

By the same author

Spring Wheat (NAC Arable Unit)

A Tropical Agriculture Handbook (Cassell)
    with D C Joy

CEREAL HUSBANDRY

E. John Wibberley

Farming Press

First published 1989

Photographs on the front cover and spine supplied by Holt Studios Photograph Library, Hungerford, Berkshire
Front **(left to right): seedling barley plants;** *Sitobion avenae* **grain aphid on wheat ears;**
discharging wheat grain from combine harvester into trailer
Spine: **combine harvesting barley**

British Library Cataloguing in Publication Data
Wibberley, E. John
    Cereal husbandry.
    1. Great Britain. Cereals. Cultivation
    I. Title

ISBN 0-85236-124-6

Published by Farming Press Books
4 Friars Courtyard, 30–32 Princes Street
Ipswich IP1 1RJ, United Kingdom

Distributed in North America
by Diamond Farm Enterprises,
Box 537, Alexandria Bay, NY 13607, USA

Phototypeset by MHL Typesetting Ltd, Coventry
Printed and bound in Great Britain by Butler & Tanner Ltd, Frome, Somerset

Contents

Acknowledgements

Many farmers and agronomists have contributed directly or indirectly to the present book and I am extremely grateful for their influence, though any shortcomings are my own. Comments on omissions or errors will be welcome, as will suggestions for improvements.

I am especially indebted to Mick Conway and the late Helen Paterson and to all the members of the Cirencester Cereal Study Group, particularly Tim Morris, who kindly read through the manuscript, and to Alasdair Barron, Stephen Bond, Robert Henly, Rex Jenkinson, Michael Limb, Peter Lippiatt, Pat Morris and David Ursell.

My colleagues Jim Lockhart and Tony Wiseman encouraged me and advice on specific chapters was generously given by Dai Barling, Harry Catling, Ray Churchill, Peter Glanfield, Alison Samuel and Bill Heatherington.

I am pleased that Dr Graham Milbourn, Director of the National Institute of Agricultural Botany, agreed to write the foreword.

Roger Smith of Farming Press deserves a special accolade for his patience and Julanne Arnold for her careful attention to detail in preparing for publication. Every effort has been made to acknowledge material quoted and to seek necessary approval. Any oversight is unintentional.

My wife Jane and sons David and Mark provided the much-appreciated family support needed during such a venture.

Finally, 'The earth is the Lord's and everything in it' (Psalm 24:1; I Corinthians 10:26). It was Jesus Christ Who likened His own death and resurrection to a grain of wheat 'falling into the ground and dying' in order to multiply. The wise husbandman will acknowledge that the ultimate power to produce and reproduce cereals lies beyond man, and we are cast in the role of responsible stewards of God's gifts.

Introduction

Cereals are the dominant arable crops of British and of world agriculture.

The advent of EC grain surpluses and the consequent decline in potential cereal farming fortunes has increased the need for good innovative husbandry of cereals and of the cropping sequences in which they feature.

The arrival of the computer on farms has brought an aid, not a substitute, for good husbandry. Indeed, its over-use can distract the unwary from adequate field observation, from proper husbandry.

Husbandry is both an age-old craft and a constantly changing, imprecise applied science. It integrates the various branches of science, economics and geography in an attempt to provide informed, caring management — in this case of cereal crops. It is balanced also in that it involves controlled use of resources yet requires stewardship to conserve them. A maxim of the ancients was that the 'land must be weaker than the husbandry'; I prefer to put it positively that the husbandry must be stronger than the land, i.e. if you cannot master boy's land loams, do not try clays! Important elements in the husbandman's (and husband-woman's!) character are perseverance and patience.

The ultimate goal of husbandry is human welfare because cereal husbandry exists with the same objectives (expected outcomes) as agriculture as a whole, namely:

- To supply *food* for an increasing world population, already 5,000 million people
- To supply *raw materials* for processing
- To provide *creative* employment in farming
- To *protect* not only the means of production, notably the soil, but the whole ecosystem for future generations of crops and people
- To generate *profit* for the farm population, without which husbandry can be deemed to have failed

Cereal husbandry consists of:

(i) understanding the cereal crop
(ii) selection, timing and efficiency of all operations designed to supply the crop's needs and to satisfy market requirements for the end-products

There has been a tremendous surge of interest in cereal crops throughout Britain and elsewhere during the past two decades and it is my belief that many farmers and students feel overwhelmed by the mass of technical literature available. I share this feeling and it is the purpose of this present work to try to help us to focus on essentials and thus sift detailed literature with greater comprehension and discrimination.

This book is therefore written with *all* cereal growers in mind, not just specialists, and for *all* students of agriculture; they are likely to need an understanding of cereals for most farming systems and districts they may encounter and it is for *all* courses involving agriculture.

It is intended as a guide through the jungle of information by machetes rather than microscope — not that the use of a microscope is at all inappropriate for the judicious cereal-grower!

The intentional emphasis of the book is on understanding and appreciating the principles and on their practical application. It is hoped that these principles will be of interest to those outside Britain and, indeed, outside agriculture altogether (either seeking general insights or studying a related subject such as biology, economics or geography).

The aim has been to present a panoramic view, painting an overall picture of the cereal scene. There is a place for both the detailed specialist document and the overview such as is attempted here. It is not clever to be unnecessarily complicated! It has been difficult to decide on what to include and what must be omitted. Advice on omissions and faults will be gratefully received. If the book raises issues in the reader's mind and stimulates closer observation and attention to husbandry detail, it will have achieved its objective. It is divided and written so that each chapter is more or less self-contained — each providing a daily reading ration for a fortnight or so, with Sundays off for rest!

A list of further reading and references appears in Appendix 1.

Minchinhampton
May 1989

Foreword

If cereal production is to remain profitable today it must be based on the application of the latest scientific principles to crop production. In *Cereal Husbandry*, John Wibberley has produced a clear and up-to-date coverage of the practical background for modern techniques in all the field operations for cereal production as well as a clear analysis of the newest scientific principles that determine the correct use of inputs.

This is a very readable and well-illustrated book which will provide farmers and students with an extremely practical guide to the latest scientific principles involved in profitable cereal management today.

DR GRAHAM MILBOURN
Director
National Institute of Agricultural Botany
Cambridge

26 April 1989

Part One

Preparation

The geographical, environmental and botanical framework

Chapter 1

BACKGROUND TO CEREAL CULTIVATION

The chief temperate cereals are wheat, barley, oats and rye plus durum wheat and triticale, whilst those of the sub-tropical and tropical areas are chiefly rice, sorghum, maize and millets. Cereals include several locally vital but internationally insignificant grains such as acha (*Digitaria exilis*).

Cereals belong to the *Gramineae*, a large mono-cotyledonous family of some 600 genera and around 10,000 species. They are named after the goddess of Roman mythology, Ceres, giver of grain. Annual 'first-fruits' offerings were made to her in harvest thanksgiving for barley and wheat which became known as 'cerealia'. The cereals are members of the grass family with relatively large edible grains. The grain is strictly a one-seeded fruit called a caryopsis in which the pericarp (fruit-coat) is thin, translucent and fused to the seed-coat or testa. Perennial cereals such as *Agrotricum*, which is a cross between wheat and couch grass, exist, but only annuals are commercially important.

Each cereal exists in many varieties. Rice, for instance, has some 7000 varieties of which only a few are cultivated (cultivars).

HISTORICAL PERSPECTIVE

It is generally agreed that cereal cultivation started some 6000 years ago in the fertile crescent of the Middle East, probably around Jericho, Israel and in Egypt (Genesis 42: 1−3: Gompertz, 1927), barley preceding wheat in this respect. Indeed it was barley which featured in the Biblical case of Ruth gleaning in the fields of Boaz and the miraculous feeding of the five thousand by Jesus using the two fish and five barley loaves of the boy. Barley (*Hordeum sativum*) remains well adapted to less fertile land than wheat.

Oats (*Avena sativa*) were probably first noticed as weeds in barley and wheat crops and cultivated later. Indeed, in Britain, Dr Samuel Johnson (1709−1784) defined oats in his famous dictionary as 'a grain which in England is generally given to horses, but in Scotland supports the people'.

The wheat first cultivated widely was emmer (*Triticum dicoccum*), a tetraploid (having double the normal number of chromosomes carrying genetic information in each cell) derived from the wild diploid spelt wheat (*T. aegilopoides*). These had grains which did not thresh out easily. Thus it was a breakthrough when Rivet wheat was selected as a freely threshing species (*T. turgidum*). However, the majority of the wheat now cultivated as the world's leading cereal of trade is the hexaploid *T. aestivum* (breadwheat). Wheat shares with rice the distinction of being separately classified by the UN Food and Agriculture Organization (FAO), whilst all the other cereals are termed coarse grains.

Rye cultivation extended habitation northwards into Europe and the USSR. Maize (*Zea mais*) originated in the New World and was adopted as the staple cereal of Indians in the Americas, hence its being called Indian corn or just corn in North America, whilst corn in Britain is an all-embracing term for cereals. Rice became the staple cereal of the Asian and humid tropics. It is cultivated as paddy (padi) in wetland, puddled fields. Dryland (upland) rice is grown in the drier tropics. Sorghum and millets of various genera fulfil the central dietary role for the dry tropics.

In about 1895 in Scotland, rye was crossed with wheat (usually *T. durum* now, i.e. durum, pasta or

3

macaroni wheat) to produce triticale. This was further developed in North America after World War II and has been adopted in Europe and the UK more latterly, where Polish cultivars are most popular. Indeed, *Triticum durum* itself has been grown as a minor specialist cereal in the UK since about 1980.

Grain is strictly a wider term than cereal because it includes legumes and other edible seeds such as beans. Oilseed crops are *not* cereals; neither is buckwheat (family *Polygonaceae*), though it is sometimes loosely considered so when grown for its grain to produce flour (Sarrasin or blé noir in Brittany).

Evidence of breadwheat cultivation in Britain dates back some 5000 years. Julius Caesar's men cut corn following their invasion in 55 BC. One of the earliest rotations practised in Britain was autumn-sown cereal (usually wheat), spring-sown cereal (usually barley) and then fallow.

Cereals occupied 50 per cent of the four-course rotation introduced in the seventeenth century by Lord Townshend on his light land at Raynham, Norfolk. In the mid-nineteenth century, Rothamsted Experimental Station in Hertfordshire began continuous cereal-growing on the same field. Similar trials have been started since and commercial practice has also been to grow continuous cereals as well as those in rotation. Paddy rice has been continuously cultivated on the same terraces for three or four thousand years in the Far East, for example, Java and the Philippines.

Early in the twentieth century, Sir George Stapledon, who worked at the Royal Agricultural College before starting the Welsh plant breeding station, advocated the greater integration of grassland and cereal production in the system of alternate husbandry or ley farming. Typically, in this system three-year leys alternate with three years of corn crops.

Wheat yields in Britain trebled between 1950 and 1985, having remained more or less static from 1800 to World War II. These yield improvements arose from:

- better varieties, notably with a greater grain/straw ratio (improved harvest index)
- better husbandry, both the promotion of positive factors such as soil potential and fertilisers and the protection from negative factors (weeds, pests, diseases) (Figure 1.1)

IMPORTANCE OF CEREAL CROPS

The importance of cereals in both world and UK agriculture is great. Whilst quantitative terms such as

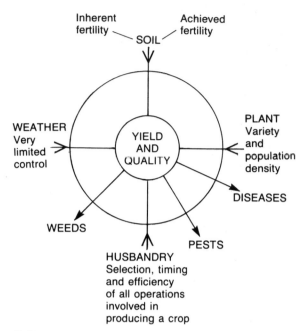

FIGURE 1.1 Factors affecting cereal yield and quality.

area, yield and production indicate the extent of cereal importance, reasons for this must be sought in terms of uses, distribution and production characteristics:

1. Uses

- Cereals are *multipurpose*, providing human diet, livestock feeds and a wide range of farm and industrial raw materials.
- Cereals are *demanded* consistently throughout the world, being a concentrated carbohydrate source with useful protein, fat, mineral, vitamin and fibre content. They have proved fairly stable-priced commodities to trade.
- They are easily *storable*.
- They are *transferable* both in terms of ease of transportation and in terms of convertibility to various end-products for different markets. This contrasts with a commodity like butter which can only be sensibly used as such.

2. Distribution

- Cereals are *adaptable*, the different species and varieties tolerating a wide range of soil, climatic and agronomic conditions.
- They are an *integral* part of most arable systems

of cropping. It is difficult to devise sensible arable rotations in the UK which omit cereals.

- Location of supplies has become an important factor in *world politics*.

3. Production

- They are relatively *easy* to produce in that a total failure of crop or markets is very unlikely by contrast with many crops. The labour requirement is fairly low per tonne produced.
- Cereals have proved fairly *straightforward* to mechanise. This arises partly from the harvest fraction being above-ground and also from their widespread cultivation which has justified the development costs of improved harvesters.
- Relatively low *capital* investment per hectare is needed for a cereal production enterprise.
- A cereal gives a relatively good ratio of *energy* output to energy input per hectare to produce it. It also gives a good yield in relation to seed planted (for instance, 40:1 or so for wheat in contrast with peas at around 15:1).

WORLD IMPORTANCE NOW

World food supplies hinge on cereal production, trade and reserve stocks of grain. Before 1940, every region except Western Europe was a net grain exporter; now Africa, Asia, Latin America, Eastern Europe and the USSR are net importers. In 1980–81 almost 90 m tonnes of wheat was traded internationally, 80 m tonnes of maize and only 12 m tonnes of rice. Never before has there been so much information about supplies or so much grain transported for trade and aid to poorer areas of the world. Twelve thousand people a day died from hunger and related diseases in 1980. In Africa, one child in three died from these causes before school age. Africa has had a rapid population growth but only a slight increase in food production over the past decade (Table 1.1).

Grain has already become arguably the most significant political weapon in the modern world. North America remains the chief exporter of cereals (she held just under 40 per cent of all world grain reserves in 1981), though the EC is increasingly significant. Whilst Russia is the world's largest wheat producer, she needs around 200 m tonnes of cereal per year and had to import 35 m tonnes in 1981. China imported 15 m tonnes for the first time in 1981.

TABLE 1.1 Comparison of population growth with food production in Africa

Year	Population in millions	% change on 5 years ago	Food index	% change on 5 years ago
1965	312	+ 13.5	105	Not available
1970	355	+ 13.8	122	+ 16.2
1975	407	+ 14.6	133	+ 9.0
1980	470	+ 15.5	146	+ 9.8
1985/86	555	+ 18.1	159	+ 9.3

NB: Between 1970 and 1975, the increase in the population overtook the increase in the food supply.

Source: Adapted from statistics contained in the *UN Estimates and Projections of Urban, Rural and City Populations 1950–2000*, published in 1982, and from FAO Production Year Books.

Africa imported a similar amount then, and both India and the ASEAN countries (Association of South-east Asian Nations) operate their own grain reserves. Indonesia increased her rice production elevenfold between 1975 and 1985, attaining self-sufficiency.

It can be strongly argued that world hunger and unemployment problems have the mutual solution of more labour-intensive agriculture. Closer crop observation, the ability to cultivate crop mixtures plus the greater recycling of nutrients in labour intensive systems can enhance output. Proper management is a prerequisite of success but unfortunately men are sometimes more recalcitrant than machines!

There are some 13,000 million hectares of land in the world and about 10.5 per cent is reckoned by the FAO to be arable, that is, 1369 M ha. Of this, some 576 M ha are occupied by cereals, excluding rice, representing about 42 per cent of the area under arable crops. World cereal yields are approximately 2 tonnes per hectare, about 30 per cent of the UK level of recent years, and total world cereal production is at 1660 M tonnes. The inclusion of an estimate for rice boosts cereals to 70 per cent of world arable land.

Trends in production and importance in area and yield terms are shown in Tables 1.2 and 1.3. The importance of cereals in the EC is indicated by Table 1.4. EC grain surpluses reached 17 million tonnes in 1986. In the UK, cereals occupy around 4 M ha, some 75 per cent of total arable land. The relative importance of cereals in UK and EC land use is indicated by Table 1.5. A century of wheat trends for Great Britain is presented in Table 1.6.

TABLE 1.2 Trends in world cereal production

A. PAST DECADES

Crop	% change during 1960s and 1970s	Mean (M tonnes) 1975–79	1979–81	1986
Cereals, total	+ 34	1449	1590	1867
Major species				
Wheat	+ 38	408	443	536
Rice (paddy)	+ 31	366	396	475
Maize	+ 39	352	422	480
Barley	+ 43	176	157	180
Sorghum	+ 42	62	64	71
Oats	+ 2	49	43	48
Millet	+ 5	40	28	31
Rye	− 1	33	25	32

Source: FAO data.

B. SHORT-TERM SUPPLY SHIFTS
(MILLION TONNES)

	Production	Consumption	Stocks
Total (all species)			
1986/7	1684.6	1657.4	457.4
1987/8	1604.5	1664.3	397.6
1988/9	1549.7	1660.8	286.5
(estimate)			
Wheat			
1986/7	530.2	523.0	175.2
1987/8	504.3	533.8	145.6
1988/9	503.2	533.5	115.3
(estimate)			

NB: Amidst much talk of EC overproduction in the late 1980s, actual cereal stocks are declining when considered globally.

Source: USDA.

TABLE 1.3 Cereals in the world, 1982 and 1987 (rounded estimates)

Cereal	Area (million ha) 1982	1987	Yield (t/ha) 1982	1987	Production (million t) 1982	1987
Wheat	239	223	2.0	2.3	478	515
Rice	143	145	2.9	3.2	415	458
Maize	131	127	3.5	3.6	458	452
Barley	78	79	2.1	2.3	164	182
Sorghum	48	44	1.4	1.4	67	62
Millet	42	39	0.7	0.7	29	28
Oats	26	26	1.7	1.8	44	48
Rye	18	16	1.7	2.0	31	32
Others (estimate)	7		1.6		11	
Total	732	706	2.3	2.5	1697	1788

Note: Total world stocks as a percentage of consumption were 19.5% in 1982 and 26.3% in 1987.

Source: Based on FAO data.

of grain supply in them is eaten directly, whereas industrially advanced countries feed 67 per cent of grain supplies to animals.

Cereals are fairly well balanced nutritionally, and whole grains are a valuable source of fibre (Table 1.8). European and North American diseases associated with constipation such as appendicitis, varicose veins and large-bowel cancer are less prevalent in societies consuming more fibre, and cereal fibre is considered particularly protective against constipation. Dietary fibre by providing bulk also protects against excessive energy intake and absorption with its resultant obesity and diabetes. Furthermore, adequate fibre protects from diseases related to cholesterol and bile acid metabolism such as gallstones and certain forms of heart disease. Western man's typical fibre intake of 20 grammes per day stands at one-third to one-sixth of the intake of rural Third World cereal diets, according to Dr D.P. Burkitt (1975). He recommends broadly for Western man a doubling of fibre and starch intake, halving of salt and sugar intake and reduction of one-third in the present fat intake. An increased consumption of whole (unrefined though physically processed) cereals can achieve much of this adjustment at once. There is reckoned to be a 10 per cent per annum growth rate in the demand for whole-grain products in the UK and even higher for those organically grown (by alternative methods of nutrient supply and weed control with minimal synthetic chemical inputs — see later).

GENERAL USES

1. Diets

Cereals consumed directly account for about 55 per cent of the human diet. Indirectly they contribute more owing to their inclusion in livestock diets producing meat, eggs, milk and dairy products (Table 1.7). EC consumption of grain is about 85 kg/head/year whilst Russian consumption is double this figure.

The proportion of diet contributed by direct cereal consumption generally increases the poorer the country, amounting to over 80 per cent of diets in the poorest areas. Less developed countries contain over 75 per cent of the world's population but produce less than 50 per cent of world grain; 86 per cent

TABLE 1.4 EC cereal production (M tonnes)*

		Belg./Lux.	Denmark	France	Greece	Ireland	Italy	Neths.	UK	West Germany	Spain	Portugal	EC 12
Common wheat	1984	1,331	2,446	32,448	1,456	602	5,439	1,131	14,970	10,197	5,550	432	76,001
	1985	1,215	1,972	28,092	1,013	495	4,610	851	12,026	9,779	4,958	367	65,378
	1986	1,325	2,171	25,564	1,124	424	4,685	940	13,845	10,286	4,038	462	64,862
	1987	1,078	2,720	26,800	1,050	309	4,872	792	12,011	9,884	5,467	445	65,427
Barley	1984	934	6,072	11,512	854	1,770	1,618	192	11,070	10,284	10,789	91	55,186
	1985	746	5,251	11,440	606	1,494	1,630	197	9,740	9,690	10,698	65	51,558
	1986	858	5,134	10,061	681	1,428	1,548	262	10,010	9,377	7,331	90	46,780
	1987	669	4,780	10,500	620	1,363	1,585	275	9,219	8,514	9,602	99	47,726
Oats	1984	118	157	1,890	54	137	438	65	549	2,973	788	152	7,321
	1985	127	171	2,206	60	106	388	58	614	3,278	719	164	7,891
	1986	58	129	1,079	75	110	413	34	435	2,727	419	146	5,625
	1987	65	115	1,112	62	93	351	52	436	2,420	503	146	5,414
Maize	1984	53	—	10,493	2,162	—	6,672	1	—	1,026	2,529	499	23,436
	1985	50	—	12,441	1,948	—	6,357	2	—	1,204	3,414	570	25,985
	1986	57	—	11,441	1,994	—	6,247	1	—	1,302	3,405	652	25,099
	1987	49	—	11,374	2,260	—	5,700	—	—	1,051	3,500	800	24,734
Total cereals†	1984	2,503	9,283	58,153	5,417	2,513	18,919	1,407	26,618	26,489	20,592	1,316	173,209
	1985	2,197	7,956	55,687	4,429	2,095	16,892	1,129	22,471	22,471	20,510	1,233	160,513
	1986	2,383	7,968	50,308	5,164	1,954	17,350	1,265	24,486	24,486	15,794	1,482	153,743
	1987	1,917	8,230	52,146	5,071	1,765	16,956	1,145	21,724	23,595	19,777	1,657	155,576

* Based on estimates at 27 October 1987. EC12 estimate for 1988 is 164.5 Mt. † Includes rye, sorghum, mixed corn and triticale.

TABLE 1.5 EC cereal areas harvested (M hectares)*

		Belg./Lux.	Denmark	France	Greece	Ireland	Italy	Neths.	UK‡	West Germany	Spain	Portugal	EC 12
Common wheat	1984	194	322	4,976	553	77	1,476	143	1,939	1,629	2,154	270	13,744
	1985	195	340	4,631	469	78	1,295	128	1,896	1,609	1,911	264	12,815
	1986	197	353	4,608	426	76	1,271	118	1,987	1,623	1,990	275	12,923
	1987	198	404	4,669	398	51	1,223	112	1,986	1,657	2,116	280	13,093
Barley	1984	152	1,181	2,108	365	304	434	34	1,979	2,006	4,023	84	12,669
	1985	135	1,104	2,255	310	298	468	39	1,966	1,949	4,246	86	12,857
	1986	145	1,088	2,090	266	283	466	42	1,917	1,947	4,334	84	12,663
	1987	139	990	1,979	265	287	452	50	1,836	1,837	4,352	90	12,277
Oats	1984	28	34	433	45	25	191	13	107	669	479	194	2,218
	1985	31	41	533	43	23	184	11	134	692	466	189	2,346
	1986	24	30	322	42	22	183	6	95	605	414	190	1,933
	1987	26	25	283	39	20	180	9	100	564	349	200	1,795
Maize	1984	8	—	1,743	231	—	961	—	—	182	440	257	3,821
	1985	7	—	1,890	221	—	923	—	—	181	526	262	4,010
	1986	3	—	1,857	218	—	833	—	—	187	525	285	3,912
	1987	7	—	1,698	226	—	770	—	—	191	526	290	3,708
Total cereals†	1984	396	1,669	9,715	1,518	406	4,897	196	4,038	4,941	7,511	958	36,244
	1985	380	1,612	9,686	1,460	400	4,636	183	4,017	4,884	7,517	945	35,719
	1986	384	1,588	9,465	1,446	380	4,642	171	4,025	4,812	7,592	976	35,479
	1987	384	1,550	9,171	1,411	358	4,524	176	3,941	4,689	7,692	1,015	34,911

* Based on estimates at 27 October 1987.   † Includes rye, sorghum, mixed corn and triticale.
‡ Provisional MAFF data for the UK at 1988 harvest are wheat 1891, barley 1895, oats 121, rye 7 and mixed corn 5.

Source of Tables 1.5 and 1.6: Eurostat, EC Commission documents MAFF and H-GCA.

TABLE 1.6 Great Britain: area yield and production of wheat, 1886–1986

'000 hectares: tonnes/hectare: '000 tonnes

Year	Area	Yield	Production	Year	Area	Yield	Production
1886	925	1.87	1,727	1936	728	2.06	1,497
1887	938	2.22	2,088	1937	741	2.06	1,529
1888	1,038	1.95	2,021	1938	778	2.56	1,991
1889	991	2.07	2,057	1939	713	2.33	1,669
1890	966	2.13	2,061	1940	727	2.27	1,654
1891	934	2.17	2,026	1941	90	2.23	2,033
1892	898	1.83	1,645	1942	1,013	2.56	2,598
1893	768	1.81	1,383	1943	1,397	2.50	3,491
1894	780	2.13	1,663	1944	1,301	2.45	3,185
1895	573	1.82	1,045	1945	919	2.40	2,209
1896	686	2.33	1,603	1946	834	2.40	1,997
1897	764	2.02	1,544	1947	875	1.93	1,693
1898	851	2.41	2,052	1948	921	2.60	2,394
1899	810	2.27	1,841	1949	794	2.82	2,238
1900	747	1.98	1,479	1950	1,002	2.64	2,646
1901	688	2.13	1,474	1951	862	2.72	2,353
1902	698	2.28	1,592	1952	821	2.85	2,342
1903	640	2.10	1,338	1953	896	3.01	2,705
1904	556	1.86	1,036	1954	993	2.85	2,826
1905	727	2.27	1,654	1955	788	3.35	2,640
1906	711	2.37	1,686	1956	927	3.11	2,888
1907	658	2.35	1,543	1957	853	3.19	2,723
1908	658	2.23	1,473	1958	892	3.09	2,751
1909	738	2.31	1,700	1959	780	3.63	2,827
1910	732	2.08	1,526	1960	849	3.58	3,037
1911	771	2.33	1,805	1961	737	3.54	2,610
1912	779	1.95	1,520	1962	911	4.35	3,968
1913	711	2.21	1,566	1963	779	3.90	3,044
1914	756	2.30	1,734	1964	892	4.24	3,790
1915	909	2.20	1,993	1965	1,024	4.07	4,166
1916	799	1.98	1,584	1966	904	3.84	3,471
1917	801	2.07	1,661	1967	932	4.18	3,899
1918	1,067	2.31	2,467	1968	977	3.54	3,466
1919	931	2.02	1,878	1969	832	4.04	3,361
1920	781	1.97	1,540	1970	1,008	4.19	4,232
1921	826	2.50	2,060	1971	1,095	4.39	4,812
1922	822	2.15	1,770	1972	1,127	4.24	4,776
1923	728	2.21	1,612	1973	1,145	4.37	4,999
1924	645	2.22	1,435	1974	1,232	4.97	6,127
1925	626	2.30	1,437	1975	1,033	4.34	4,486
1926	666	2.07	1,382	1976	1,230	3.85	4,738
1927	689	2.20	1,512	1977	1,076	4.90	5,274
1928	588	2.30	1,350	1978	1,256	5.27	6,617
1929	559	2.41	1,351	1979	1,371	5.23	7,167
1930	567	2.02	1,145	1980	1,441	5.89	8,468
1931	505	2.03	1,026	1981	1,490	5.84	8,705
1932	542	2.18	1,184	1982	1,662	6.20	10,307
1933	704	2.41	1,693	1983	1,693	6.37	10,789
1934	751	2.51	1,889	1984	1,936	7.72	14,938
1935	758	2.33	1,771	1985	1,897	6.33	12,039
				1986	1,994	6.97	13,890

Source: MAFF and H-GCA.

TABLE 1.7　Sources of man's food energy

Food		Percentage of energy supplied
Cereals		56
Rice	21	
Wheat	20	
Corn	5	
Other cereals	10	
Roots and tubers		7
Potatoes and yams	5	
Cassava	2	
Fruits, nuts and vegetables		10
Sugar		7
Fats and oils		9
Livestock products and fish		11
Total		100

Source: Brown, 1975.

TABLE 1.8　Proximate composition of cereal grains (dry matter basis)

Cereal	Protein $(N \times 6.25)*$ (%)	Fat (%)	Soluble carbo-hydrate (%	Crude fibre (%)	Mineral matter (%)
Wheat					
Manitoba	16.0	2.9	74.1†	2.6	1.8
English	10.5	2.6	78.6†	2.5	1.8
Mixed grist	15.0	2.1	78.6†	2.4	1.9
Maize					
Flint	11.1	4.9	80.2§	2.1	1.7
Dent	10.6	4.6	81.0§	2.2	1.6
Sweet	12.1	9.1	74.5§	2.2	2.0
Sorghum	12.4	3.6	79.7§	2.7	1.7
Millet	13.6	5.4	77.9§	1.3	1.8
Rye	13.8	1.4	79.7§	2.6	2.2
Barley	11.8	1.8	78.1§	5.3	3.1
Rice					
Paddy	9.1	2.2	71.2§	10.2	7.2
Brown	11.0	2.7	83.2§	1.2	1.8
Polished	9.8	0.5	88.9§	0.3	0.6
Oats					
Whole grain	11.6	5.2	69.8§	10.4	2.9
Groats	14.9	7.0	74.6§	1.3	2.1

* $N \times 5.7$ for wheat and rye; $N \times 5.95$ for rice.
† Available carbohydrates determined by hydrolysis.
§ Calculated by difference.

Source: Kent, 1983.

2. Livestock feeds

Cereals may be fed as whole grains, ground, crushed, rolled, acid-treated or caustic-soda treated. Over 40 per cent of the British oat crop is fed on the farm of origin; the corresponding figure for barley is over 20 per cent and for wheat, below 5 per cent.

The UK is self-sufficient in feed wheat and feed barley. Both the EC and the US have imported manioc (cassava) from Third World countries as a cereal substitute − a crazy policy it seems when those same areas have burgeoning food deficits.

Whole crop cereal silage may be made by cutting crops when the grains are soft cheesy-ripe. Some crops are grown deliberately for this purpose and may include other species, notably vetches or other legumes. The bulk produced may be good but cereals alone give a low protein silage by contrast with pasture grass alternatives. Heavily diseased or irregular crops may be taken for silage as a second choice, especially if patches in a crop have been filled by later-sown seeds. Green oats are made into hay in some regions such as Israel.

Cereals may be harvested as whole, near-ripe crops and fractionated industrially (Figure 1.2).

Cereal straw provides both feed and bedding, to some extent simultaneously, especially for loose-housed cows and other livestock. It may be fed fresh or ammonia- or caustic soda-treated to improve its nutritional value for ruminants. Processed straw has also been incorporated into compound feeds.

3. Industrial Uses

A range of industrial uses and the processing of cereals for eating are discussed by Kent (1983) and by Andersen (1984) (see Figure 1.3).

Apart from the physical properties which make grains a source of adhesives and fillers for various purposes, they are also sources of specific chemicals, notably starch and dextrin. As concentrated energy sources their starch can be converted to sugars and then alcohol for use as fuel, and starch can also be made into biodegradable plastics and other goods.

Straw is a potential fuel. Some 1000 M tonnes are produced annually in the world, only part of which is used for animal feeding and bedding. This could be very important since yields of 4 t/ha are equivalent to the annual increment of temperate mixed forest, and fuelwood supplies in less developed countries are increasingly critical. The problem in any industrial usage of straw is its bulk and the separation of supply from areas of demand. The wide adoption of high-density balers would help greatly.

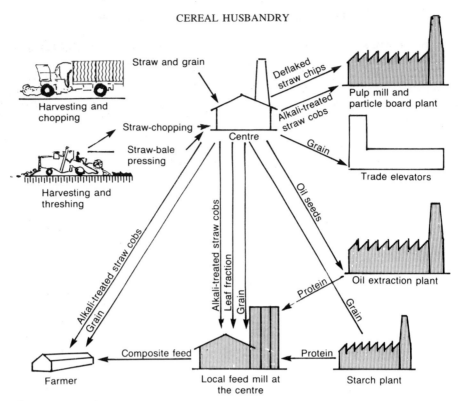

FIGURE 1.2   The agricultural refinery concept. (From Rexen and Munck, 1984)

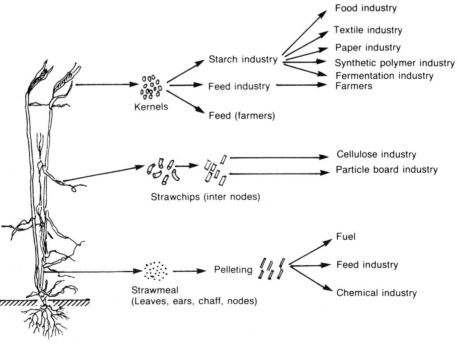

FIGURE 1.3   Transformation of a cereal crop into intermediates for use in various industries. (From Rexen and Munck, 1984)

The fibre content of straw is high and makes it a potential material for the manufacture of coarse paper, pulp, packaging, insulation and construction board. In addition, straw crafts are important rural industries overseas, not only for aesthetic items but for useful goods such as straw mats and ropes. Staniforth (1982) reviews the commercial uses of straw fibre. Uses of cereals in the EC and UK are indicated in Figures 1.4 and 1.5.

DISTRIBUTION OF CEREALS

Table 1.9 gives the major producers of cereals in the world. In order of quantity of production, wheat ranks first, closely followed by rice and maize, then barley, sorghum, oats and rye. Other cereals are of relatively minor importance overall though they may be staples in particular districts. In order of rainfall requirements in the tropics, rice requires the most, followed

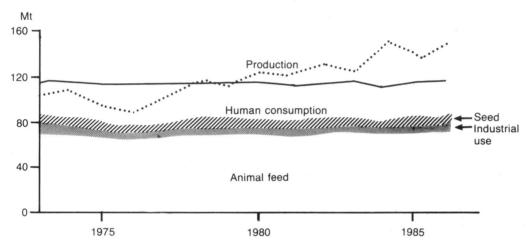

FIGURE 1.4   EC cereal production and uses.

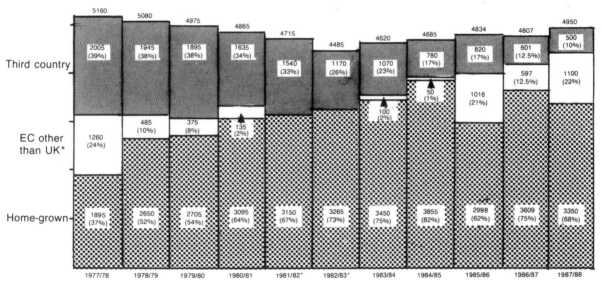

*Imports from EC in 1981/82 and 1982/83 are estimated at approximately 25,000 tonnes and 50,000 tonnes respectively.

FIGURE 1.5   Millers' usage of wheat ('000 tonnes) in the UK. (Source: NABIM)

TABLE 1.9   World production (major producing countries) of wheat, maize, sorghum, barley and rice (M tonnes)

	1983/84	1984/85	1985/86	1986/87	1987/88(a)
WHEAT (b)					
USA	65.9	70.6	66.0	56.9	57.3
Canada	26.5	21.2	24.3	31.4	26.1
Australia	22.0	18.7	16.2	16.2	13.0
Argentina	12.8	13.2	8.5	9.0	9.5
EC 12	63.8	82.9	71.6	71.7	75.2
USSR	77.5	68.6	78.1	92.3	80.5
Eastern Europe	35.4	42.1	37.1	39.3	39.8
China	81.4	87.8	85.8	90.3	88.0
India	42.8	45.5	44.1	46.9	46.0
Others	61.3	60.9	67.6	74.9	72.1
World total	489.4	511.5	499.2	528.9	507.4
MAIZE (c)					
USA	106.0	194.9	225.5	209.6	181.3
Brazil	21.0	22.0	21.0	26.5	24.0
Mexico	9.3	9.9	10.5	10.0	10.1
Argentina	9.2	11.5	12.1	9.5	11.5
South Africa	4.4	8.1	8.1	7.8	8.5
Thailand	4.0	4.4	5.4	4.1	3.0
EC 12	21.8	23.1	25.7	25.0	23.3
USSR	13.3	13.6	14.4	12.5	15.0
Eastern Europe	33.2	35.4	33.6	38.8	33.5
China	68.2	73.4	63.8	69.0	75.0
Others	56.8	62.2	63.0	62.3	61.8
World total	347.2	458.6	483.1	475.1	447.1
SORGHUM (c)					
USA	12.4	22.0	28.5	23.9	19.0
Australia	1.9	1.4	1.4	1.2	1.5
Argentina	6.9	6.2	4.2	3.0	3.2
South Africa	0.5	0.6	0.4	0.5	0.7
Thailand	0.3	0.4	0.3	0.3	0.3
Mexico	4.0	4.1	3.7	4.3	4.0
India	11.9	11.4	10.1	10.2	9.0
China	8.4	7.7	5.6	5.4	5.2
Nigeria	2.7	3.7	3.5	3.6	3.4
Sudan	1.8	1.1	3.6	3.4	3.1
Others	8.2	7.7	9.1	9.8	9.6
World total	58.9	66.2	70.5	65.6	59.0
BARLEY (c)					
USA	11.1	13.0	12.9	13.3	11.3
Canada	10.2	10.3	12.4	14.7	14.0
EC 12	42.9	54.5	50.8	46.7	48.4
USSR	50.0	41.8	46.5	53.9	62.0
China	6.8	7.3	6.2	6.1	7.0
Eastern Europe	15.4	17.1	16.4	17.0	16.4
Others	23.5	24.8	26.5	27.4	26.2
World total	164.7	174.4	176.7	182.6	188.8

(continued)

	1982/83	1983/84	1984/85	1985/86(d)	1986/87(e)
PADDY RICE					
Argentina	0.3	0.5	0.4	0.4	0.4
Australia	0.5	0.6	0.9	0.7	0.5
Bangladesh	21.3	21.8	21.9	22.6	23.1
Brazil	7.8	9.0	9.0	10.3	10.5
Burma	14.4	14.4	14.3	14.3	12.5
China (f)	161.2	168.9	178.3	168.5	171.1
EC 12	1.6	1.5	1.7	2.0	1.9
India	70.7	90.2	87.5	96.2	90.0
Indonesia	33.6	35.3	38.1	39.0	38.4
Japan	12.8	13.0	14.8	14.6	14.6
Korea (South)	7.3	7.6	8.0	7.9	7.9
Pakistan	5.2	5.0	5.0	4.4	5.2
Thailand	16.9	19.5	19.9	19.7	18.0
USA	7.0	4.5	6.3	6.1	6.1
Others	58.9	61.0	62.4	63.7	63.2
World total	419.5	452.7	468.4	470.3	463.5

(a) Estimates available as at end October 1987.
(b) July–June year.
(c) October–September year.
(d) Preliminary.
(e) Forecast.
(f) Includes Taiwan Province.

Source: United States Department of Agriculture (USDA).

by maize, sorghum and finally millets, which are drought tolerant.

Of the temperate cereals, wheat is the most southerly distributed; it is also cultivated in the high-land tropics. Wheat is important over the widest range of latitudes of any cereal (Figure 1.6). It requires greater sunshine receipts and higher soil fertility to perform well than do the other temperate cereals. Barley is grown on lighter land, whilst oats are found in more northerly latitudes and rye is the northern-most cultivated cereal.

UK DISTRIBUTION

Figure 1.7 shows the general distribution of wheat and barley in Britain. Wheat requires an early start to the growing season and plenty of summer sunshine to ripen; it is more resistant to winter frost than barley and much more so than oats. It responds more to soils having plenty of body — strong land — and richer land.

Wheat predominates in the heavy land areas of the eastern and East Midland regions of England. It is the leading cereal in Lincolnshire, Bedfordshire, Cambridgeshire, Essex, Suffolk, Kent, West Sussex, Nottinghamshire, Warwickshire and the Buchan area of Aberdeenshire. Wheat and barley are fairly balanced in Buckinghamshire, Leicestershire, Hereford and Worcester.

Elsewhere *barley* has been the leading cereal, dominating chalk and limestone districts of the south and east up to the Lincolnshire and Yorkshire wolds. However, the wheat area caught up with barley in the early 1980s in England and Wales for the first time since the recording of annual statistics began in 1866. Barley increases in western and northern areas and wherever land is lighter, including Norfolk. It is now a more popular feed grain than oats on account of higher yields and easier management.

Oats occupy only as much as 5000 hectares in Devon, Hereford and Worcester and Kent. They occupy over 6000 hectares in Wales, not far behind wheat, but still under one-tenth of the Welsh cereal crop, which in total is equivalent to 2 per cent of the English corn area.

FIGURE 1.6   Wheat producing regions of the world. (Canadian Wheat Board, 1980)

Rye approaches 2000 hectares only in the counties of Norfolk and Suffolk. Forage rye varieties are widely distributed, being used to supply early bite (the first spring grazing supply). Triticale is slightly expanding and classified with rye in statistical returns.

Mixed corn is generally oats and barley but may include beans or peas. The mixture can outyield pure stands of the constituents but cannot be more intensively managed. It is called dredge in the South-west and mashlum (usually including legumes) in the North. Its area has exceeded oats and wheat added together in Cornwall, covering some 6000 hectares, almost 15 per cent of the total cereal crop in that county. Otherwise it has only just exceeded 1000 hectares in Devon and Dyfed. Mixed cropping is a very beneficial practice worldwide but complicates crop management. It is declining fast for UK corn.

Maize for grain is of very minor occurrence on a field scale, being confined to the eastern part of the South coast of England where sunshine receipts and growing season length are maximal, though maize for silage is more widespread, especially now that earlier-making forage varieties are available. Sweetcorn varieties are of minor horticultural importance in the UK.

SIGNIFICANT TRENDS IN CEREAL PRODUCTION

Ten-year periods are the relevant ones on which to base changes or decisions about farming policy, not what happened last harvest, which may have been atypical for one's farm. The decade 1975 to 1985 will now be considered. It was one of great change and expansion in the UK:

● Area under cereals up slightly, especially due to an increase in wheat planting, which showed a swing towards milling wheats.
● Great expansion in the area under winter barley to around 55 per cent of the total barley area, at the expense of spring varieties. The main reasons

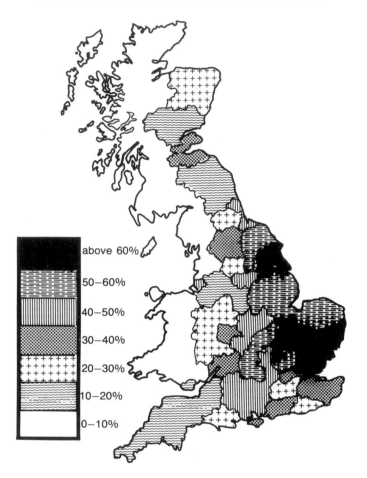

FIGURE 1.7 Total wheat and barley area as a percentage of total land area on agricultural holdings. (From Food from Britain)

for this were the increasing adoption of systemic fungicides and the use of improved two-row and some useful six-row varieties (i.e. six rows of grains in each ear). The pendulum can swing back towards spring varieties!

● Adoption of longer runs of consecutive cereal crops and high inputs to sustain them, though some return to rotations towards the end of the decade.

● Adoption on a wider scale of reduced cultivation systems, including direct drilling, though now a detectable return to ploughing if only on a one year in three or five basis.

● Great expansion of oilseed rape production, chiefly winter varieties, usually following winter barley and preceding winter wheat. This alternative crop is combine-harvested. It occupied 25,000 hectares at the start of the decade and 300,000 hectares at the end, going on to almost 400,000 hectares.

● Yields greatly increased: wheat by 45 per cent, barley by 35 per cent and oats by over 25 per cent (see Figure 1.8).

● Exports of UK cereals went up and were set to increase further.

The decade 1975−85 was marked by increasing famine, especially in Africa, and yet countries like India succeeded in maintaining adequate reserves through the green revolution (new varieties and concomitant management) coupled with a strategic reserve policy. Overall, world supplies kept pace with the population increase (Figures 1.9 and 1.10).

As far as Europe is concerned, the problem of expanding cereal surpluses has become of huge

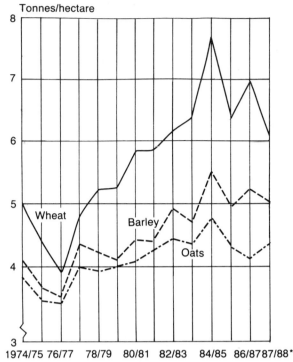

*Provisional (deduced from end-September MAFF production estimates)

FIGURE 1.8   Recent trends in United Kingdom yields of wheat, barley and oats. (Source: MAFF)

political and agricultural significance. Having been encouraged to produce and having succeeded in doing so, farmers are now seen by some as villains. However, Jonathan Swift (1667−1745) wrote the now-famous lines in *Gulliver's Travels*, 'whoever could make two ears of corn or two blades of grass to grow upon a spot of ground where only one grew before, would deserve better of mankind and do more essential service to his country than the whole race of politicians put together.' Certainly the challenge of good husbandry has never been greater for farmers and the challenge of guiding overall production has never been greater politically.

CEREAL POTENTIAL

Austin (1978) quoted the potential yield of winter wheat as 12.9 t/ha. Record crops have now achieved this performance. The UK national average yield of winter wheat was 7.7 t/ha in 1984. Yields changed little between 1900 and 1945 but have trebled since. World wheat yield still averages only 2 t/ha.

Irrigated rice crops (padi) using short-strawed, quick-maturing (90-day) varieties can produce four crops per year with a total yield in excess of 20 t/ha. Three crops per year are quite common. However, only one-quarter of Asia's rice farmers have access to such irrigation. Whilst the reduction in real prices

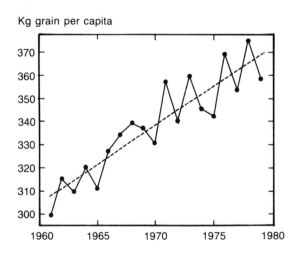

FIGURE 1.9   Average supply of cereal grain in kilogram per world inhabitant over the period 1961−79. (From McKey, 1981)

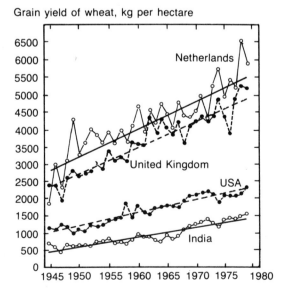

FIGURE 1.10   Increase in grain yield in selected countries. (From McKey, 1981)

of rice resulting from production increases in countries like Indonesia and the Philippines has benefited the landless and urban poor, it has worsened the plight of the smaller rice farmer who lacks irrigation.

The Centre for World Food Studies at Wageningen in the Netherlands is using mathematical models to calculate the maximum biological production of crops by applying systems analysis and crop production simulation techniques. It is assumed that a good farmer can produce about 70 per cent of the maximum biological yield — this figure is called the crop production ability (CPA); then the actual yield expressed as a percentage of CPA gives the crop performance index (CPI). The CPI indicates the scope for improvement in yield of any particular crop grown at a specific site. For the UK wheat crop, CPI was around 30 per cent in 1950 whereas it rose to around 80 per cent by the late 1980s.

BIOTECHNOLOGY AND CEREAL CROPS

Biotechnology is a wide, multi-disciplinary, new field of science dealing with the use of micro-organisms, genes and biochemical components of cells to produce goods or provide services. It includes:

- *genetic engineering*, for instance, incorporating genes for improved gluten quality in wheat and threonine (an amino acid) content in barley and using recombinant DNA to give more durable disease resistance.
- *tissue culture*, for example, immature wheat embryos have been cultured at Rothamsted to produce hundreds of plantlets which can subsequently be assessed in normal field trials. These plants show considerable variation from which new cultivars of wheat may emerge, so diversifying and accelerating the plant breeding process.
- *fermentation*, for instance, at the Plant Breeding Institute in Cambridge, the enzyme alpha-amylase from wheat, which hydrolyses starch (and makes for low Hagberg numbers in wheat), has been incorporated into yeast by transferring the gene which controls this enzyme (more genetic engineering!). The yeasts may then be used to digest starch from cereals in the commercial production of ethyl alcohol (ethanol, an industrial alcohol).
- the perhaps more distant prospect of *nitrogen-fixing* nodules on cereals is being further investigated; certainly pearl millet can fix nitrogen symbiotically when a *Spirillum* bacterium is there.

Some commentators predict that this sort of work will equal or exceed the impact of the green revolution based on the dwarfing (reduced height or Rht) genes of the Japanese wheat cultivar Norin 10 which affected wheat and rice yields so much from the 1960s onwards. Others equate its potential impact with that of the computer and micro-electronics revolution.

Chapter 2

THE CONTEXT FOR CEREAL CROPS

This chapter considers the two great environmental aspects of climate and land as they affect cereal husbandry.

CLIMATE AND CEREALS

Climate affects cereal cultivation in several ways both long term and short term (weather variations):

- Restricts species and, indeed, cultivars of cereal which can be chosen.
- Directly influences their establishment, e.g. a wet, closed autumn limits the planting of the target winter wheat area on heavier land and necessitates some alternatives including spring wheat.
- Determines their development and subsequent performance in both yield and quality aspects.
- Affects the incidence and severity of problems — weeds, pests, diseases. The microclimate within areas of a crop can provide a focal zone for disease development (Plate 2.1).
- May directly damage crops, e.g. hail, drought, blind grains through wetness at flowering, lodging by wind and rain.
- Alters the responsiveness of crops to applied treatments both to promote growth (such as fertilisers and PGRs) and to protect from problems (the biocides).

Historically, the seasonal climate has greatly influenced British cereal prices, and national prosperity has fluctuated with the wheat price of the previous harvest. For instance, the wheat price at Exeter in 1608 was 50 per cent above the average for the decade 1600–1609 owing to a bad season. Now with EEC surplus stocks and pricing mechanisms the average person is insulated from the realities of seasonal weather fluctuations to considerable disadvantage as far as an understanding of farming is concerned.

Strategies to combat weather problems include:

PLATE 2.1 *Microclimate and disease probability: persistence of wetness in a wheat crop.* (Courtesy of IACR Long Ashton Research Station, University of Bristol)

- Choice of appropriate variety.
- Timely crop establishment and timely application of all treatments.
- Adequate drainage and subsoil management to maximise the rooting potential.
- Irrigation. This is especially vital on sands such as those of Nottinghamshire and East Anglian Breckland. Moisture supply can critically affect establishment of spring cereals on such land and profoundly limit the growth rate, particularly from GS 30−59. Strategic use later can prolong green leaf area which is vital to maximise grain-filling. Awned cereals (those possessing bristled heads) tend to tolerate drought better than non-awned ones and taller varieties often have a correspondingly deeper root system.

Figure 2.1 indicates climatic trends over a decade in south-east England. It gives the overseas reader a very general guide to the background conditions for UK cereal production. Obviously, increasing latitude limits the length of the growing season, though it is compensated by higher summer light

intensity. The Gulf Stream protects the west of Britain from late frost (and in some places May frost is rare), whilst easterly winds in the winter can be bitter along the eastern side of the country and on exposed land, especially when cereal crops are not protected by snow cover.

Altitude has a profound effect on the growing season, shortening it by about two weeks for every 100 metres up. Local undulations of the land creating winter frost pockets or humid zones in summer can be influential. Aspect of slopes is also important, a southerly one obviously favouring greater growth in the northern hemisphere.

As a minimum, the diligent cereal farmer now monitors rainfall and soil temperature at 10 cm as guides to his expectations when inspecting crops and to the timing of treatments.

The excellent series of six regional books published in 1984 by the Soil Survey of England and Wales entitled *Soils and Their Use* seeks to integrate the effects of soils and climate on land work (Figure 2.2), as well as cropping.

Of global concern is the 'greenhouse effect' due

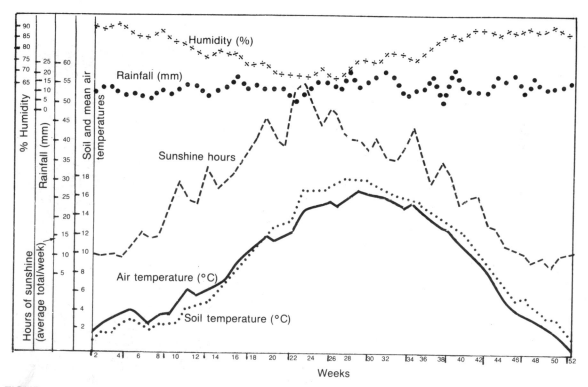

FIGURE 2.1   Annual weather trends (mean of a decade showing weekly averages), near Reading, South-east England.

Soil series	Soil*	Year	MWDs	Autumn / Winter / Spring (SEP OCT NOV DEC JAN FEB MAR APR)	MWDs
Sherborne	a	Normal	66		14
Sherborne	a	Wet	40		0
Moreton	ab	Normal	56	CIRENCESTER 825 mm annual rainfall	9
Moreton	ab	Wet	30		0
Evesham & Haselor	d	Normal	16		0
Evesham & Haselor	d	Wet	0		0
Sherborne	a	Normal	88		27
Sherborne	a	Wet	65		0
Moreton	ab	Normal	78	SOMERTON 725 mm annual rainfall	22
Moreton	ab	Wet	55		0
Evesham & Haselor	c	Normal	48		12
Evesham & Haselor	c	Wet	25		0

MWDs: Number of good machinery work days during the period indicated

Frequent opportunities for autumn landwork   Frequent opportunities for spring landwork   Little opportunity for landwork

*Soil assessment = workability in relation to wetness: b = average, a and ab above average, c and d below average

FIGURE 2.2   The effects of soil and climate on landwork, Sherborne association. (Source: Soil Survey of England and Wales, 1984)

to carbon dioxide accumulation following combustion of fossil fuels and net loss of forest cover, plus gases such as chlorofluorocarbons (CFCs) from propellants. The result is a rise in mean global temperature of around 0.5°C and more erratic weather during this century. Some zones, notably those nearer the tropics, are suffering more frequent droughts whilst others, such as in Northern Europe, may find conditions more conducive to crop production.

LAND AND CEREALS

It is of fundamental importance for any farmer to *know* his soil, including the often considerable variation within fields. Part of the regular kit in any farm vehicle should be a lightweight spade. A small trowel and hand-auger are also useful. A bare hand and sensitive boot are paramount, hence the old saw 'The best doung (manure) is the farmer's boot.' Most problems in husbandry have underlying causes in the soil, and failure to investigate such causes can lead to a 'cure symptoms' approach to cropping which is unsatisfactory and never-ending. Deal with *causes* as far as possible first.

There are three vital points anyone should check when new to farming a block of land:

1. Where are the *boundaries* and what is their condition? It has been known for new staff to combine the neighbour's crop by mistake!

2. Does the land need *draining*? This is the most fundamental *physical* improvement that can be made to the soil where it is needed.

3. Does the soil need *liming*? Acidic soil conditions are frequently encountered in Britain. Raising the soil pH by applying lime is the most fundamental *chemical* improvement that can be made where it is needed. It is possible, especially on lighter textured soils, to overlime and jeopardise the crop. An optimum pH of around 6.5 to 7.5 is desirable for cereals. Good farming frequently consists in balancing extremes.

UNDERSTANDING SOILS

The cereal farmer will 'know' his soil by doing the following:

A.   Determining its type (texture)
B.   Examining the profile (vertical section to one metre plus)

C. Testing it (or, more usually, having it analysed)
D. Working it
E. Classifying it as to land capability, soil series, the individual's own criteria

A. Determining Soil Type (Texture)

All the practically important properties of soil can be related to its texture, which is determined by the proportions of sand, silt and clay particles present in it. The distinction between these particles is based on their diameter, as agreed in 1927 by the International Soil Science Society and subsequently revised (Figure 2.3).

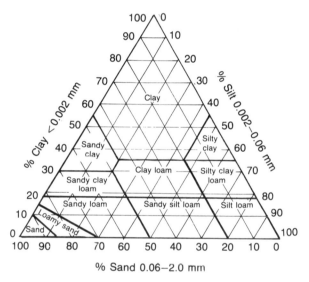

FIGURE 2.3   The soil textural triangle.

A normally moist sample of soil in the field should be handled to assess the broad description:

Gritty, sharp, sounds abrasive	= Sandy
Smooth, soapy, silky	= Silty
Sticky, mouldable, takes a shine	= Clay
No particular one of the above	= Loam

Note: Higher organic matter may give a *greasy* feel to soil; high calcium carbonate in a calcareous soil *softens* the basic texture; high moisture content accentuates the clay fraction. A given proportion of clay exerts a bigger influence over soil behaviour than the same proportion of sand because of its smaller, more reactive particles.

A *loam* is a mixture of sand, silt and clay. The description of a loam can be further qualified according to the dominant particle type where one is prominent: clay loam, silt loam, sandy loam. Further sophistication in assessment may be possible with practice: sandy clay loam, etc. More elaborate procedures exist for the measurement of soil texture but their relevance depends on how representative the sample is of the soil area to be managed.

Knowing the soil type is fundamental for determining the following:

1. Most usual names of the soil.
2. Physical conditions – structure and tilth that can be attained.
3. Ease of working – heavy, medium or light – and selection of cultivation system.
4. Water-holding and conducting behaviour.
5. Nutrient-holding behaviour and requirements.
6. Activity of living things – the range and rate of biological activity both beneficial and harmful, for example, decomposition rate and nature of weed flora.
7. Behaviour towards applied chemicals, especially residual (soil-acting) herbicides.
8. Cropping choice. Different cereals have their own preferences and tolerances. Wheat on the whole prefers heavier soils richer in nutrients and is less sensitive to physical conditions than barley. Barley tolerates rather more alkaline, somewhat less rich soils but reacts more to poor physical conditions. In fact barley withstands high salts in general better than the other cereals. Oats tolerate poorer conditions than either wheat or barley as far as nutrients and lower pH are concerned. However, winter oats are extremely sensitive to frost-heavage unless a well-established root system has developed. Oat responds to the richer lowland environment normally afforded its cousins and can yield significantly above its national average on such soils. Rye is the cereal with the reputation for the most frugal soil demands, partly owing to its very exploratory root system. Acid sands and exposed sites may be tolerated but, like oats, when given the better land normally allotted to wheat, rye can yield well over its rather misleading national average.

B. Examining the Profile

Whilst the majority of cereal roots occur in the top 50 cm of soil, the wheat crop sends some roots down to 1—1.5 m, sometimes 2 m. It should not therefore be beyond our interest to examine the top metre of soil. Apart from waiting for a pipeline to cross one's land or a drainage scheme to commence, it is fairly straightforward to examine 1 m cores using a post-hole digging auger on the back of the tractor. A few trial pits may also be dug in strategic spots.

When one considers the far-reaching significance of the soil conditions on farming operations and performance, time spent furthering one's practical understanding of the soil is well invested. It should be a priority before time spent juggling on the computer with figures concerning inputs applied to the surface. Indeed the quantity and type of such inputs are often directly related to what has or has not been done to improve soil conditions to depth. It is my impression that a disproportionately low amount of management time is spent examining soils, the need for which is not rated perhaps highly enough by many cereal growers. Part of the problem lies in the wide tolerance of cereals to variations in soil conditions before crop failure ensues. Most other crops are far fussier. There is considerable sub-clinical yield loss in deteriorating soil conditions before symptoms become obvious.

What features should be considered when examining a soil profile for cereal growing?

Depth   The depth of topsoil on much good cereal-growing land — notably calcareous downland of chalk and limestone districts — is very limited. The depths of the various layers (horizons) which occur in many soils are significant.

Type   The texture of the different layers should be observed, along with the percentage and nature of stones, which exert a considerable influence on cultivations and performance of corn on brashy (stony) land, including positive benefit in aiding moisture retention.

Structure   This is a profile feature of great consequence. It refers to the way in which mineral particles (sand, silt and clay) are grouped to form aggregates (Figure 2.4).

The structure achieved in a soil is limited by the texture of the materials available to build into aggregates. However, within this limitation, the extent to which an optimum structure is attained depends on soil management. Soil structure is much more easily damaged than it is improved. Mellow tilths are less easily damaged, being less extreme in texture than raw tilths. The latter include non-calcareous clays and

silts which turn to slurry when wet and rocks or powder when dry.

Historically, a *biological* emphasis was given to the concept of soil fertility. Then since the advent of pioneering work on fertilisers and soil chemistry by Lawes and Gilbert in the mid-nineteenth century, a *chemical* emphasis has been associated with soil fertility. During the past quarter-century, the increasing size and weight of machines used in cereal production has posed compaction problems and thus a *physical* emphasis has been associated with soil fertility, which has increasingly been equated with 'good structure'. In reality, soil fertility is a comprehensive concept referring to the soil's ability to produce and go on producing useful crop yields. In order to achieve this the soil must provide:

- Space
- Water
- Air
- Anchorage
- Balanced supply of all essential mineral nutrients
- Favourable pH (around 6.5—7.5)
- Warmth
- Absence of toxins and restrictions

The first four of these requirements are clearly closely related to soil structure. As mentioned previously, structure can be described according to the degree and type of aggregates present and their size, shape and stability (to wetting, drying, traffic and cultivations). Structure can also be considered in terms of the pattern of pore spaces created by the aggregates. This pattern determines the balance between air and water held and the ease of penetration for cereal roots.

Compaction is a frequent occurrence in modern cereal farming. Its severity and location in the profile need to be monitored. A spit of soil can be dug when cereal roots are well established (say April to June for winter corn) and bumped on the ground just enough to see whether fissures (cracks) develop

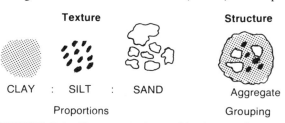

FIGURE 2.4   Representations of texture and structure (not to relative scale, which is impossible to portray).

vertically as desired or horizontally indicating compaction. It is also possible to over-loosen soil!

Colour   This is principally an indicator of the drainage status of the soil (Figure 2.5). Mottled soil colours are a common indication of some degree of impedance of drainage.

Accumulations of organic matter   Are dead roots and crop residues decomposing at a steady, acceptable rate? The *quantity* of organic matter in the soil is often quoted as if it were the only consideration. A stagnant pond contains a lot of organic matter but this is scarcely beneficial! Also a wet, heavy field is likely to be oversupplied with organic matter. Equally, at levels of organic matter often accepted as ideal for cereal growing — say 3 to 5 per cent in mineral soils — the question must be asked, 'What is the *quality* of this organic matter?' Sawdust is organic matter; so is rich garden compost. Just as a business analysis of cereal production must consider the question of the rate of return on capital invested, one must ask what is the rate of decomposition/replacement of the organic matter in a soil. Particularly with intensive winter cereals, it is vital that residues should decompose rapidly. This process not only releases elements which can be used again by the next crop (mineralisation) but also removes a potential reservoir for trash-borne diseases and pests. The end result of decomposition is the much more slowly dissipated *humus* which has twice the capacity of the best clay and 20 times of the worst clay to hold onto nutrients, apart from its role in aggregating sands and aiding retention of plant-available water.

It is obvious that the narrower the spectrum of ingredients contributing to soil organic matter, the less diverse it is going to be. Just as we are keen to give our farm livestock comprehensive diets with a complementary range of ingredients, the same principle would seem applicable to the soil for its micro-organisms. Much more research is needed on the impact of soil organic matter quality and the rate of turnover on cereal performance. There is a close correlation between the soil's ability to support a large and diverse soil population and its suitability as a root environment. Thus the nature of its organic matter is some evidence of its suitability for cropping. Earthworm populations are perhaps the easiest to observe and measure. Their population size depends greatly on the degree of soil disturbance as well as these other factors.

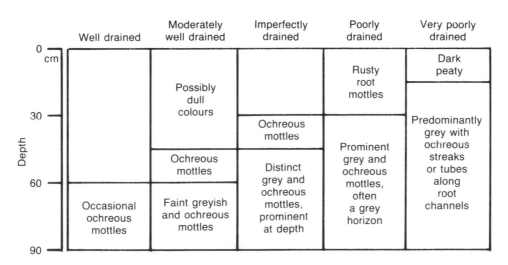

Note: Ochreous varies from yellow to rust in colour.

FIGURE 2.5 Soil colour and drainage class as recognised by the Soil Survey. During the winter or early spring, poor drainage is often self evident but during dry periods, careful profile examination is needed to detect it. (ADAS diagram adapted from Soil Survey of England and Wales)

C. Testing Soil or Having it Analysed

It is possible to purchase colorimetric kits to assess the chemical constituents of the soil. Rapidly registering probes for pH and moisture status are also obtainable. Commercial firms and ADAS offer a laboratory analytical service for soil samples. Some farmers have obtained computerised, detailed analyses for all essential elements, both major and minor (trace), from American laboratories amongst others.

Whatever the degree of sophistication of the testing, it remains axiomatic that the usefulness of the results depends on how sensibly the sample was taken. There are three commonsense considerations of good sampling (which apply in principle to any bulky commodity, including the grain harvested):

1. Representative   The sample should truly represent what is to be treated as a result of the analysis. Areas of obviously contrasting soil type within a field should be dealt with as separate blocks. Headlands, gateways and any obviously atypical spots should be excluded from the sampling or dealt with separately.

2. Randomised   The samples should be taken without bias from scattered spots throughout the whole area to be treated. Fields may be walked as shown in Figure 2.6, but in practice many are not square! The scatter of samples should be taken evenly throughout the field or block of land to 15 cm depth for routine topsoil analysis purposes.

3. Replicated   The accuracy of the sampling or at least its correlation with reality is increased by the number of samples taken, which should never be less than 10 and preferably 25. It is statistically better to have many small samples than a few large ones. Normally the individual samples are bulked, mixed and then subsampled for analysis. Although analysis may be conducted on several replicate subsamples, each of these is often a very small quantity of soil.

When one considers that the top 15 cm of 1 hectare of topsoil weighs around 2500 tonnes (1000 tons per acre) and analysis is conducted on a few grammes,

it is clear how vital it is to sample sensibly. Digital print-outs to numerous decimal places are routinely possible nowadays with the sensitivity of analytical equipment (which can detect literally millionths of one millionth of a gramme of chemicals!). However, all this is futile unless the original sampling fulfils the above three Rs of good procedure.

D. Working the Soil

Farmers around the world who practise hand cultivation have an intimate knowledge of local soil variations and problems. The closeness and detail of this knowledge is clearly diminished from the perspective of an air-conditioned cab serviced by radio or stereo headphones! Nevertheless it is vital for the manager of cereals to have worked the land himself and thus have gained a clear insight into the realities encountered by those he directs in cultivating or otherwise treating the land.

The art of husbandry very much depends on dealing with the soil under as near optimum conditions as possible. No more than broad principles can be enunciated in a textbook, as these lessons have to be learned in the field (Figure 2.7). Traditionally they have been handed down as part of the local rural lore. Sometimes the view held of a soil's limitations may be outdated by new techniques but generally the farmer is wise to heed the advice of older countrymen on how the local soil behaves. The farmer of a raw tilth, such as a marine clay, will need to be geared with sufficient, more powerful tackle to handle the land during the much-restricted period suitable for cultivations than that required by a mellow tilth such as a calcareous loam.

E. Classifying the Soil

Soil series   In 1983 and 1984 the Soil Survey of England and Wales completely mapped the soil series of the country on six sheets, scale 1:250,000, with corresponding descriptive books. Whilst the soil series descriptions give more detail of pedological interest than the practical farmer requires, a study of the notes will increase his understanding of his own land.

A soil series is a group of soils whose members have similar profiles (vertical sections of 1 metre) and differ only in the texture (type) of the topsoil. They are generally given local names after the area where the profile type was first described, e.g. Andover, Hanslope, Denchworth.

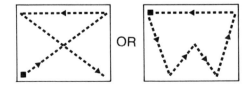

FIGURE 2.6   Field walking for soil sampling.

FIGURE 2.7 Entertainments at which we have never assisted: a ploughing lesson at an agricultural college. (From *Punch*, 1928)

Land capability In 1969 Bibby and Mackney published a seven-grade land-capability classification system for Scotland, and in 1974 England and Wales caught up with an agricultural land classification (ALC) scheme. These systems are concerned with evaluating those features which are largely unalterable physical limitations to cropping such as altitude, slope, depth of soil, wetness of site. They do not take account of the present level of management (for example, phosphate index, degree of weed infestation).

The ALC is really a system for assessing the *flexibility* of cropping possible on the land. Land is in five grades, grade 1 being superior and having little or no restrictions on cropping, whilst grade 2 has minor limitations. Grade 3 land makes up almost 50 per cent of the total, and it was sub-classified in 1976 into 3A (fine texture), 3B (shallow, stony) and 3C (fine texture, lies wet). Grades 4 and 5 have severe and very severe limitations respectively, making them largely unsuitable for cereals.

Most countries have their own system of assessing land capability. So do some farmers.

FIELD DRAINAGE FOR CEREALS

Only about one-third of UK land is naturally well drained. Poor drainage is caused by one or more of the following:

- *Site*   Low-lying or basin land
  High rainfall area (even 750 mm of annual rainfall delivers 7500 t/ha (3000 tons/acre) of water for the land to cope with)
  Spring lines
- *Soil profile*   Impermeable material, such as high clay fraction
  High water-table

The detrimental consequences of excess water in the soil arise largely from its displacement of oxygen (which diffuses some 10,000 times slower through water than through soil air). The results are poor establishment of the crop and reduced rooting depth, volume and activity. All biological activity is impaired in anaerobic conditions but some micro-organisms continue to thrive and produce toxins such as ethene (ethylene), hydrogen sulphide and excess ions of aluminium and manganese, all of which can inhibit roots. The slower, festering decomposition of organic matter from previous crop residues also means that seedlings of the next crop are hindered and more likely to contract diseases from the still-lingering trash.

The *benefits* of improved field drainage for cereal growing are, therefore:

● Better soil structure can be achieved.
● Rooting is thereby encouraged.
● Nutrient uptake is consequently greater.
● Decomposition of trash is accelerated and so conditions are healthier.
● Cultivating periods are increased by some 3.5 weeks extra on average in both autumn and spring for all appropriate soil series in the UK. This allows greater probability of drilling cereals at the target time in nearer optimum conditions followed by more rapid passage through the vulnerable seedling phase of the crop's life.
● Yield potential from the land is therefore substantially improved. (I recall one site where the farmer drained just the worst-affected part of a field only to find the following year's wheat crop yielded exactly double the supposedly acceptable part of the field — which was subsequently drained as well but incurred double tackle-delivery and other overhead charges!)
● Land value is increased in real usefulness terms and financial value is soon likely to offset the gross drainage costs (often still near only one-fifth of the market value of the land in the UK).
● There is greater flexibility in day-to-day management and in possible cropping choice for the field.

In order to decide on an appropriate drainage scheme, the procedure should be:

1. First seek the advice of one or more specialists — the Ministry of Agriculture, drainage contractors, land agents or other private consultants — to identify the causes and to recommend cures.

2. Research any old maps in the farm archives and study the site itself for evidence of previous under-drainage and other underground services including water supplies for drinking troughs.

3. Commission or conduct your own site survey, starting at the lowest point to ascertain adequacy or otherwise of the departing watercourse, including consequences and wayleave through lower neighbouring land; to determine outfalls into ditches, marking

these by coloured posts when installed to ease later location for maintenance; to check positions for main drains (often now 80—160 mm perforated plastic pipes, though clay pipes are still made) along lines of least fall (often 1 in 100 to 1 in 400) followed by lateral pipes (usually 50—60 mm perforated plastic, unless the run is so long that some 80 mm pipe may be needed to link into the main). Lateral falls are greater than the mains, usually between 1 in 50 and 1 in 200. (See Table 2.1.)

4. Consider the main variables in scheme design, which involve:

● Pipe spacing — typically 20 m.
● Pipe depth — typically 75—90 cm, not less than 60 cm.
● Pipe diameter (possibly filter-wrapped on very fine sandy soils).
● Use of stones as permeable backfill over pipes to within at least 37 cm of the soil surface. This can almost double scheme costs. Some schemes now use close-spaced (3.5—5 m) plastic pipes at normal depths but without backfill to achieve greater effects at similar cost, it is claimed. Backfill is generally needed where poor permeability is the problem (Plate 2.2).
● Use of secondary treatments. Moling is the preferred one usually done at intervals of about 2.5—3 m and depths of 52 cm. Alternatively, subsoiling is used on stony sites shallower (30—35 cm) and closer spaced (1.5 m). Sometimes subsoiling is justified in addition to moling on very dense clays.

5. Get proper plans drawn up, taking the opportunity to modify the field layout if appropriate and to tie in with existing schemes where these remain effective.

6. Notify the contractor in plenty of time as to when you want the work done. Programme the work over several seasons where drainage demand involves a large area and high capital outlay in order to suit cash flow and to minimise interference with cropping programmes. Consider draining through a standing cereal crop once land is dry enough in spring and before stem extension is too far advanced. These measures spread the working season for drainage contractors, increase their confidence to invest in better equipment and enable more work to be done when soils are less liable to be damaged.

7. Ensure regular maintenance of ditches, outfalls and renewal of moling and/or subsoiling as conditions may dictate (usually within five years for moling and often three for subsoiling). See Plate 2.3.

PLATE 2.2   Plastic-pipe underdrainage with permeable backfill being applied. Narrow trencher drainage machine with laser-plane grading. (Courtesy of D.W. Clark Ltd)

IRRIGATION

Soil moisture supplies in late April and May are vital in laying adequate foundations for a good harvest. Later on, moisture availability is often the limiting factor in securing adequate grain filling and thus both TGW and hectolitre weight. When considered alongside the many sophisticated inputs cereal growers are prepared to use now, the supply of extra water does seem an obvious treatment to contemplate!

Yield responses vary hugely according to conditions and species; 10—15 per cent can be expected for dry areas. In real droughts such areas may show doubling to quadrupling of yields with irrigation. Best responses to irrigation of cereals in both trial work and commercial practice are associated with:

- Sandy soils (e.g. Gleadthorpe, Notts. EHF trials).

- Application during late stem extension, especially just before ear emergence.
- Ensuring that SMD (soil moisture deficit) does not drop below 25 to 50 mm. Normally total water use of up to 50 mm is strategic except in a real drought; 25 mm per dose is sensible.
- N supplies being correspondingly good.
- Late disease protection for the consequent susceptibility to colonisation by ear diseases after irrigation.
- Use of straw shortening and/or stiffening treatments.

Mobile irrigators have made irrigation more practically feasible for cereals, though lodging is highly likely in spite of growth regulator use. Irrigation facilities are normally only justified for more respon-

TABLE 2.1 Example of soil series and drainage design for cereals

Series	Topsoil type	Lateral pipe spacing (m)	Pipe depth (m)	Backfill need	Secondary treatment (M = moling S = subsoiling)
Adventurers	Fen peat	15	> 1	No, but pipes often need filter-wrapping	No
Denchworth	Calcareous clayey to very fine loamy (poorly permeable)	20–30	0.75–0.90	Yes	M (sometimes plus S)
Hodnet	Reddish fine to coarse loamy (subsoil slowly permeable)	10–20	0.75–0.90	Preferable	S (and for pans which often arise)
Whimple	Reddish fine, often silty loam over clay	20–30	0.75–0.90	Yes	S

Source: Soil Survey of England and Wales data.

sive crops like vining peas, fruit, potatoes, vegetables and grassland.

Management and capital costs take some justifying for most systems at present, but the possibility should be kept under review. Light land rotational cereal growers in the Eastern counties are much more likely to have irrigation facilities and to justify the extra

PLATE 2.3 *Failure to maintain secondary treatment — mole channels in this case — revealed by aerial visibility of pipe-drain lines.* (Courtesy of D.W. Clark Ltd)

capacity which will allow cereal irrigation when necessary.

CULTIVATIONS

Cultivations include all soil tillage processes carried out before, during and after the growing of a crop which can influence its performance: they can be considered longer term (directed to the *soil*, e.g. subsoiling, subsoil mixing) and shorter term (directed to the *crop*, i.e. its tilth requirements).

The variable factors which determine the choice of an appropriate cultivation system for cereals include its objectives, cereal species, soil types, sites, seasonality, weed problems, power sources and equipment available.

The possible *objectives* of a cultivation treatment for a cereal crop may be:

- *Soil structural adjustment*, i.e. aggregate size control.
  Seedbed preparation — to ensure maximum contact between seed and soil for moisture imbibition and rootlet development and to provide ample consolidation to ensure good control over drilling depth.
  Rootbed preparation — adequate treatment to depth for the root system to establish unhindered.

- *Soil structural improvement*, i.e. air/water balance optimised and

Removal of caps and crusts
Pan busting
Increasing permeability generally (decreasing bulk density by aerating).

- *Soil conservation*
Erosion control by contour cultivations and use of bunds and benches where appropriate overseas. Erosion should be considered as a potential British problem on some land.
Water conservation often goes with good erosion control, e.g. tie-ridges overseas; overcultivating loses moisture, which is critical on light soils in spring.

- *Incorporation of fertilisers*
BOM (bulky organic manures) need ploughing in. Certain inorganics, e.g. urea, ammonia and lime; placement of phosphates, especially for maize.

- *Disposal of crop residues*
Disposal is important if residues are diseased, but often it is done for convenience and appearance (complete burial unmixed is often undesirable). Straw incorporation is now a topical issue. Chopping on the combine and then ploughing is currently the most used method and usually avoids yield depression.

- *Weed control*
Creation of a false seedbed − that is, soil disturbance leading to annuals germinating.
Burial of existing growth.
Repeated fragmentation to starve perennials after stimulating growth (especially couch, but glyphosate is now used, often pre-harvest).
Removal of couch and ground elder for burning.

- *Pest control*
Exposure of grubs for birds.
Consolidation to suppress wireworms, etc., was advocated in the UK during World War II!
Avoidance of clods to reduce shelter for slugs.

Indeed, a passionate concern about soil erosion in the tropical world needs to be applied with increasing measure to the temperate zone. Erosion (loss of topsoil) occurs by the agencies of both water and wind: soil conservation includes all means of prevention. In the UK, the Cranfield team of the Soil Survey and Land Resource Centre has measured annual topsoil loss rates as some 2 t/ha on clays, 15 t/ha on silts and over 45 t/ha on some sands, whilst the rate of formation of new topsoil approximates to 1 t/ha/year.

Predisposing field factors for erosion, apart from heavy rainfall and/or high winds, include the following:

- sandier soils such as found under cereals throughout the UK from north Norfolk to south Devon and Somerset, from the West Midlands to Nottinghamshire and Bedfordshire
- sloping land which really needs tree or grass cover at least in periodic belts across the slope
- large fields and exposed sites without windbreaks and wash-stops
- cultivations and tramlines up and down slopes, rather than following the contours
- fine, loose tilths
- declining organic matter status
- intensive arable cropping

Soil is indeed the farmer's basic asset, and any civilisation loses it at its peril. Long-term land care is at the heart of true husbandry. Land should be treated like a new-born baby − kept in its place, its face kept clean and its bottom dry!

Seedbed Preparation

Art and science both contribute here! It is sound discipline to check from depths upwards (wheat roots to 2 m), i.e. drainage, subsoil management and then topsoil condition.

The seedbed needs to be *clean, rich, moist, fine enough* and *deep enough*. The traditional method of achieving a seedbed is by progressive trimming: e.g. plough, disc or rotavate, harrow.

Appearance can be deceptive, and zeal to make it clean may

- dilute nutrients and organic material from surface layers
- deplete moisture
- damage structure, causing compaction and/or panning because so many treatments were gone through

The choice of seedbed preparation method depends on:

- soil type
- tilth needed for a particular cereal
- time available: shorter time may necessitate larger equipment pulled by a more powerful tractor. There is a trend towards 'multiple' implements to give fewer passes, e.g. two passes − plough plus power harrow with a bridge-link to drill with a light tine bar behind for silts on same day. Why

PLATE 2.4   *Close-up of 'paraplow' unit and soil disturbance effect.* (Courtesy of Howard Rotavator Co.)

PLATE 2.5   *Surface cracking effect of Howard 'paraplow' through stubble.*

PLATE 2.6 Subsoil improvement to 35 cm depth using Howard 'paraplow', after stubble burning.

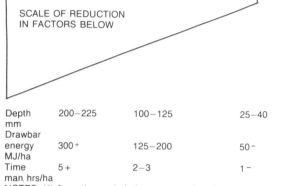

	TRADITIONAL e.g. Plough, disc, tyne × 2, harrow, roll, DRILL	REDUCED (MINIMAL) e.g. Shallow plough, press or tyne, rotary, etc. DRILL	DIRECT DRILLING Spray only DRILL
Depth mm	200–225	100–125	25–40
Drawbar energy MJ/ha	300+	125–200	50–
Time man. hrs/ha	5+	2–3	1–

NOTES: (1) A continuum is being compared and contrasted: various data used.
(2) Soil types/seasons/site conditions all vary.
(3) Periodic deep soil treatments must enter long-term comparisons.

FIGURE 2.8 Cereal cultivation systems contrasted.

not the same for other soils except clays, on which plough, disc (set straight) followed by Vaderstadt roll may well be the best system.

Traditionally the procedure was to treat topsoil (invert and burst), mix and then firm and make fine enough.

General rules of thumb

- Deep for maize
- Fine for spring corn
- Cloddy (but firm) for autumn corn; needs to be finer for residual herbicides to work
- Level for precision seeders
- Fine, firm, flat for very small-seeded minor species

Reduced cultivations and direct drilling have been popularised (see later text, Figure 2.8 and the excellent book by Dr Harry Allen, 1981). There is a discernible trend towards more subsoiling (Plates 2.4–2.6) and/or rotational ploughing, i.e. part of the farm each year, often one-third to one-fifth of the cereal area (Plates 2.7–2.11). However, pure direct drilling has produced problems of carbon and weed control (see later).

Choice of Cereal Cultivation System

Plough if drainage is poor, surface structure is damaged, crop trash is excessive, cereals are to be drilled after October

Shallow cultivate if land needs levelling, false seedbed needs to be created to encourage volunteer cereals to grow, burnt ash needs statutory scratch to clear it

Direct drill if soil type is suitable (notably calcareous and clay loams), soil structure is reasonably good, a good burn is obtained of straw and stubble, site is more or less free from perennial weeds

The purposes of cultivations after drilling are:

1. To control weeds — by hoe (e.g. inter-row for maize and sorghum) and/or spray
2. To keep surface loose—aids infiltration of water, breaks cap for aeration, reduces run-off damage, 'dust mulch' restricts evaporation
3. To firm soil around plants, e.g. after heaving over-winter, in order to reduce lodging later
4. To level the surface and push in stones (to protect harvesting machines later)
5. To earth up, e.g. ridged maize and sorghum
6. To thin crops

TABLE 2.2 Some alternative autumn cultivation systems for cereals on a range of soils

	Lighter land	Loams	Loams	Heavy land	Calcareous/ heavy loams	Silts/silty clay loams
	Spring tine	Spring tine	Spring tine	Plough	(Pre-direct -drilling)	Spring tine
	Plough	(? Leave)	(? Leave)	Disc	Spring tine	(Leave)
	Roll	Plough	Cultivate	(Leave)	(Leave)	Plough
	(Leave)	Power-harrow	Cultivate	Roll	Spray	Power-harrow/ drill (bridge-link)
	Spring tine	(Leave)	Spring tine			
	Spring tine	Roll				
		(Leave)				
		Spring tine				
Relative cost (100 ≡ £20 in 1987–88)	225	275	150	235	100	210

Notes:

(1) Timing of treatments is a critical art to minimise risks/maximise effectiveness of tilth and to protect against weed seeds/shed corn. Timeliness includes autumn ploughing to allow for frost tilth ahead of spring-sown crops.

(2) The initial spring tine treatment may be a statutory scratch after straw/stubble burning.

(3) Subsoiling may be additional at similar cost to ploughing on from 10–60 per cent of fields in any one season (depending on the land type), especially if no ploughing is done.

(4) Reasons for ploughing including grass weeds (especially barren brome and blackgrass); trash after poor burns; stubble of rape or of cereals after baling; compaction, rutting or panning impeding drainage; for wheat to be planted from November onwards in the UK.

(5) Moisture loss *must* be minimised through avoidance of recreational/protracted tillage.

(6) Many growers will also roll after the drill to improve seed/soil contact.

(7) Costs must not be cut at the expense of seedbed quality but the power harrow is complex as well as costly though it is effective on clods.

(8) Circumstances alter cases, and ploughing, for instance, may be the last thing to attempt on some heavy soils in some seasons!

(Note: the trend is away from rolling/harrowing after emergence of most cereals, though this was once the prevalent practice. Start with a good tilth and promote rapid crop cover by timely sowing whenever possible to obviate the need for after-cultivations.)

Effects of cultivations which may be adverse include the following:

- Powdering if too dry
- Loss of pores within aggregates and smearing if too wet
- Loss of moisture — overaeration
- General compaction by heavy implements/local pans (Figure 2.9)
- Loss of uniformity — physically (stones and lumps up) and biochemically (infertile subsoil up)
- Dilution of organic matter — warm, moist conditions accelerate breakdown and deep cultivations will mix limited OM to depth
- Decline of earthworms — arable soils under the plough typically have one-third of the populations of equivalent soils under direct drilling or under grassland

PLATE 2.7   Cross-ploughing of scuffled stubble using Eberhardt split-body reversible plough to give cultivator/mixing effect.

PLATE 2.8   Stubble to seed bed on light loam using six-furrow Kverneland variable furrow width, semi-mounted plough: only needs levelling and consolidation before drilling.

PLATE 2.9 *Ploughing out stubble to form a near-fit seed bed on loam: push/pull ploughing using Ransomes three-furrow reversibles.* (Courtesy of Ransomes, Sims and Jefferies Ltd)

PLATE 2.10 *Consolidation and levelling of ploughed land by trailed soil-packer/furrow-press.*

PLATE 2.11   Clay loam requires more cultivation once ploughed: it is liable to greater risks from extreme seasons, wet or dry. (Courtesy of Colchester Tillage Ltd)

Recording the Cultivation Systems for Cereals: Checkpoints

1. Background information

- Obtain soil map/zones of variation (perhaps by aerial photographs); remote sensing techniques promise to provide more widely affordable information
- Produce a cultivation equipment inventory
- Summarise current cultivations policy

2. Practice to date

- Identify underdrainage/land improvement schemes in place
- Outline cultivations sequences (ref. soil types) in current use for the different cereals

3. Results achieved

- Record tilth descriptions achieved (good and bad; use photographs to educate all staff involved)
- Record rates of work (typical/actual) — see Table 2.3

Improvements

- Isolate problems to treat in priority order
- Specify current issues for staff opinion/possible change

TABLE 2.3   Cultivations: typical work rates

Operation	Typical work rate (ha/hr)
Plough (Rev 3 furrow mounted 55 KW Tractor)	0.45
Plough (Rev 6 furrow semi-mounted 110 KW Tractor)	0.8
Heavy disc (2.5m; 10 cm deep)	0.9
Spring tine (3.5m; 10 cm deep)	1.9
Harrows (4.5m)	2.3
Rotavator (1.5m; 15 cm deep)	
On grass, etc.	0.4
On plough	0.6
Rolls	
Heavy (2.4m)	0.7
Light, 3-Gang (7.3m)	2.6

Source: Scottish Agricultural Colleges, *Farm Management Handbook.*

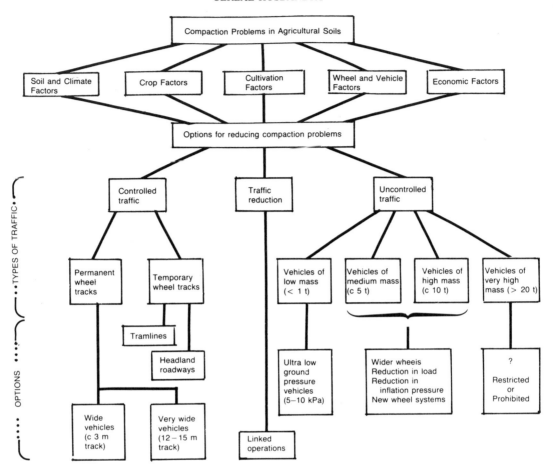

FIGURE 2.9   A simplified diagrammatic representation of some of the options available for reducing compaction in relation to the factors affecting the cultivation system as a whole. (From Soane, 1982)

Chapter 3

CEREAL CHARACTERISTICS

The various parts of the different cereal species must be identified, their functioning (physiology) considered and their growth stages through the life cycle described if one is to understand the basis of yield and quality determination and thus employ sound crop husbandry.

Space does not permit exhaustive diagrams and photographs of all cereals, but all key parts are covered in this chapter (see Plates 3.1—3.5 and Figures 3.1—3.3):

- Grains
- The vegetative plant
- Heads

A hand lens is vital equipment when examining cereals, and a number of enthusiastic UK cereal farmers either carry a pocket microscope or, more likely, have one in the farm office. *Attention to detail* is the watchword of good husbandry, which starts with close observation of the crop.

THE CEREAL PLANT AT WORK

It is essential to have a basic grasp of how a living thing functions (of its physiology) in order to manage it effectively. The basic process upon which all green plant production, including cereals, rests is photosynthesis, represented simply by:

$$6CO_2 \; + \; 6H_2O \xrightarrow[\text{Chlorophyll}]{\text{Light}} C_6H_{12}O_6 \; + \; 6O_2$$

Carbon dioxide   Water   Chlorophyll   Glucose   Oxygen

This process not only provides the majority of the food consumed by man and livestock, since carbon dioxide derivatives provide over 90 per cent of cereal yield, but also it replenishes oxygen depleted by respiration.

There is no evidence that maximum photosynthesis per unit area of crop has increased at all since cereals were domesticated (Evans, 1975). Thus total *biological yield* or biomass (all plant parts) has advanced little if any in modern cereal varieties. What has improved is *economic yield* through:

- Breeding for increased grain:straw ratios and disease resistance. Grain:straw can now be as favourable as 2:1 and is often 1:1; only recently it was 1:2 and it still remains as poor as 1:5 in some traditional varieties of sorghum and millet in Africa and Asia. Harvest index (percentage of above-ground dry matter yield which is grain

  measured as $\dfrac{\text{DM in grain}}{\text{Total DM}} \times 100$) varies in

practice between 30 and 55 for current wheat and barley cultivars. Dwarf varieties based on the Rht2 (reduced height) genes have contributed significantly to the so-called 'green revolution' in both wheat and rice yields.

- Better crop management (the theme of this book).

There is considerable variation in net photosynthesis in the field owing to:

1. *Low and variable efficiency of light interception* Commonly only 1 and up to 3 per cent of total incident solar radiation is actually trapped by the crop, though this could be trebled in the field by timely development of an optimum-density leaf canopy and by selecting the most efficient cultivars.

(text continued on page 41)

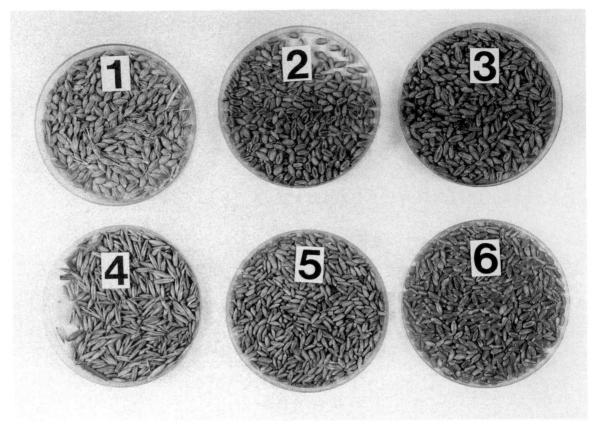

PLATE 3.1   Cereal grains: 1=barley; 2=wheat; 3=triticale; 4=oats; 5=rye; 6=durum.

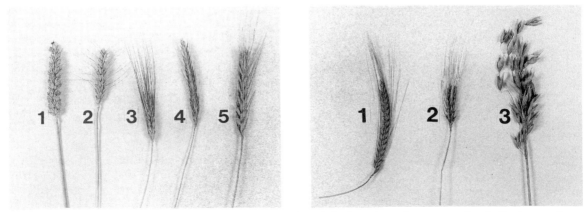

PLATE 3.2   Cereal heads: (Left) 1=common wheat; 2=bearded wheat (awned common wheat); 3=durum wheat; 4=rye; 5=triticale. (Right) 1=two-row barley; 2=six-row barley; 3=oats.

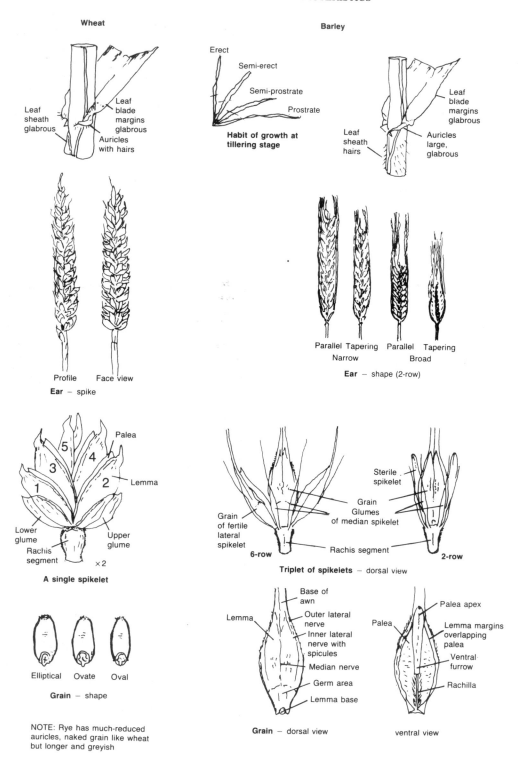

Wheat

Leaf sheath glabrous

Leaf blade margins glabrous

Auricles with hairs

Barley

Erect

Semi-erect

Semi-prostrate

Prostrate

Habit of growth at tillering stage

Leaf blade margins glabrous

Leaf sheath hairs

Auricles large, glabrous

Profile   Face view

Ear − spike

Parallel Narrow   Tapering   Parallel Broad   Tapering

Ear − shape (2-row)

Palea

5   4

3

1   2

Lemma

Lower glume

Rachis segment   ×2

Upper glume

A single spikelet

Grain of fertile lateral spikelet

Sterile spikelet

Grain
Glumes
of median spikelet

Rachis segment

6-row   2-row

Triplet of spikelets − dorsal view

Elliptical   Ovate   Oval

Grain − shape

NOTE: Rye has much-reduced auricles, naked grain like wheat but longer and greyish

Lemma

Base of awn

Outer lateral nerve

Inner lateral nerve with spicules

Median nerve

Germ area

Lemma base

Palea

Palea apex

Lemma margins overlapping palea

Ventral furrow

Rachilla

Grain − dorsal view   ventral view

FIGURE 3.1   Characteristics of wheat and barley plants. (Courtesy of NIAB)

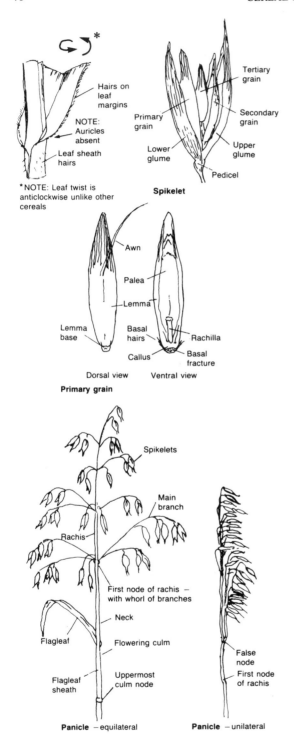

Hairs on leaf margins

NOTE: Auricles absent

Leaf sheath hairs

*NOTE: Leaf twist is anticlockwise unlike other cereals

Tertiary grain

Primary grain

Secondary grain

Lower glume

Upper glume

Pedicel

Spikelet

Awn

Palea

Lemma

Lemma base

Basal hairs

Rachilla

Callus

Basal fracture

Dorsal view   Ventral view

Primary grain

Spikelets

Main branch

Rachis

First node of rachis — with whorl of branches

Neck

Flagleaf

Flowering culm

Flagleaf sheath

Uppermost culm node

False node

First node of rachis

Panicle – equilateral   **Panicle** – unilateral

FIGURE 3.2   Characteristics of oats. (Courtesy of NIAB)

PLATE 3.3   *Spring barley cv Triumph.* (Courtesy of C.T. Miles, DSIR, New Zealand)

PLATE 3.4   *Triticale cv Lasko.* (Courtesy of Semundo Ltd)

PLATE 3.5   *Harvesting forage maize. A 2m Krone precision chopper taking three rows simultaneously and capable of up to 90t/hour.*

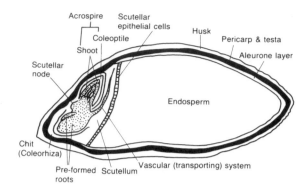

FIGURE 3.3   Longitudinal section of an ungerminated barley grain. (From IOB)

2. *Variation in photosynthetic rate*   The biggest distinction identified relates to the speed of the biochemical pathway between the one-carbon (C1) molecule of carbon dioxide and the six-carbon (C6) glucose; those which rapidly act to produce a C4 molecule are maize, sorghum and millet, whilst the less efficient C3 cereals — wheat, barley, oats, rye and rice — use a slower biochemical pathway.

3. *Different environmental responses*   C4 cereals respond up to double the light intensity of C3 cereals, tolerate higher temperatures and use water twice as efficiently (transpiration ratios — kilograms of water used per kilogram of dry matter yield — are 300—350 for C4 by contrast with 500—700 for C3). In addition, C3 cereals are actually inhibited by normal atmospheric oxygen content at 21 per cent.

Considering the cereal crop as analogous to a business, the fundamental question is, 'By how much does the weight gain by photosynthesis exceed all the processes of loss, notably respiration?' See Figures 3.4 and 3.5.

A business analysis of the cereal crop might include the following:

1. *Capital value = total crop dry matter*
2. *Productive capital = leaf area index (LAI)*, i.e. area of leaf per unit area of ground (around 7.5 seems ideal for wheat on good land)
3. *Factory production span = leaf area duration (LAD)*, i.e. days of green leaf area survival between sowing and harvest. Varieties differ significantly in leaf production characteristics, and extra nutrients, especially nitrogen, can influence leaf area and colour (Table 3.1 and Figure 3.6).

4. *Net production = crop growth rate (CGR)*, i.e. dry weight increase over time
5. *Investment programme = dry matter distribution*, i.e. to GRAIN for the next generation of the crop
6. *Efficiency measure = net assimilation rate (NAR)* = grams extra dry matter per gram of leaf dry matter, i.e. $NAR = \dfrac{CGR}{LAI}$

In the analysis of fully replicated seedling wheat trials over eight weeks at 20°C with ample moisture at three densities with and without nitrogen $\equiv$ 125 kg/ha, CGR at near-optimum density was increased by 43 per cent with nitrogen. Leaf to root ratio at this density was increased by nitrogen from just under 1:1 to almost 2:1 but dry matter percentage was depressed from a mean of 15.7 to 13.5 per cent and NAR was cut to only 56.5 per cent of the efficiency without nitrogen. Nevertheless, nitrogen application improved yield. These data (Wibberley, 1974, unpublished) illustrate some of the complex responses of a cereal crop.

Maize and other C4 cereals have a photosynthetic rate some 55 per cent greater than wheat (C3), double the translocation rate (movement of products of photosynthesis to grains) and some 60 per cent greater CGR than wheat — exceeding the photosynthetic rate differential because C4 cereals do not suffer photorespiration (loss of carbohydrate by respiration in daylight) as do C3 plants, and thus NAR is higher for C4 cereals. This greater efficiency of maize explains the agronomist's desire to have varieties suitable for grain production in the UK.

TABLE 3.1   Grain yield and leaf-area indices in winter wheat cultivars at two fertility levels grown at Cambridge in 1977—78

Cultivar	Low-fertility yield		High fertility yield	
	kg/ha	LAI	kg/ha	LAI
Little Joss	3300	3.6	5220	6.9
Holdfast	2910	2.7	4960	6.6
Cappelle-Desprez	3740	3.1	5860	7.4
Maris Widgeon	3570	3.1	5680	8.1
Maris Huntsman	4040	3.1	6540	8.2
Hobbit	4630	3.0	7300	6.3
Mardler	4070	3.3	6210	7.6
Armada	4240	3.1	6860	7.6
Mean	3812	3.1	6079	7.4

Source: Data from Austin *et al.*, 1980.

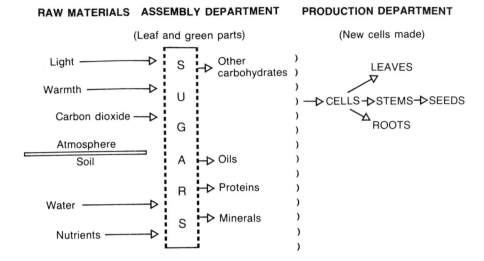

RAW MATERIALS   ASSEMBLY DEPARTMENT   PRODUCTION DEPARTMENT

(Leaf and green parts)   (New cells made)

FIGURE 3.4   The cereal crop as a business.

W	=	P_E	−	R_D	−	R_N	+	S
Final grain weight	=	carbohydrate from ear photosynthesis	−	ear respiration by day	−	ear respiration by night	+	shoot assimilation (including flag leaf and sheath and upper stem)
Wheat	100	=	24	−	28	−	11	+	115
Barley	100	=	79	−	24	−	10	+	55

NOTE: These data emphasise the importance of ear photosynthesis for barley and of the flag leaf for wheat.

FIGURE 3.5   Balance sheet. (After Thorne, 1965)

N treatment to date	Relative green leaf area	Mean plant height (mm)
Nil	▨	185
+ November	▨	195
+ February	▨	200
+ November + February	▨	215

FIGURE 3.6   Winter barley: effect of early N on relative green leaf area and plant height in April. (From Jenkinson and Wibberley, 1986)

CEREAL GROWTH AND DEVELOPMENT

The cereal plant is described as determinate; that is, it has a vegetative phase of root and leaf production giving way to a reproductive phase which ends with production of ripe grain. It uses its accumulating dry matter (growth) to enable it to pass through the various stages of its life cycle (development). These stages are accompanied by recognisable external and internal changes in the plant, though the two do not necessarily correspond in different cultivars. The temperate cereals develop flowers in response to increasing day length (that is, they are photoperiodic); what actually triggers them is the duration of darkness rather than light. Tropical maize cultivars are sensitive to short days but temperate ones are bred to be less sensitive. Development in response to environment (phenology) is affected by factors such as temperature, e.g. the stress of high temperature accelerates flowering but with lower resultant yield, whilst true winter varieties of cereal need to experience a prior cold period to trigger ultimate flowering and grain formation (this is called the vernalisation requirement).

The Rothamsted Experimental Station is using a model to represent these aspects of growth and development (Figure 3.7). However, growth does not proceed in exact step with calendar dates from season to season and in any case varies with sowing date and soil conditions. An appropriate frame of reference was needed. A descriptive scale of growth for cereals was first produced by Feekes and Large in 1954 for the purpose of defining disease severity in relation to recognisable stages of the plant. This was adopted in British farming by the early 1970s when systemic fungicides, hormone weedkillers and plant growth regulators were becoming more widely used and effects varied with stage of crop at application time. However, the scale was awkward and imprecise, running from 1–11.4. In 1974, Zadoks *et al.* introduced a more precise decimal key to 100 growth stages — GS 0–99 — for the purposes of analysing both weed competition and responses to weed control at different stages, and it is internationally recognised for all cereals and grasses (Figures 3.8–3.10). This hundred point decimal scale divides neatly into ten phases of the life cycle (with secondary numbers to represent relevant recognisable stages within each) as follows.

0 = Germination

These ten initial stages enable the researchers on seed physiology to describe precise points in water imbibition, hormone and enzyme action and activation of the germ (embryo plant) within the seed (Figure 3.11). Germination rate is environmentally dependent, notably on temperature, oxygen and moisture supply. Evidence that germination has taken place is the emergence at GS 05 of the young root (radicle) closely followed at GS 07 by the coleoptile (thin sheath which protects the young shoot or plumule and which disintegrates by about the four-leaf stage).

1 = Seedling Growth

These stages describe the early development of the main shoot as it produces its leaves and moves from dependence on 'deposit account' reserves of food from the seed to 'current account' production from its own leaves.

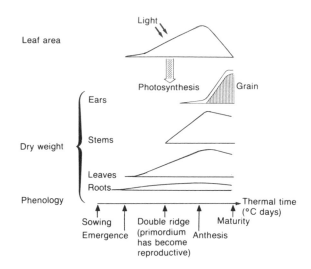

FIGURE 3.7 Overview of changes in the cereal plant over time — as used in the modelling of crops at Rothamsted Experimental Station.

Winter barley initiates its ear primordium once the second leaf has unfolded and completes laying down all its grain sites by the six-leaf stage. Winter wheat begins the same process only at the four-leaf stage and completes it by the second node stage (GS 32). In March-sown spring barley in England, rate of leaf appearance is steady at about one every five days, each successive one being larger than its predecessor (until developing ear competition takes effect later on). Cereal leaves grow from their bases so the tips are older, making the plant less susceptible to canopy surface damage.

The fibrous root system is developing critically during this vulnerable, juvenile phase (Figure 3.12). The code is logical, e.g. 13 = seedling with main shoot and three leaves unfolded (i.e. with ligule visible at the base of the leaf blade), 14 = with four leaves, etc. Younger leaves emerge like a telescope from within the sheath of the next older one below. The value of the scale is now becoming more apparent for the vigilant farmer who monitors the rate of progress of his different crops. If seed is too deeply sown, for instance, development rate will be slower.

2 = Tillering

Tillering is the production of extra side shoots. These arise from buds in the axils where leaf sheaths join

(text continued on page 46)

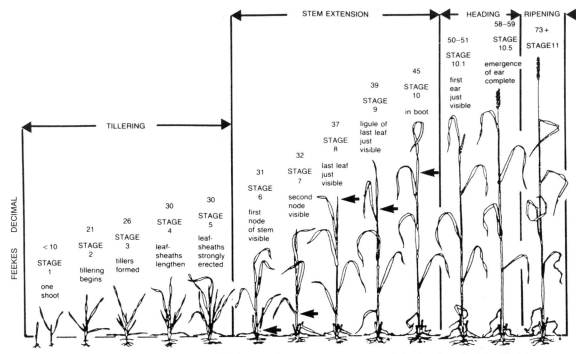

FIGURE 3.8   Growth stages in cereals: Feekes scale (after Large, E.C.) and decimal scale (after Zadoks, J.C., Chang, T.T. and Konzak, C.F.)

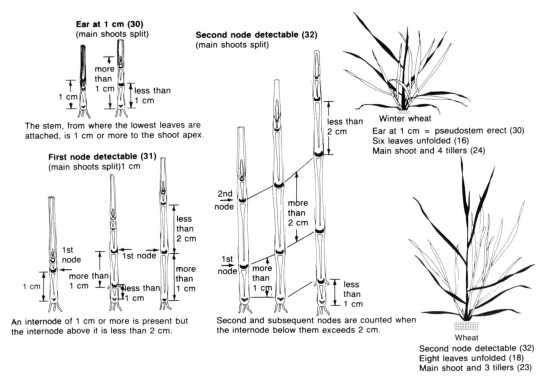

FIGURE 3.9   Stem elongation and internal progress. (From D.R. Tottman, 1987; drawings by Hilary Broad)

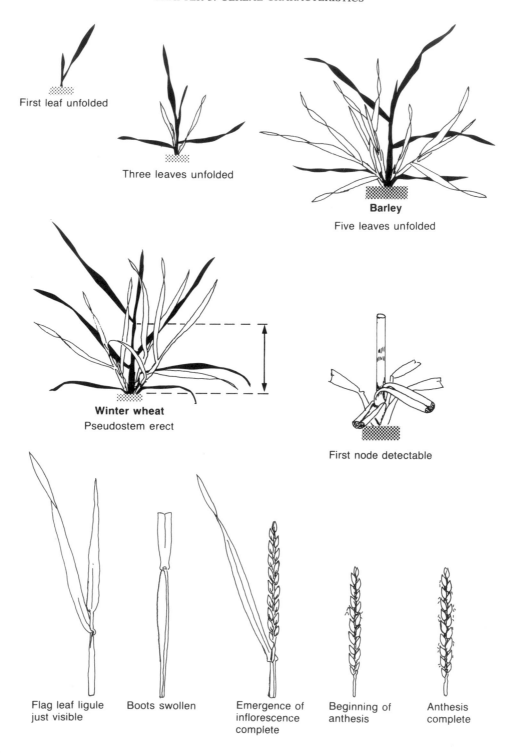

First leaf unfolded

Three leaves unfolded

Barley
Five leaves unfolded

Winter wheat
Pseudostem erect

First node detectable

Flag leaf ligule
just visible

Boots swollen

Emergence of
inflorescence
complete

Beginning of
anthesis

Anthesis
complete

FIGURE 3.10 Summary of some important stages of cereal growth described in accordance with the decimal code. (From Tottman and Makepeace, 1979)

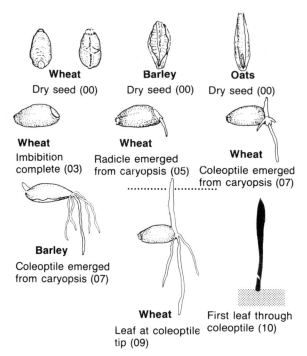

FIGURE 3.11   Germination and seedling growth.
(From D.R. Tottman, 1987)

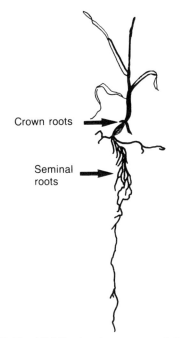

FIGURE 3.12   Mid-September sown winter barley
seedling (drawn to half natural size) shown at the
time when later crops are only just sown!

the stem at the base of the cereal plant at ground level (the crown). This process is also known as stooling or suckering in some countries and is the basis for sometimes ratooning rice and sorghum crops in the wet tropics (i.e. cutting successive harvests from only one initial sowing).

Some cereal breeders have thought tillering undesirable and, indeed, most maize cultivars scarcely produce tillers at all, whereas original types gave ten or twelve. A uniculm (main shoot only) variety of barley exists but has not proved commercially successful since it lacks standing power and the capacity to compensate for poor conditions that its tillering cousins possess and that results in higher, more reliable yields for them.

The amount of tillering depends on cultivar and growing conditions such as timing of nitrogen fertiliser (Figures 3.13 and 3.14). Given no competition from surrounding plants, a single barley seed, of cv Igri for instance, can produce thirty or more tillers, but in a crop with 325 plants per square metre, it may produce only five.

The proportion of tillers which actually go on to develop their own associated root systems and to bear ears is critical. Generally, tillering capacity is greater in winter rather than spring varieties, in two-row rather than six-row barleys and in dwarf rather than tall wheats. Leaf number per tiller is more or less constant for any particular variety so controlling tiller density to achieve an optimum ear population is the critical factor. Excessive tillering will not only prove competitive (even parasitic!) against ear-bearing tillers but also creates a dense crop in which a favourable micro-climate exists for many diseases. On the other hand, inadequate ear density limits yield, so a compromise is sought as so often in husbandry decisions. Wheat compensates more effectively than barley for low ear density. In any one variety, few tillers die if fast dry matter accumulation per plant is sustained from the time of maximum tiller production up to anthesis. Early nitrogen application during the tillering phase encourages greater tiller numbers to form, and at maximum tiller stage, nitrogen and other inputs to maintain growth rate will encourage tiller survival.

Varieties which have limited tillering ability very often have strong apical dominance (overpowering influence of the main shoot exerted through its own internal hormone concentration). This apical dominance can be lost either if the main shoot is damaged by pest attack or affected by early lodging (stem collapse) or if plant growth regulators (PGRs) suppress it deliberately or by accident. A late phase of secondary tillering in such circumstances can greatly inconvenience harvesting since there is usually a wide differential in ripening between oldest and youngest ear-bearing tillers; however, the process does compensate somewhat in yield terms.

A plant can now be described as at GS 14, 22 (main shoot with four leaves and two tillers), i.e. given a double, full description (Plate 3.6). The description GS 24 is logical, i.e. main shoot plus four tillers; GS 25 plus five tillers, etc. Until the end of tillering, the ear primordium (growing point) stays below soil surface level, where it is protected.

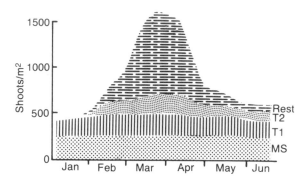

FIGURE 3.13   Number of different categories of tillers: winter wheat sown in October. MS = main shoot; T1 = first tiller, etc.

3 = Stem Elongation

The plant is now fully tillered. Tillering ceases sometime in early April for early October-sown winter wheat in southern Britain. This is triggered by the internally developing ears, which now begin to compete in earnest for their share of the plant's resources (Figure 3.15).

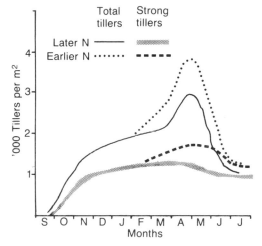

FIGURE 3.14   Effect of N timing on tillering pattern of winter barley. (From Jenkinson and Wibberley, 1986)

PLATE 3.6   *Wheat seedling at Growth Stage 14,22 (main shoot and two tillers). 1=seminal roots; 2=epicotyl; 3=crown or nodal roots.*

First the plant assumes a more upright posture loosely known as pseudo-stem erect stage or more precisely from the Weed Research Organisation's work as 5 cm length of main shoot leaf sheath in winter wheat, about 7 cm in barley. Then the first node (joint) is detectable (GS 31), quite soon followed by the second (GS 32) and so on to GS 35 usually. Nodes are 'roundabouts in the traffic-flow system of the plant', as well as possessing the capacity to help re-erect early-lodged crops as long as the node is green and active. Stems, apart from those of maize and sorghum, are hollow except at the nodes.

Plants must be dissected to examine ear development because it can vary some tenfold in size at this

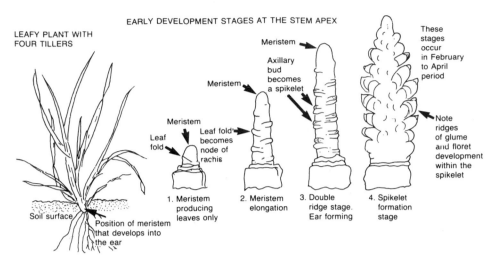

FIGURE 3.15   Simplified diagrams of an ear forming from the tiller apex of winter wheat (after D.M. Barling). The apical meristem changes from leaf production to ear formation following cold treatment (vernalisation) and increasing day length (photo period).

stage even though varieties appear at the same GS externally (Figure 3.16). PGR applications are used in an attempt to restrict excessive stem extension and so have more photosynthates for the grain as well as enabling the stem cells to remain more fortified, thus providing a stronger stem against any threats to lodge (knock over) the crop. Hormone weedkillers applied after GS 30−31 can damage ears and induce shrivelled grain. Leaf area of the plants increases greatly from this time (in the winter wheat crop cited above

to a June peak target of around LAI 7.5). Ear development is proceeding and grain numbers retained per ear are determined (Plates 3.7 and 3.8). The crop normally grows very quickly during this phase.

4 = Booting

GS 45 marks the stage where the developing ear is swelling visibly inside the leaf sheath of the flag (top)

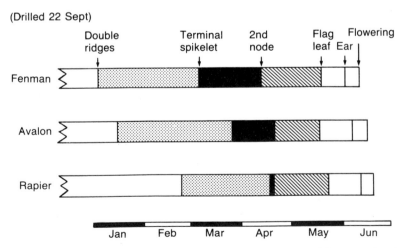

FIGURE 3.16   Wheat: timing of development stages 1984 − varietal differences. (ICI/Broom's barn work)

leaf. It is important to continue to protect the crop carefully from diseases during this phase. It is a period of continuing death of superfluous tillers, and the last-formed spikelets and florets also abort to leave a number which the crop can, hopefully, sustain through to harvest. The pattern of primordia production varies considerably between species and according to sowing dates (Figure 3.17).

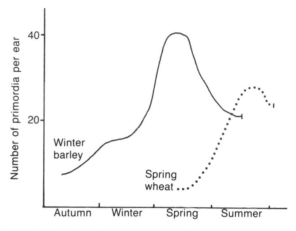

FIGURE 3.17 Contrasting development patterns of the earliest- and latest-harvested British cereals (note that varietal and sowing date differences can also be substantial.)

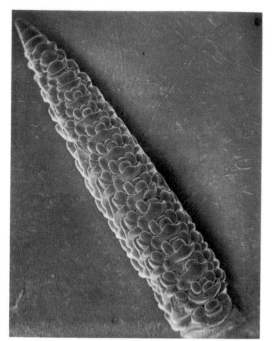

PLATE 3.7 Ear of two-row winter barley cv Igri at awn primordium stage in April. (Courtesy of RHM Lord Rank Research Centre)

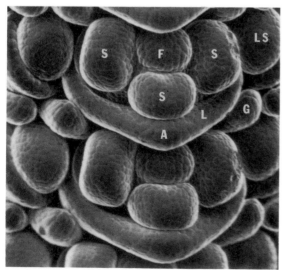

PLATE 3.8 Winter barley (cv Igri) electronmicrograph of developing ear — close-up of spikelets at awn primordium stage in April. A=awn; L=lemma; G=glume; S=stamen; F=floret; LS=lateral spikelet. (Courtesy of RHM Lord Rank Research Centre)

5 = Ear Emergence

Ears emerge in response to shortened nights. This stage marks the end of leaf expansion and the onset of leaf senescence except for the flag leaf and leaf two. Stem sugar content decreases rapidly. Ear emergence results typically in a main shoot plus two or three ear-bearing tillers in two-row barley, main shoot panicle plus one or two others in oats and main shoot plus maybe one (occasionally two or three) in wheat. Six-row barley behaves much more like wheat. The two-row ears of the most widely cultivated British barleys arise from the infertility of the side spikelet rows so that only the central spikelet row is fertile on opposite sides of the rachis (inflorescence central stalk). Two-row barley will abort between 30 and 50 per cent of its potential grain sites but retain some 95 per cent of remaining fertile florets to produce grains. Wheat similarly loses around 40 per cent of its floret initials but only produces grain from some 80 per cent of its retained florets, giving about 3.5 grains per spikelet on average at harvest. Thus, ears/m² varies more for barley than for wheat, whilst grains/ear varies more for wheat.

Everything now hinges on grain growth, the grain itself acting as a 'sink' in physiologists' language to receive carbohydrate from the various 'sources' in the plant. Stoy (1966) represented this diagrammatically (Figure 3.18). Most of the carbohydrate stored in cereal grains is produced by photosynthesis after ear emergence. The percentage of assimilates actually deposited in the grain in the various stages from ear emergence can be reckoned:

Ear ⟶ Flowering ⟶ Milk ⟶ Yellow ⟶ Ripe
emergence   (cheese)
         5%   25%   53% ripe   17%

Stem sugar content decreases rapidly during senescence though barley typically retains reserves which can be mobilised to supply up to half the ultimate grain carbohydrate, thus buffering it well in drought.

6 = Anthesis (Flowering)

Self-pollination leading to self-fertilisation is usual in barley, wheat, oats, rice and triticale. However, they can all be induced to cross for breeding purposes. Protogyny (development of the ovaries much earlier than the anthers) ensures the need for cross-pollination in pearl millet. Maize, sorghum and rye are also normally cross-pollinated though they can be selfed for breeding purposes. The lodicules swell to open the pales to allow extrusion of the anthers to release pollen onto the wind (Figure 3.19).

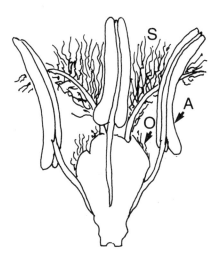

FIGURE 3.19   The flower of oat: O = ovary, S = stigma, A = anther. (From Nelson, 1946)

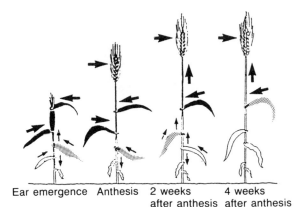

Ear emergence   Anthesis   2 weeks   4 weeks
                          after anthesis after anthesis

FIGURE 3.18   Photosynthesis and distribution of assimilates in a maturing cereal plant. Depth of colour indicates increasing rate of photosynthesis, size of arrows indicates magnitude of assimilate transport (after Stoy, 1966).

Wheat, for example, exhibits anthesis once the ear has emerged so that its base is some 5 cm clear of the flag leaf ligule. An individual ear can complete the visible process amazingly quickly, often early in the morning. Stamens dangle, anthers split, pollen is released and takes but five minutes to germinate on the feathery stigma of the 'home' ovary and about an hour to send down a pollen tube into it. This contains three nuclei, two of which fuse within the day with ovary nuclei to produce, respectively, the embryo or young plant and the endosperm or carbohydrate store of the grain.

The whole crop may take around a week to complete anthesis if it is uniform. In the north of Britain rather than the south, and in a crop of rye rather than wheat, the process takes more than twice as long to ensure maximum chance of cross-pollination; this explains the prevalence of ergot disease in rye ears exposed to risk for a longer period.

7 = Milk Development

This is an absolutely critical phase of stowing away the grain carbohydrate. Therefore it is important to monitor the potential raids of late aphids which can dramatically reduce yield (the ears will already have received a protective fungicide as necessary).

8 = Dough Development

Ample time to mobilise all possible reserves into the

grain is now critical to yield. Early death of the plant through drought or disease can be very damaging.

9 = Ripening

In the UK, grains will fill to between 25 mg for the thinnest barley and up to 65 mg for the plumpest wheat. Some millets are as low as 6 mg whilst maize may be 360 mg. In a survey of 38 dryland rice crops in Nigeria (Wibberley, 1975, unpublished) the mean grain weight was 22.9 mg in a range from 15 to 46 mg, with 85 per cent of the sample in the 20s. For sorghum see Table 3.2.

TABLE 3.2 The influence of weeding on yield and grain weight of tall Farafara (4 m) and short Kaura (2.5 m), sorghum cultivars, Plateau State, Nigeria

	Cultivar	Yield (t/ha)	Thousand grain weight (g)
Weeded	Tall Farafara	2.31	25.9
(three times)	Short Kaura	2.61	27.0
Not weeded	Tall Farafara	1.21	24.5
	Short Kaura	1.88	26.5

Source: Wibberley, 1976, unpublished.

CEREAL ROOTS

These merit a special section because shoots demand a disproportionate amount of our attention and the Zadoks scale inevitably omits root progress. Letcombe Laboratory did some excellent work on roots (Figure 3.20).

- Cereal roots grow actively from their tips, by contrast with cereal leaves which grow from their bases. They can remain active as far as half a metre behind the tip.
- Seminal or primary roots are the first 'seed' roots to emerge; three to six of them support the plant during the first month or so of life. They develop first-, second- and third-order lateral branches and penetrate to considerable depth. They may amount to a total length of 5 metres by the time a winter wheat crop is a month old, though they never occupy more than 5–10 per cent of the total root volume of a fully grown crop.
- The secondary root system develops usually one to two months after germination. These thicker roots (300–700 μ or 0.3–0.7 mm; cf. seminal roots at 220–400 μ) arise at the crown and are also called nodal or adventitious roots.
- Lateral roots develop from these to produce a fibrous system overall with abundant root hairs. They

FIGURE 3.20 Root and shoot systems of winter wheat, one, two and three months after sowing in mid October. (Ellis, F.B. & Barnes, B.T., ARC Letcombe Laboratory, unpublished)

are finer (100−200 μ) and tend to occupy a greater proportion of the total system if compacted soils restrict the larger roots. They are vulnerable to local soil nutrient deficiencies whereas the main roots survive unless the whole plant is deficient. They normally spread sideways up to 1 metre.

Maize also develops prop roots from just above the crown which help to support it.

- Roots may occupy as little as under 1 per cent of the total soil volume to 15 cm depth and no more than 5 per cent.
- Most roots are concentrated in the top 25 cm of the soil profile. In October-sown winter wheat, depth of rooting extends to 0.5 m by February, to 1 metre by the five-leaf stage/end of tillering and to as much as 2 metres at peak root development, coinciding with full ear emergence (GS 59) during June when total root length may exceed 80 metres.
- Root growth rate accelerates from 5 mm per day over winter to 25 mm per day in late spring.
- Root development is obviously correlated with seasons and soil depth (Figure 3.21 and Table 3.3). Root production is more consistent than shoot production. Variations in soil depth and therefore in rooting can make considerable visible differences to the uniformity of ripening on shallow soils such as chalk downland and limestone wolds.
- Root functions for the cereal crop include:

Anchorage and support.
Production of hormones to control growth patterns.
Production of exudates. These are organic substances which leak out of active roots continuously, attracting beneficial micro-organisms which help to protect roots from pathogens as well as assisting in crop nutrition. These exudates may account for as much as 10 per cent of the total loss of photosynthates generated by a crop.
Water absorption − drawn from as far as 100 mm away from the root surface (rhizoplane).
Nutrient absorption − closely allied with water absorption. Whilst very mobile ions like nitrate (NO_3^-) can reach a root from 100 mm away, the least mobile like phosphate ($H_2PO_4^-$) need to be within 1 mm.

- Both root formation and activity are restricted by oxygen deficiency in the soil and by extremes of pH.

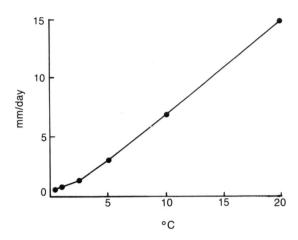

FIGURE 3.21   Effect of temperature on rate of root elongation of cereals. Mean values for wheat, barley and rye. (From Valovich and Grif, 1974)

TABLE 3.3   Relative wheat yields according to seasons and soil depths: silt loam near Reading

Soil depths (cm above gravel)	Intermediate (1974)	Seasons Dry (1975,1976)	Wet (1977,1978)
< 70	83	63	87
70−90	90	80	99
> 90	100	100	100

Source: Data of F.B. Ellis.

YIELD ANALYSIS

Yield is built up from a series of components:

Yield = plants/hectare × tillers/plant × proportion of ear-bearing tillers × grains/ear × weight/grain (mg)

Crop structure in practice can compensate to an extent for lowering of one component by raising of another (Table 3.4). In an Essex trial, four wheat varieties at 312 plants/m² had 34 grains per ear but at only 77 plants/m² they had 44 grains per ear.

The contributions of various plant parts to final grain yield have been variously estimated. Ranges for wheat and barley are shown to be wide. In maize, it is reckoned that the net contribution of the ear photosynthesis is nil or even negative, as is indeed the case for some wheat data (Table 3.5). Half the maize grain carbohydrate is reckoned to come from the top third of the crop canopy.

The factors determining grain yield include:

- Crop density
- Light interception, dependent on leaf angle, leaf area index and standing power of the crop

TABLE 3.4 Wheat: ear population, ear size, TGW and yield

Population (ears/m²)	Grains per ear	Yields in t/ha at TGW (1000 grain weights in grammes)			
		35g	40g	45g	50g
300	30	3.2	3.6	4.1	4.5
	35	3.7	4.2	4.7	5.3
	40	4.2	4.8	5.4	6.0
400	30	4.2	4.8	5.4	6.0
	35	4.9	5.6	6.3	7.0
	40	5.6	6.4	7.2	8.0
500	30	5.3	6.0	6.8	7.5
	35	6.1	7.0	7.9	8.8
	40	7.0	8.0	9.0	10.0
600	30	6.3	7.2	8.1	9.0
	35	7.4	8.4	9.5	10.5
	40	8.4	9.6	10.8	12.0

TABLE 3.5 Sources of grain carbohydrate (percentage) in wheat and barley

	Wheat	Barley
The ear itself	0—10	25—45
Flag leaf, its sheath and stem above	60—85	30—35
Lower parts of plant	10—20	20—45

TABLE 3.6 Target head populations

Cereal		Heads/m²
Winter wheat		500—600
Spring wheat		700—800
Durum wheat		650—750
Winter barley	Two-row	1000—1200
	Six-row	500—650
Spring barley	Two-row	1000—1200
Winter oats		650—750
Spring oats		700—800
Rye		700—800
Triticale		600—700

Note: In general, richer growing conditions can sustain higher densities and it must be remembered that crops can compensate by grain number and/or grain size for varied head populations. Varieties differ in density optima somewhat.

- Duration of green leaf survival after ear-emergence (LAD)
- Flag leaf size, especially in wheat
- Ear size as a 'sink' and for photosynthesis (especially in barley, including its awns)
- Temperature and duration of grain-filling period

Figure 3.22 presents an overview whilst Table 3.6 suggests possible target head populations for cereal species.

Cereal crop husbandry is about the understanding of crop behaviour, selection and timing of all operations in order to exploit this in an effective management system.

Any one of the following variables can be calculated if the others are known:

$$\text{Yield t/ha} = \frac{\text{ears/m}^2 \times \text{grains ear} \times \text{TGW (g)}}{100,000}$$

or

$$\frac{\text{grains/m}^2 \times \text{mean grain weight}}{100,000}$$

Table 3.7 gives an example of crop structure, yield and quality relationships for some winter wheat crops.

TABLE 3.7 Winter wheat cv Avalon: crop structure, yield and quality of seventeen Cotswold crops

Season	1982	1983	1984
Seeds/m²	390	440	420
Plants/m²	328	374	346
Establishment %	84.1	85.0	82.4
Ears/plant	1.29	1.37	1.42
Ears/m²	424	512	493
Grains/ear	35	42	33
Wt/ear (g)	1.56	1.39	1.55
Actual yield (t/ha)	6.63	7.10	7.66
TGW (g)	45.97	34.00	47.70
Sp wt (kg/hl)	79.79	79.50	80.83
Protein (%)	11.08	12.20	12.15

Note: TGW reached 51.4g in one sample whilst specific weight attained 82.5 kg/hl and was not recorded at below 76 kg/hl. Protein did, however, fall below 10% in one sample in 1982 but climbed to 12.7% in another in 1984. Measured Hagbergs all fell within an acceptable range from 252—482.

Source: Wibberley, 1985.

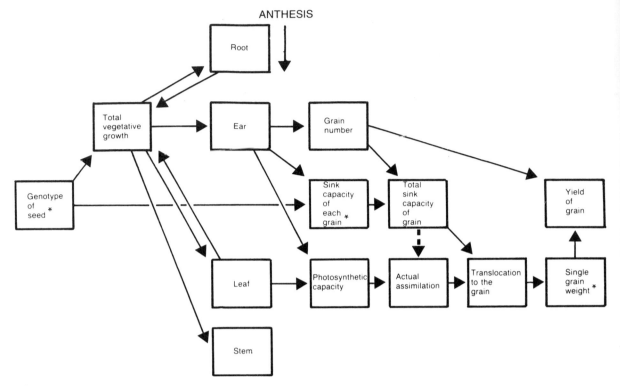

FIGURE 3.22   Relationships of the principal physiological characters which determine yield. (From Bingham, Agric. Progress **44**)

CEREAL CROP QUALITY

Apart from yield alone, farmers need to heed quality requirements of their cereal markets, especially in times of oversupply. Quality refers to suitability for end-use. Standards are liable to change either to control market supply (relaxed in times of shortages; tightened in times of surpluses) or to satisfy novel markets yet to be discovered. Many aspects of quality are subjective and locality- or market-specific. Always check before growing. Seed sample and intervention (EC bulk-bought surplus stock) standards are detailed in Appendices 3 and 4 and quality is generally dependent on appropriate choice of variety and good husbandry — especially lodging prevention, timely harvesting and careful cleaning, drying and storage.

General Physical Measurements Applicable to all Cereals

- An evaluation of *purity* (degree of freedom from all matter other than the grain concerned) — percentage by weight and named contaminants.
- *Entirety* of grains (i.e. proportion of broken and sprouted grains present).
- *Colour* of sample (still a lot of mystique about this in the malting barley trade!)
- Grain *size*. The proportion of small grains (2 mm screenings) is relevant. A high figure not only shows a poor sample but probably many shed in the field too.
- Grain *weight* (absolute weight) adjusted to a specified moisture content. (Beware: yields and other

results vary considerably between dry weight, 14 or 15 per cent moisture or 'as combined' moistures!) The weight is usually expressed as that of one thousand grains (TGW) in grams. It can be depressed significantly by high levels of late-applied nitrogen (Figure 3.23).

- Grain *density*. This was based on the Anglo-Saxon bushel measure (roughly 8 gallons or 36 litres). Pounds per bushel vary widely for the different cereal species; add to that the variation in accepted weight of wheat meant from district to district and it becomes clear that 'bushels per acre' is an unsatisfactory *quantitative* (yield) measure. Yet it is still used as such in North America! It is a very useful *qualitative* measure and has now been metricated into SI units, viz. kilograms per hecto-litre (loosely called kph but correctly abbreviated kg/hl).

Instead of using large weights and hundred-litre measures the figures are actually derived from grams occupying half-litre measures. Grain that is not well filled will not pack tightly in the measure so will weigh less. The term 'specific weight' is used now for kg/hl and in some countries 'test weight' is more usual. It is sometimes also abbreviated HLW for 'hectolitre weight'.

In practice, many markets require a minimum of 76 kg/hl for wheats, and for barleys, 64 kg/hl is now the accepted minimum target.

General Chemical Measurements Applicable to all Cereals

- Moisture content (mc) is the first consideration because by difference it indicates the true amount of the more valuable dry matter present and it affects the keeping and processing qualities of the grain. The normal level required is 14−15 per cent, and low levels often attract a premium.
- Other chemical constituents vary not only according to species (Table 1.8) but also with special market requirements within each species. Note the high protein content of the top-quality Canadian (Manitoba) wheat in contrast with the low level for paddy rice, the high fat content of oats and the high fibre and mineral content of barley.
- More detailed chemical information often relates to an assessment of protein quality, viz. the proportions of the various constituent amino acids present (Table 3.8). There are cultivar differences and breeders can select for some very specific chemical components.

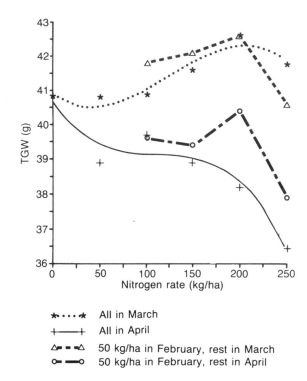

★ ‧ ‧ ‧ ★   All in March

+————+   All in April

△‑ ‑ ‑△   50 kg/ha in February, rest in March

○‑ ‑ ‑○   50 kg/ha in February, rest in April

FIGURE 3.23   Effect of N rate and timing on TGW (1000 grain weight) @ 15% mc(g.), Gerbel six-row, Eastleach, 1985.

Specific Quality Requirements of Different Cereals for Particular Uses

Wheat (*Triticum aestivum*)

The wheat grain consists of some 82 per cent endosperm, 3 per cent germ and 15 per cent 'skins'. It must be noted that a good milling wheat is one which is hard, i.e. separates clearly into a sizeable flour fraction (70−75 per cent); the remainder is bran or outer skins (some 6 per cent), weatings (inner skins, inseparable starch grains and nutrient-rich aleurone layer) and germ. For breadmaking purposes, wheat may need many additional features (see Gedye *et al.*, 1981, and Flour Milling and Baking Research Association annual leaflets for details). Grow milling varieties as first wheats after a break crop.

The most important features are briefly considered here:

Protein content   Crude protein (i.e. all nitrogenous compounds including non-proteins) is taken as percentage of N × 6.25 (on the basis that proteins contain an average of 16 per cent N). True protein

TABLE 3.8 Percentages of essential and other key amino acids present in cereal grains (means of cultivars)

	Wheat	Barley	Oats	Rye	Triticale	Maize
Arginine	0.60	0.45	0.70	0.50	0.72	0.34
Cystine	0.20	0.20	0.20	0.20	0.36	0.16
Histidine	0.20	0.30	0.20	0.30	0.30	0.23
Isoleucine	0.60	0.60	0.60	0.60	0.41	0.31
Leucine	1.00	0.90	Nil	0.80	0.74	1.24
Lysine	0.40	0.20	0.40	0.40	0.62	0.15
Methionine	0.18	0.14	0.18	0.16	0.14	0.12
Phenylalanine	0.70	0.70	0.70	0.70	0.53	0.31
Threonine	0.40	0.40	0.40	0.40	0.37	1.60
Trytophane	0.14	0.13	0.15	0.10	0.22	0.07
Tyrosine	0.50	0.40	0.40	0.30	0.39	0.33
Valine	0.60	0.70	0.70	0.70	0.59	0.43

Source: USA data courtesy of J. Arthur.

is taken as N × 5.7 to take account of the proportion of the total nitrogenous compounds that are actually proteins. High protein content is indicative of a good extensible grain gluten level which is essential to make good breadmaking dough as opposed to biscuit dough (Figure 3.24). Millers seek 11 per cent protein or more at 14 per cent mc (the rest of the EC expresses protein on a percentage of dry matter basis!). Premia are usually payable to the grower. Protein content is largely dependent on variety. Spring cultivars generally outdo winter ones (Figure 3.25). Extra nitrogen just before ear emergence can boost grain protein (see later), as can urea given just before green leaf disappears. It is vital to dry grain carefully.

Hagberg falling number This test indicates the alpha-amylase enzyme content of the grain. This enzyme breaks down starch to sugar in the grain, thus reducing the strength of the crumb structure in bread which is also sweeter and dark-crusted. The test involves heating a suspension of ground wheat for 60 seconds and then dropping in a plunger and recording the seconds it takes to reach the bottom of the mixture. Results vary from just over 60 to above 400; 220 plus is desirable. Low Hagbergs are related to particular varieties and delayed harvesting, especially in wet conditions (1985 in most of the UK! – note Table 3.9).

Dough machinability This relates to the stickiness of the dough, since those varieties which produce doughs that adhere to processing machinery are clearly uneconomic to use. The test simulates commercial dough-mixing equipment.

Zeleny score This test assesses baking quality of the flour. It measures the sedimentation rate of a flour suspended in a lactic acid solution. This further indicates high gluten level and high gluten quality. Thus it distinguishes a strong (high gluten) flour from a weak one (suitable for biscuits). Values can range from less than 10 to above 75. Above 35 is desirable and 20 is minimal. Variety chosen is crucial; so is avoidance of overheating during drying.

Appendix 3 indicates present and proposed standards for breadmaking wheat in the UK. Some five million tonnes of wheat are milled in the UK during most years, about 70 per cent of which is for bread-making (Canadian and other hard wheats are

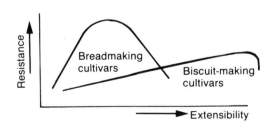

FIGURE 3.24 Extensometer test results for the doughs of typical breadmaking and biscuit-making wheat cultivars. (After Thompson & Whitehouse, 1962)

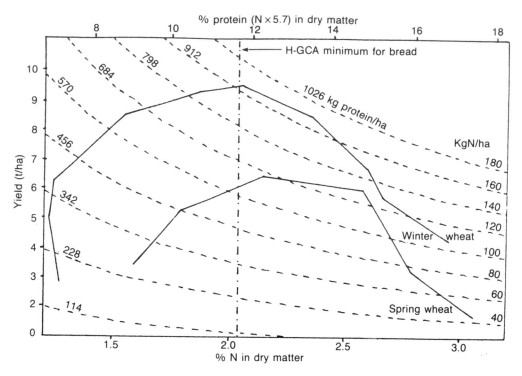

FIGURE 3.25   Yield and protein in relation to N uptake in wheat. (After Benzian & Lane, 1979).

TABLE 3.9   Grain quality survey − historical data for twelve seasons of British wheat and barley crops

	GB WHEAT (average values)											
	1976	*1977*	*1978*	*1979*	*1980*	*1981*	*1982*	*1983*	*1984*	*1985*	*1986*	*1987**
Specific weight (kg/hl)	76.2	74.7	74.8	74.4	76.2	76.3	76.4	76.2	78.2	72.9	76.8	73.4
Small grains (% < 2.0 mm)	3.2	1.1	1.1	1.8	1.5	2.2	2.0	3.0	2.9	2.5	2.1	2.2
Hagberg falling number (HFN)	297	127	203	219	252	263	313	305	273	162	222	175
% of sample > 180 HFN	94	23	64	65	82	85	93	91	79	39	72	47
Protein content (14% mc)	12.0	10.3	9.8	9.5	10.4	9.8	11.1	10.4	10.3	10.9	10.5	11.2
% of samples > 11.5% PC	64	8	11	5	15	5	36	17	19	26	18	41
1000 grain weight (grams)†	35.3	44.1	49.2	44.9	46.2	44.5	45.3	42.8	48.0	45.0	46.7	44.8
	UK BARLEY (average values)											
Specific weight (kg/hl)	63.7	66.2	65.0	67.2	64.9	66.4	67.3	68.5	68.8	65.0	67.3	64.7
Small grains (% < 2.2 mm)	27.2	7.4	7.3	5.6	8.0	9.5	5.8	7.8	10.0	5.1	4.0	5.4
Nitrogen content (% dm)	2.19	1.84	1.87	1.71	1.94	1.86	1.96	1.93	1.97	1.91	1.84	1.77
% of sample < 1.8% NC	14	40	30	60	34	45	30	34	30	38	39	60
Fibre content (% dm)	5.6	4.8	4.8	4.8	5.4	5.3	5.2	4.8	5.0	5.0	4.6	−
1000 grain weight (grams)†	28.3	34.0	40.9	42.3	40.3	40.4	44.1	40.1	41.8	42.0	46.6	41.4

* Provisional results.

† 1976−77 g on dry matter basis; 1978−1987 g at 15% mc.

Source: H-GCA.

blended to boost quality of the overall grist), 12 per cent for biscuits and the remainder for all other products.

Durum wheat (Triticum durum or pasta wheat)

- It must have 70 per cent plus of vitreous (hard, flinty, amber-translucent) grains, *not* mitadine (soft opaque) grains; early harvesting and careful, slow drying produce this vitreous grain.
- It needs Hagbergs above 220, specific weight above 78 kg/hl; protein levels above 12.5 per cent are sought.
- Premia of 50−60 per cent are obtainable over prices for milling common wheats but only if the grain meets the above specifications and is neither heat-damaged nor sprouted (yields are about 60−70 per cent of common wheats in the UK).

Barley (Hordeum sativum)

Feed barley does not have any specific criteria attached to it though high protein content is desirable. For *malting*, barley must meet certain criteria. Maltsters like pale yellow, well-filled grains in addition to the following:

- High *germination* percentage, because the malting process involves steeping grain to 45 per cent mc and sprouting it to allow enzymes to convert

starches to malt sugar. As near 100 per cent as possible is therefore needed.
- Low *nitrogen* content. High levels slow down the malting process and trouble both brewing and distilling processes and can cause cloudiness and poor keeping-quality in beer. A percentage of 1.5 N is sought but higher levels are tolerated when supplies are short.
- High *starch* content. This is the converse of high protein. It ensures a high extractable malt yield, which should be 20−24 per cent of the grain weight submitted for malting. The residual brewers' grains is a most useful feed selling wet for some 20 per cent of the price of whole grain feed barley.
- High *activity* of precisely the enzymes which impair breadwheat quality.

There is a useful export market for British malting barley, especially to West Germany. Rye is also malted and distilled to make whisky, especially in Canada and the United States.

Achievement of high malting quality rests on:

1. Choice of variety
2. Avoiding lodging, and harvesting the crop in a ripe state
3. Avoiding a late N dosage, i.e. not after GS 30−31 in any quantity; total N level and the season, however, are the critical factors (Figure 3.26)

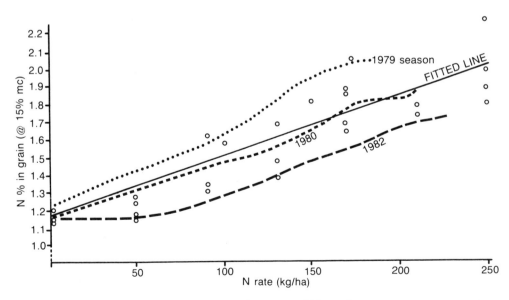

FIGURE 3.26   Winter barley: N rate and per cent N in grain over seven seasons in Eastleach, Glos. (Fitted line = average, and extreme years are shown.)

4. Careful drying (basically treat as for seed corn)

The best soils are light to medium land and the east of England is favoured over the western climate in likelihood of securing malting samples. Malting barley normally accounts for some 20−25 per cent of UK barley production. Premiums for malting over feed barley prices have been consistently higher than those for quality over feed wheat in the UK; they are around 20−25 per cent for the best samples down to 10 per cent for inferior but acceptable ones. For this reason, many UK farmers are opportunist growers of it. Table 3.9 indicates UK barley quality.

Oats (Avena sativa)

- Some 25 per cent of the UK supply is used for human consumption in porridge, muesli and other oatmeal products; thin-skinned varieties like Peniarth are preferred for oatmeal production, giving 55 per cent plus of groat, i.e. kernel, extraction rate rather than 40 per cent for high husk types. Naked spring oats enjoy premia up to 40 per cent and are concentrated sources of nutrients (but use dust masks when handling them!).
- Millers like to contract to take all of it at harvest and store at around 10 per cent mc.
- Oats once suffered from the action of lipase, an enzyme that degrades its rather high content of fats and oils, so turning it rancid; this is now controlled.
- At least 50 kg/hl is the expected specific weight.
- There is no EC intervention buying arrangement to date so prices fluctuate.
- There is sometimes a lucrative market in sales to horse owners.
- There is a risk of making or masking a wild oat problem on the farm.

Rye (Secale cereale)

- Rye can command milling wheat prices in the UK, especially under contract for Ryvita crispbread biscuits; it typically has a weak flour and is liable to sprout in the ear.
- It is second only to wheat as a bread grain in the north temperate zone. The bread stays moist and is often deliberately soured as in the black bread (smörbröd) of Scandinavia.
- Rye can be malted.
- Requirements for these markets parallel those for milling wheats (more like biscuit-making varieties) and malting barleys, respectively.
- Above all, rye must be protected from the toxic, grain-infecting fungus called ergot; and ergoted grains must be separated from any grain for consumption either by passing through indented cylinders or by flotation.

Triticale

- This product of a wheat (usually *Triticum durum*) mother crossed with a rye (*Secale cereale*) father is of local importance in highland Africa (e.g. Ethiopia, Kenya highlands) but especially in Eastern Europe, notably Poland, and North America.
- It offers not only useful livestock feed potential at a cost-saving per tonne of 3−5 per cent over other feed grain for compounding, but also milling potential − both of these markets owing to its higher content of certain essential amino acids (especially lysine) than other cereals (Table 3.8). In view of this, there is also some market interest from the health-food trade.

Information on the quality requirements for tropically produced cereals is unfortunately outside the scope of this book (but see Kent, 1983).

In processing cereals for human consumption, food manufacturers often blend cereals of different species, cultivars or field lots in order to obtain their desired qualities. This blending is also practised to an extent by export shippers to satisfy such criteria as minimum specific weight expectations.

Chapter 4

SEED AND VARIETIES

CHOICE OF CEREAL SEED

It is not wise to set low standards for the vital initial decision on choice of cereal seed. On the other hand, useful economies can be achieved by sensible judgements, e.g. strategic use of carefully grown, farm-saved seed.

Quality

Important seed features are:

1. Germination capacity This is the percentage by *number* of the cereal seeds which show the radicle (young root) during the period of the germination test. Almost 100 per cent is possible, 95–98 per cent common; below 85 per cent is quite unacceptable. Ideal conditions of moisture, aeration and temperature (20°C) are provided. There is a general correlation with likely field establishment (Figure 4.1).

2. Purity This is the percentage by *weight* of seed of the species and even variety which the sample is supposed to contain. Impurities include seeds of other crops and weeds, broken grains, dust and soil. Over 99 per cent purity is sought; below 98 per cent is unacceptable.

3. Health Seed can be tested for such deep-seated diseases as loose smut by the Official Seed Testing Station (OSTS) at Cambridge. Loose smut occurs to a variable degree in many commercial and farm-saved seed samples. It is not infrequent to find 5–10 per cent of seeds infected. Ergot in the sample is also to

be rigorously avoided, being most troublesome in ryes and some wheats.

4. Vigour Rothwell Plant Breeders have worked extensively on the physiology and implications of seed vigour. RHM (Rank Hovis McDougall) have also pioneered research on this facet of seed and seedling quality. This work has been centred at the Lord Rank Research Station in High Wycombe (see Hayward,

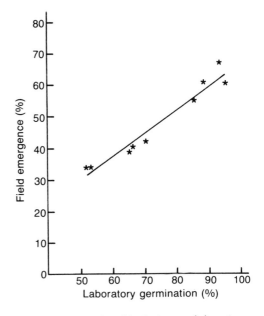

FIGURE 4.1 Relationship between laboratory germination percentage and field emergence percentage for spring wheat.

60

1978). Whilst the germination test assesses the seed's capacity to sprout under ideal laboratory conditions, the vigour test aims to screen seed samples to ensure they have adequate and uniform capacity to 'get up and grow' sturdily in the more testing conditions of the field. Ability to develop shoot and root system to predetermined extents within the period and conditions of a laboratory test is used to give a pass or fail result for the question 'Is it a high vigour lot?' Seed which has passed these tests has been grown in many sites and conditions nationwide alongside low vigour samples. RHM believe that some 10−15 per cent of cereal seed used is of unacceptable vigour. They have found that high versus low vigour samples give:

- Better establishment: around 20 per cent improvement in initial plant population on average over many farm conditions
- Yield improvement of up to 10 per cent owing to greater ear numbers and/or ear size
- Advantage at all seedrates, but especially at low ones.

Some results from the Scottish Crops Research Institute appear in Figure 4.2.

5. TGW (thousand grain weight)   This is the weight of 1000 grains in grams (g). Unfortunately, within reasonable limits, there is no clear correlation between seed size and performance in the field in either establishment or yield terms. However, the boldness (size and plumpness, therefore weight) of seeds can vary considerably − more than twofold in fact − so information on this feature is useful in calculating seedrates and total seed purchases necessary to plant a given area (Plate 4.1). In addition, grading of seed to a uniform size helps uniformity of the crop.

6. Variety   This must be suitable for the intended site and purpose (market). Statutory cereal seed standards are enforced by seed-crop inspection, official seed testing and a network of registered processors and registered merchants. They are summarised in Appendix 4. It is wise to retain and store carefully a small sample of any seed lots sown so that it may be submitted for tests in support of any claims that may be pending following poor field performance.

Sources

A cereal grower must buy certified seed or save his own seed from last year's crop. It is illegal to buy

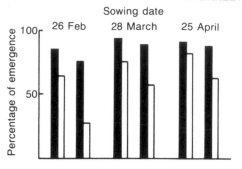

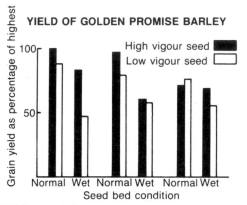

FIGURE 4.2   Influence of seed vigour on emergence and yield of spring barley in normal and wet seed beds. (Based on data of D.A. Perry, SCRI)

seed from another farmer unless that farmer is registered as a seed merchant. Merchants can only sell certified seed. (Appendix 5 shows recent certifications.)

Certification

The EC scheme, known as the Restricted Generation System, of cereal seed multiplication is used. Breeders of cereal varieties maintain stocks, generally looking after the first three or four years of the normal six-year seed production cycle (note that this is not to be confused with the initial development of a new variety!):

Year 1   Progeny rows
Year 2   Progeny rows
Year 3   Plots → Pre-basic seed (white label + violet stripe)
Year 4   Plots → Basic seed (white label)

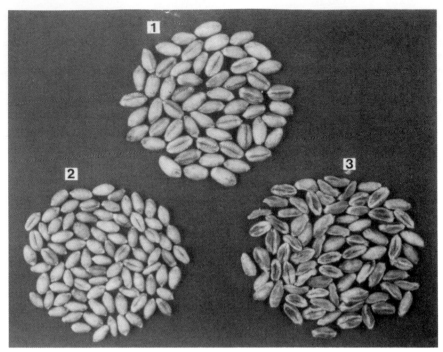

PLATE 4.1   Variation in thousand grain weight. 1 Large plump grains (1000 grain weight of 63 g); 2 Small plump grains (1000 grain weight of 33 g); 3 Large shrivelled grains (1000 grain weight of 40 g). (Source: Gedye, 1981)

Year 5   First certified generation, C1 seed (blue labelled sacks)

Year 6   Second certified generation, C2 seed (red labelled sacks)

(Note: Retain labels with the field records: seed growers must display labels in the field.)

This system, coupled with the standards for each of the two grades, simplifies choice.

Rigorous selection is needed in all stages of multiplication to ensure varietal purity. One faulty ear of wheat reproducing itself over five seasons could produce around 75–80 tonnes of faulty grain! Rye seed has to be imported to avoid such change since it is cross pollinated. The highest standards apply to the earliest stages of multiplication because deviations here will obviously be multiplied later. A few growers with very high standards of field hygiene and management may be sub-contracted by seed-houses to multiply pre-basic seed, others to multiply basic seed and most to multiply C1 seed to give the principal commercial grade, C2. The only circumstances in which the expensive C1 seed is likely to be bought is with such multiplication in view and in the early years after

a new promising variety has been launched when C2 seed is unavailable or scarce.

Some cereal farmers rate the importance of seed quality so highly that they use a preponderance of C1 seed in order to increase the chances of uniformity and success. The value of this is debatable. HVS (higher voluntary standard) seed of the C2 generation caters for a demand, being freer of impurities.

Growing for Seed

The attractions of growing corn for seed are:

1. Premiums are attainable, often of 20 per cent over commercial uses, depending on generation of seed involved and scarcity of the variety.

2. The higher standards of husbandry it imposes are a challenge and welcome discipline to some growers.

3. Seed produced for home-use will be of a higher standard than average for farm-saved seed.

Points to bear in mind in producing cereal seed are as follows:

1. Clean land   This applies especially to grass weeds (notably wild oats) and cereal volunteers. Use roguing (hand pulling) on populations of these up to about 250−300/ha. The WRO-developed chemical glove is not suitable for seed crops: only total weed extraction and disposal is good enough. Walk the tramlines in pairs armed with plastic fertiliser bags into which to stuff the culprits! Of the non-grass weeds, cleavers, wild onion and wild radish can be particularly troublesome in seed crops.

2. Previous cropping   The previous *two* years must have seen neither another cereal species, other varieties of the same species nor uncertified seed of the same variety if one is producing basic seed.

For C1 seed production the year before must have been the same species sown to certified seed of the same variety: if the year before that was the same variety it must have been sown with certified (not farm-saved, later generation) seed.

For C2 seed production no other variety in the preceding year is acceptable, though another species is. The same variety can have been grown on that field but it must have been sown with certified seed.

3. Isolation   There must be a least 2 metres separation between cereal seed crops (either fallow, hedge, ditch, track or another non-cereal crop); 50 metres separation is needed between two-row and six-row barley varieties and the same separation from possible loose smut sources for wheat as well as barley-seed crops.

4. Uniformity   Choose sheltered fields as far as possible. The crop must emerge, develop and stand well. Although lodged crops are not disqualified by EEC standards, HVS requires that not more than one-third of the field be lodged. Not only is a lodged crop difficult to inspect but also it is more difficult to harvest and often costlier to clean. Avoid risky applications of late herbicides or materials which may distort growth (check varietal susceptibility to herbicides and insecticides); avoid excessive nitrogen and use straw-shorteners/stiffeners if necessary.

5. Inspection   It is necessary for qualified crop inspectors to check that statutory requirements have been met. For home-saved seed crops it is advisable to have a similar inspection. Fields must be labelled either into the crop or on gateposts with a label from a bag of the seed sown.

6. Clean equipment   Ensure that all equipment is properly cleaned of any sources of contamination, biological, physical or chemical. This includes seed drill, combine harvester, trailers, drying and handling equipment and storage silos. Minimise handling because risk of both contamination and broken grains increases with each handling.

7. Harvesting   Pre-harvest glyphosate may not be used because it may impair germination. If there is a clash of fields ready to harvest, obviously give priority to seed crops. There is a correlation between harvesting weather plus seed's optimum readiness and its subsequent performance when sown.

It is advisable to discard from seed purposes at least the first pass around the headland. Where weeds have ingressed further, more discards may be necessary. Apart from avoiding a higher-risk zone for weeds, such a procedure allows some chance for any uncleaned previous crop residues to pass through the combine.

Combining needs to proceed with rather gentler thrashing than is suitable for feed crops. It is easy to develop hairline cracks in the grain which predispose it to damage to germination capacity, especially following subsequent chemical dressing.

8. Drying   Grain for seed is ideally precleaned and then dried with care; 43°C is a safe temperature. However, the wetter the grain coming off the combine, the lower the temperature should be:

Grain moisture content (mc) off combine	Drying temperature
30%	43°C
24%	53°C
18%	65°C

The Scottish Colleges give more cautious advice:

49°C maximum if grain is *under* 24% mc
41°C maximum if grain is *over* 24% mc

High temperatures in which grain cooks in its own juices damage germination capacity! Target moisture content for long-term storage is 14−15 per cent.

9. Liaison   Develop a co-operative dialogue with the merchant, who should be seen not as an adversary but rather an ally in achieving good marketing standards!

10. Specialisation Although there are a number of specialist seed-growing farms, it is advisable for the average cereal grower to concentrate on producing properly only one or two varieties for seed. Obviously, the number which can be separately handled with neither confusion nor excessive cleaning-time for equipment depends on the system, but restriction is wise. (See Appendix 4 for standards.)

Seed Processing

Clearly, systems vary somewhat in the sophistication of equipment and extent of sampling done along the line, but the normal stages in a modern plant are:

a) Arrival at seed warehouse (sampled)
b) Pre-cleaner (to exclude especially straw, stones, soil)
c) De-awner (to exclude awns and chaff)
d) Cleaner (aspirator to exclude dust and very light grains; screens to exclude oversized and thin grains plus weed seeds)
e) Indented cylinders
f) Specific-gravity separator (to exclude other cereal seeds and damaged grains). Samples are important at this stage
g) Bulk storage
h) Chemical treatment
i) Packaging in 50 kg four-ply paper sacks

Seed processors must be officially registered to preserve standards in the farmer's interests. Home-saved seed on farms does not have to be so processed. Indeed, its weaknesses may be linked to the fact that it less often gets such a degree of treatment.

Using Home-saved Seed

The attractions of this are:

1. There is significant cost saving (around 70 per cent of the price of C2 seed after processing and treatment).
2. The grower can check for himself that processing and treatment have been done to his satisfaction. Mobile seed-cleaning plants can do a good job (and some a bad one!). Particular points to check are weed-seed removal (notably wild oat) and uniform application of seed-dressing chemicals.
3. The seed is available on site when needed for timely sowing. This is especially relevant for earlier-sown autumn corn when merchants often cannot process all requirements fast enough.

4. It can be argued that home-saved seed which has been multiplied on that farm over several generations develops adaptation to conditions — a limited eco-type. Up to six generations are quite safe if carefully grown, handled and tested.

The problems associated with home-saving are:

1. Overcooking during drying, thus damaging germination capacity.
2. Weed-seed contamination is generally much greater than allowed for in merchant's seed. An indication of the contrast is given by data collected in a National Institute of Agricultural Botany survey of seed at the drill in spring 1970. Of 620 collected samples (of 3.18 kg), 378 were tested at OSTS, giving the following results:

Merchant's seed	36% of samples had weed seeds
	4% of samples had over ten weed seeds
Farm-saved seed	89% of samples had weed seeds
	74% of samples had over ten weed seeds
	18% of samples had over one thousand weed seeds!

In all, 90 species of weeds were recorded but more than half these occurred in fewer than 1% of samples. Most frequent weed seeds were:

black bindweed	24% of samples
cleavers (goosegrass)	21% of samples
spring wild oats	19% of samples
knotgrass	18% of samples
Persicaria	18% of samples
couch	15% of samples (although chiefly rhizome-spread!)

In addition to contamination, farm-saved seed is less likely to receive such general care as given to a seed-contract crop. In order to successfully benefit from farm-saved seed a grower needs to impose upon himself the same disciplines and standards as needed for seed-contract fields.

A recent survey showed that over three-quarters of farm-saved seed failed EC minimum standards. The OSTS can test for:

● Germination — a rapid seed viability (RSV) staining test can give results within 2 days as to whether seed passes the EC minimum of 85 per cent germination capacity. A full germination guidance test on 2 replicates of 100 seeds gives results in 7 to 11 days of precise germination percentages.

- Purity — weed seeds and other contaminants are analysed.
- Seed-borne diseases — these can be assessed by microscopic examination.

Cereal Seed Dressing

The majority of the cereal hectarage of the UK is now sown with chemically treated seed. The objectives of the chemicals are seed and seedling protection from diseases and pests. Single purpose dressing (SPD) is for common seed-borne diseases but does not protect from pests, including birds. Dual purpose dressing (DPD) also includes HCH (Hexachlorohexane), which controls wireworm and deters wheat bulb fly and birds.

Seed dressing has several aspects:

1. It is economically preferable to spraying the crop — less chemical is used and application is easily put on target. Some dressings interfere with the smooth flow of seed through corn drills and may increase wear of drill parts.

2. It is ecologically preferable to spraying except for seed-eating birds and their food chains. It does not seem to control pigeons much, though, and there is also the risk of inducing pathogen resistance through the repeated easy treatment of seed. Oat loose smut has already developed such resistance to mercurial compounds.

3. It is capable of giving wide-spectrum protection for the crucial seedling stage. The spectrum obviously depends on the materials used: HCH covers wireworms and other soil pests, whilst organomercurial compounds cover many superficial fungi.

4. It is capable of giving specific protection against particular anticipated problems, e.g. wheat bulb fly control using carbofenothion or chlorfenvinphos; cereal mildew control using ethirimol or triforine; loose smut control using carboxin mixtures.

5. It may be capable of giving more persistent (several weeks') protection, e.g. for seedling and foliar diseases Baytan (triadimenol, which is closely related to triadimefon) and Ferrax (flutriafol, ethirimol and thiabendazole).

It is important that the following precautions are observed:

- Dressing is evenly applied to seed.

- Treated seed is dyed and labelled. (NB. The *dye* is not the chemical, which penetrates much deeper. There have been accidents when treated seed has been consumed; e.g. some years ago in Iran, when dressed, dyed and labelled seed arrived too late for the planting season, it was washed by some consumers and then eaten. Washing removed only the dye; mercurial compounds produce irreversible brain damage.)
- Only dry enough seed is treated (14—16 per cent mc). In both wet and over-dry, cracked seed, excessive amounts of chemical can penetrate and damage the germ. This leads to failure or distortion of seedlings.
- Hands are washed thoroughly after handling treated seed.
- Treated seed is not fed to livestock.

Hot-water treatment remains an option for seed-borne disease control on a limited scale. It is tedious, risky and of dubious efficacy. The water temperature for effective check of the pathogens is only around 1—1.5°C below the temperature above which germination capacity is impaired. Thus facilities for accurate temperature control, accurate exposure time and careful drying are needed. The seed is neither protected from immediate colonisation by soil-borne fungi nor from soil pests.

Cereal seed dressings have improved the protection of crops at their most vulnerable stage of survival.

VARIETIES (CULTIVARS)

A recommended varieties booklet is published early each year by the National Institute of Agricultural Botany (NIAB) in Cambridge, based on all their regional trials, which generally produce yields some 15—20 per cent above national averages. Characteristics are scored on a scale from 1 (low degree of expression) to 9 (high). The majority of cereals grown are varieties from the NIAB booklet. However, at present farmers are free to choose unapproved varieties (e.g. Golden Promise spring barley) provided they appear in the current National List or in the Common Catalogue of the EC member states. The reason for non-approval by the NIAB has usually been a particular disease-susceptibility of the variety, but its agronomic advantages may still persuade farmers to grow it with disease protection. The Scottish Colleges publish a useful recommended list every year for Scotland.

CHOICE

Farmers' considerations when choosing a variety include:

1. Purpose Whether barleys are for malting or for livestock feed is largely a matter of variety. The same is true whether wheats are for milling or for livestock feed ('quality' wheats or feed-wheats).

2. Locality There are regional variations in performance of varieties. Soil type and exposure of the particular field is important: a weaker-strawed variety should not be given rich land exposed to the wind. Varieties may be chosen to suit the particular blend of conditions on a farm.

3. Disease resistance Resistance to expected diseases on a particular farm reduces dependence on fungicides and other means of disease control. Particular disease pressure depends on the climate of the district and on the farming system and methods adopted.

4. Yield This has been the predominant consideration for farmers, and the breeder has sought to combine as many other desirable features as possible. Yield differentials shown on the NIAB list are usually magnified under high-input field conditions.

5. Experience Farmers need to be more reluctant than some are to abandon a well-tried variety which they have learnt to grow properly. This learning cannot be done solely from trials conducted elsewhere. A sudden expansion into a new variety is unwise until a farmer has proved it on a smaller scale on his/her own farm.

6. Diversity It is unwise to grow too large an area of one variety on the principle of risk spreading. The chief risks are from disease and lodging. The NIAB have categorised varieties into 'diversification groups' (Figure 4.3) according to the particular races of a pathogen for which they have resistance. The probability of breakdown to disease is reduced by such diversification. It is also less likely that a spread of varieties would be equally affected by other adverse factors such as weather damage and market weakness.

7. Timing It is wise to have a sensible balance between winter- and spring-sown varieties and between earlier and later maturing varieties of each if cereals are grown on any scale. This spreads sowing and harvesting dates and has other advantages related to field hygiene and soil management.

	Apollo	Avalon	Brigand	Brimstone	Brock	Fenman	Galahad	Hornet	Longbow	Mercia	Norman	Parade	Rendezvous	Slejpner
Apollo	m	m	m	m	+	+	m	m	m	+	m	+	m	m
Avalon	m	ym	ym	m	+	+	m	m	m	+	ym	+	m	+
Brigand	m	ym	ym	m	+	+	m	m	m	+	ym	+	m	+
Brimstone	m	m	m	ym	+	+	m	m	m	+	ym	+	m	+
Brock	+	+	+	+	+	+	+	+	+	+	+	+	+	+
Fenman	+	+	+	+	+	+	+	+	+	+	+	+	+	+
Galahad	m	m	m	m	+	+	ym	m	m	+	m	+	m	+
Hornet	m	m	m	m	+	+	m	m	m	+	ym	+	+	m
Longbow	m	m	m	m	+	+	m	m	ym	+	ym	+	m	+
Mercia	+	+	+	+	+	+	+	+	+	+	+	+	+	+
Norman	m	ym	ym	ym	+	+	m	ym	ym	+	ym	+	m	+
Parade	+	+	+	+	+	+	+	+	+	+	+	+	+	+
Rendezvous	m	m	m	m	+	+	m	+	m	+	m	+	m	+
Slejpner	m	+	+	+	+	+	+	m	+	+	+	+	+	ym

+ = Good combination; low risk of spread of yellow rust or mildew
y = Risk of spread of yellow rust
m = Risk of spread of mildew

FIGURE 4.3 Varietal diversification scheme to reduce spread of yellow rust and mildew in winter wheat shows the risks involved in growing different combinations of varieties on the farm. Whenever possible, choose low risk combinations of varieties to grow adjacent to each other. (Source: NIAB, 1988)

DESIRABLE FEATURES OF CEREAL VARIETIES

These are placed in two categories by the NIAB:

Agricultural characters These include yield, standing power, shortness of straw, earliness of ripening, latest safe sowing date for winter varieties, winter hardiness (oats and barley), resistance to sprouting in the ear, seed-shedding resistance, resistance to ear loss (barleys) and resistance to diseases and pests. Other factors include tolerance of herbicide and other chemical treatments and of trace element deficiencies.

Grain quality characters These include thousand grain weight, specific weight (kg/hl), milling and bread/biscuit-making features (wheats), kernel content (oats) and malting grade (barleys).

THE CONTRIBUTION OF NEW CEREAL VARIETIES

This has been reviewed by Silvey (1978) and Austin *et al.* (1980). Both attribute some 50 per cent of the yield improvements for wheats to better varieties and 50 per cent to better husbandry. Silvey deals with wheats and barleys from 1947 to 1977 whilst Austin *et al.* cover wheats from 1900 to 1980. Silvey calculated a 1 per cent per annum contribution to yield improvement by new barley varieties whilst in the decade 1967–77 new wheat varieties contributed a 3 per cent per annum improvement in wheat yield. Riggs *et al.* (1981) reviewed a century of barley varieties (Table 4.1).

The NIAB estimated the cost to produce new varieties as a percentage of their extra value to growers during the period from 1964 to the late 1970s: barleys cost 5 per cent of their extra value whilst wheats cost only 2 per cent. The introduction of plant breeders' rights to royalties with the Plant Varieties and Seeds Act of 1964 has provided a considerable stimulus to commercial cereal breeding. In 1967 the National Seed Development Organization (NSDO) was set up to achieve the best commercial availability of new varieties consistent with high standards. The NSDO also promotes UK-bred varieties abroad by sponsoring them in official foreign trials and by marketing efforts overseas.

Annual charts of varietal popularity are published — a sort of cereals' 'Top of the Pops'. The MAFF have collected data from 1969, whilst seed-testing records from OSTS and from agricultural merchants are further sources of information on popularity before that date. The Plant Royalty Bureau now provides this information.

Appendix 5 shows the 1988 varietal picture. UK cereal seed certification data show wheat and barley varietal popularity and proportions of spring varieties.

TRENDS IN VARIETIES

Popularity fluctuations for wheat varieties are shown in Figure 4.4. These change quite rapidly since the size of cereal seed markets justifies their continuous development. Competition is also encouraged. Farmers are advised neither to abandon proven varieties too readily nor to get out of date on developments and awareness of new varieties: subscribe to the NIAB and receive annual lists and regular information. Information on individual varieties dates rapidly and students are advised to familiarise themselves with the principles of variety development and choice and with key features of the most prevalent varieties.

BREEDING NEW CEREAL VARIETIES

Wheat, barley and oats are naturally self-pollinated and thus described as inbreeding crops. They breed true to type and pure lines are easily established but cross pollination to develop new varieties has to be done generally by hand methods. It normally takes about six generations by pedigree breeding to establish a new uniform variety which can then be submitted to breeder's trials, taking a total of about fourteen generations to supply commercial quantities of a new variety (see Figure 4.5). Spring varieties may be bred more rapidly by having two generations per year, one of them grown in the southern hemisphere, usually New Zealand.

Other methods of accelerating the development of new varieties include the use of radiation to induce mutations. Most mutants are unfavourable but some have led to the introduction of useful varieties, e.g. Golden Promise spring barley. Biotechnology promises great acceleration.

Where one particular character is sought, such as resistance to a specific disease, a high-yielding variety may be first crossed with a wild species of grass or

TABLE 4.1   Spring barley: varietal changes 1880–1980

Variety	Year introduced (approx.)	Yield (t/ha)	Height (cm)	Harvest index (%)
Goldthorpe	1880	4.82	102	39
Kenia	1932	5.09	96	40
Spratt Archer	1933	5.19	92	41
Plumage Archer	1935	5.44	98	40
Rika	1951	5.42	85	45
Proctor	1953	5.65	84	42
Vada	1960	5.70	89	42
Zephyr	1966	5.95	79	46
Golden Promise	1966	5.51	65	48
Julia	1968	6.20	77	45
Georgie	1976	6.30	72	50
Triumph	1980	6.68	70	50

Note: Data of Austin *et al.* show similar trends for winter wheats, i.e. harvest index 35 → 50%, plant height 127 → < 70 cm.

Source: Data of Riggs *et al.*, PBI, Cambridge, for trials conducted 1978–80.

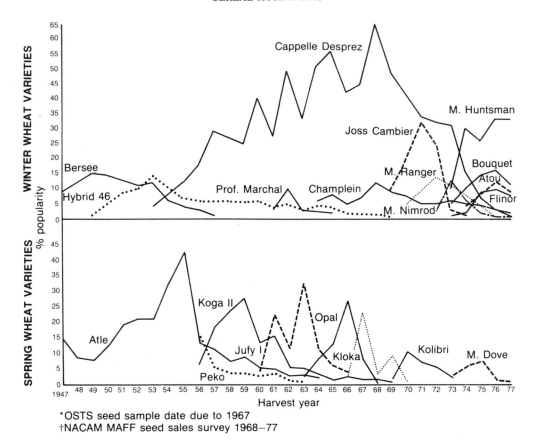

FIGURE 4.4   Popularity of winter and spring wheat varieties 1947−77 (as per cent total winter and spring wheat seed samples* and sales†). Note that a new range of varieties has now replaced these. (From Silvey, NIAB, 1978)

cereal and the progeny then backcrossed to the original cereal parent. The problem then is to select offspring which incorporate only the desired new character from the wild source and not also its many undesirable features.

In view of the time taken to produce new varieties there is scope for progressive farmers to identify promising new varieties a few years before their wider commercial availability, by which time they may have learnt to grow them and may be in a position to grow good seed crops too. There is clearly an important element of risk and of judgement in anticipating a new variety's likely popularity! Such anticipation has to be made for seed multiplication purposes when the following quantities may be involved:

Year 1   Single plants of new variety
Year 2   Ear rows
Year 3   Breeder's seed: 400 kg
Year 4   Pre-basic seed: 10 tonnes
Year 5   Basic seed: 200 tonnes
Year 6   C1 seed: 6000 tonnes (enough for 35−40,000 hectares)

Hybrids

The use of F1 (first filial generation) hybrids has greatly contributed to improvement of maize yield worldwide. Sorghum, millet and rice hybrids exist. However, hybrids have not proved popular for temperate cereals to date owing to difficulties of commercial seed production. Wheat and barley hybrids have been tried but not developed, whilst wheat and rye (triticale) are of limited UK interest at present but the subject of substantial development work, notably in Canada.

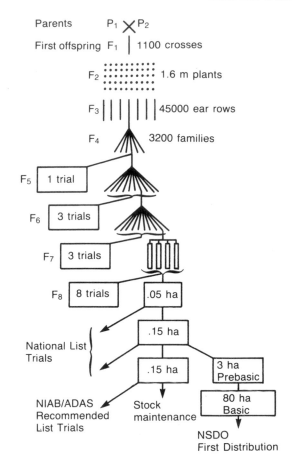

Parents $P_1 \times P_2$

First offspring F_1 | 1100 crosses

F_2 .... 1.6 m plants

F_3 | | | | | | | 45000 ear rows

F_4 3200 families

F_5 | 1 trial |

F_6 | 3 trials |

F_7 | 3 trials |

F_8 | 8 trials | | .05 ha |

National List
Trials | .15 ha |

                | .15 ha | | 3 ha Prebasic |

NIAB/ADAS   Stock | 80 ha Basic |
Recommended maintenance
List Trials

NSDO
First Distribution

FIGURE 4.5   Winter wheat selection and multiplication system. (From PBI)

Wheat has been crossed with couch to produce *Agrotricum* or perennial wheat. Its yield has been disappointing and it is somewhat reluctant to depart when one wants to change crop! Hybrids are very expensive and could lose their advantage to other new varieties quite rapidly.

HERITABILITY OF DESIRABLE CEREAL FEATURES

1. Yield

Total dry matter production in old and new varieties of cereal differs little in a given set of field conditions. Yield differences have been achieved by increasing the proportion of total dry matter which is grain (harvest index). The harvest index for old varieties was around one-third whilst current varieties average

50–55 per cent (Figure 4.6). Total dry matter production can only be improved if photosynthesis is greater. More efficient light interception by more vertical leaves plus higher intrinsic rates of photosynthesis per unit of leaf may be possible. Leaf area duration, i.e. length of life of the ideal green leaf area, may be extended more easily than these factors to boost yield. Harvest index may be further improved, though breeders consider an average of 60 per cent to be the ceiling for wheats and perhaps a little less for barleys. Semi-dwarf wheats were bred incorporating genes from the Japanese variety Norin 10 which were included in successful varieties such as Hobbit bred by Dr Francis Lupton at the PBI, Cambridge. Semi-dwarf wheats now account for over 40 per cent of all winter wheat seed sales. They need optimum conditions to realise their potential whereas taller varieties often adapt to lower fertility; perhaps they have more stem base reserves to help them to withstand stress. Not all semi-dwarfs have stiff straw, e.g. Hobbit benefits from straw stiffeners. The characters associated with higher yield potential such as shorter straw are highly heritable; so also are field characters such as resistance to sprouting in the ear.

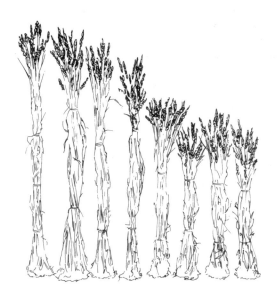

FIGURE 4.6   Eight successively introduced winter wheat varieties showing the progressive reduction in height associated with increase in yield. Left to right: Little Joss (1908), Yeoman (1916), Holdfast (1935), Maris Widgeon (1964), Maris Huntsman (1972), Hobbit (1977), Avalon (1980), and Norman (1981). (From PBI)

2. Disease Resistance

There are two main bases for disease resistance:

Single gene This shows great uniformity in level of disease resistance from site to site but carries a high risk of breakdown since it is more likely that a pathogen can develop a physiologic race capable of over-riding a single gene.

Polygenic resistance Here several genes act together to confer resistance. There may be a less predictable degree of resistance shown from site to site but less chance of breakdown. However, such varieties may also act as carriers of the pathogen by being only 'field tolerant' and thus pose a threat to nearby susceptible varieties.

Especially in connection with disease resistance there is concern amongst some breeders that the parental basis of modern cereal varieties is from a relatively narrow gene-pool. Most improved varieties are closely related to existing varieties. This is not particularly surprising but does emphasise the importance of world collections of all cereal varieties and cereal relatives. Such genetic conservation and the establishment of gene banks is vital since it leaves the widest possible options for future cross-breeding programmes. It is also important that some varieties should be selected for ability to exploit less favourable conditions including lower chemical inputs.

3. Quality Characteristics

In wheats, good milling quality (i.e. clean separation and high flour extraction rate) is very much a matter of variety. So also is breadmaking quality, which is dependent upon such key factors as protein quality and alpha-amylase enzyme activity (Table 4.2). High protein content depends less on variety than on field conditions but the highest yielding varieties tend to be poor in both protein content and quality. Malting quality is very much a matter of variety of barley.

Most desirable features of cereals are separately inherited and so it is possible, though not necessarily easy, for breeders to assemble different combinations of them in new varieties.

VARIETY/HUSBANDRY INTERACTIONS

The biggest development of the past 20 years influencing choice of variety has been the use of systemic fungicides since their introduction in 1969 and

TABLE 4.2 H-GCA classification of varieties of wheat

CLASS I Preferred Breadmaking Varieties
Includes those varieties most suitable for inclusion in breadmaking grists:

Winter − *Avalon, Mercia*
Spring − *Alexandria, Tonic*

CLASS II Acceptable Breadmaking Varieties

Minaret (spring), *Brimstone, Parade, Rendezvous*

CLASS III Non-breadmaking varieties

Includes those varieties most favoured for use in biscuit grists and those varieties predominantly used in animal feed, but which may be incorporated in certain grists:

Brigand, Brock, Fenman, Galahad, Hornet, Longbow, Norman, *Slejpner*

Note: (i) Varieties appearing in italic type have hard endosperm texture; other varieties listed have soft endosperm texture.
(ii) This is regularly revised.

commercial expansion from 1972 onwards. Most cereal growers now select varieties for yield first, quality second and disease resistance last of all. The NIAB have responded to this trend by screening varieties under full fungicide protection as well as without fungicides. Full fungicide protection in this context may mean levels beyond the economic optimum. Reliance on fungicides rather than varietal resistance is of some concern in view of the risk of inducing pathogen resistance to the chemical.

Varieties which tolerate a wide range of sowing dates are attractive. Fenman is such a wheat variety which may be sown as a true winter variety or indeed as a spring crop. It is clearly more useful to have winter varieties which tolerate spring planting so that failure to sow in autumn does not necessitate discard of seed or storage until next season. Spring varieties hardy enough to withstand autumn sowing can be useful if they provide a quality niche in the market, e.g. Triumph spring barley for malting; however, survival rates over a really hard winter have been disappointing.

Generally speaking, the higher the grain yield potential of a variety, the higher the state of fertility needed to support it. Furthermore, although under ideal conditions two varieties may support similar surviving tiller populations, the higher tillering variety will be better equipped to compensate for less suitable conditions. Thus Armada winter wheat (low tillering) was more efficient where fertility was good whilst Huntsman was a safer choice for poorer soil conditions. Amongst winter wheats Maris Huntsman has been a generally 'forgiving' variety, achieving quite high yields relatively easily in a wide range of conditions: this tolerance and flexibility accounts for the dominance of its relative Cappelle Desprez for some fifteen years from 1953 and of Proctor spring barley over the same period. Although not officially recommended by the NIAB owing to its mildew susceptibility, Golden Promise spring barley retains significant importance on the basis of its flexibility and early maturity.

The testing of varieties under commercial conditions with various husbandry treatments is increasing throughout the UK, work having been pioneered by such farmer groups as the Cotswold Cereal Centre based at the Royal Agricultural College in Cirencester (now the headquarters of Arable Research Centres Ltd with nationwide centres).

Apart from national data and regional trials there is really no easy substitute for experience of varieties on each farm. This emphasises the value of ample field records, noting reactions to applied treatments of different varieties, especially growth regulators, nitrogen and herbicides.

BLENDS OF VARIETIES

The admixture of two or more varieties, usually three, has been advocated most consistently by Sinclair McGill Seeds Ltd. Farmers have practised this for some years, though not widely. The arguments advanced in favour of it are:

- Greater aggregate exploitation of the environment by the mixture compared with sole crops of constituent varieties.
- Much more diverse resistance to disease and thus better yields and possibly reduced dependence on fungicides. Sinclair McGill have recorded reductions of up to 50 per cent of sole variety mildew levels by using a three-way blend of spring barleys. In severe mildew situations such blends have given up to 10 per cent greater yields. Less evidence exists to date for either *Rhynchosporium* or net blotch.

Merchants, maltsters and distillers do not generally favour such blends and there can be difficulties of differential maturity between component varieties from the farmer's point of view.

Part Two   Promotion

Positive aspects of growing cereals

Chapter 5

SOWING CEREALS

The standards attained at this foundational stage of the crop are crucial. Five important aspects will now be considered:

- Seed rate (quantity)
- Sowing depth (vertical distribution)
- Spacing (horizontal distribution)
- Sowing date
- Sowing method

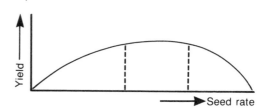

FIGURE 5.1   Yield and seed rate of cereals.

SEED RATE

In practice the relationship between seed rate and grain yield is far from precise (Figure 5.1):

Why should yield be maximised over such a wide range of seed rates?

1. The original plant population established varies for the same seed rate because

- Viability of seed samples varies. A small variation in laboratory germination capacity is magnified in the field (Figure 5.2). Under favourable conditions germination starts in four or five days.
- Seedling vigour varies, i.e. inherent rate and strength of early development vary.
- Thousand grain weight (TGW) varies. The range for wheat grains is 30−65 grams (world average around 36g; UK average around 48g); for barley 25−55 grams. In other words, there can be more than a twofold variation in the number of seeds sown per square metre at the same seed weight (kg/ha).
- Establishment percentage varies even though germination percentage and TGW might be identical

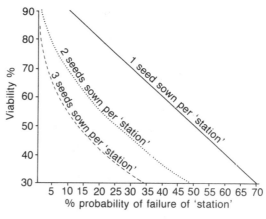

FIGURE 5.2   Seed viability, sowing density and probability of failure.

for two lots of seed. This is known as the field factor. It depends on seedbed conditions including tilth, trash level and moisture status, weather damage, early pest and disease incidence (Table 5.1; Plate 5.1). The field factor is the basis for

75

TABLE 5.1 Effect of methods of straw residue management on the number and size of winter wheat seedlings (sown on 29 September) after shallow tillage (7 cm) or by direct drilling on a clay soil

Straw treatment	Direct drilled		Shallow cultivated	
	No. of plants m^{-2}	Mean plant wt(g)	No. of plants m^{-2}	Mean plant wt(g)
9 November sampling				
Burnt	336	0.034	295	0.034
Stubble only	205	0.027	241	0.026
Stubble with straw chopped and spread	177	0.024	194	0.023
19 April sampling				
Burnt	243	0.96	285	0.84
Stubble only	169	0.74	207	0.70
Stubble with straw chopped and spread	133	0.52	193	0.63

Source: Data of Christian D.G. and Miller D.P., ARC Letcombe Laboratory.

the old rhyme on seed rates, 'One for the pigeon, one for the crow, one to rot and one to grow'!

2. Cereal plants have considerable capacity for compensatory growth. Yield depends not only on plant numbers/hectare but also on tillers per plant, the proportion of these which will bear ears, grain numbers per ear and the weight of each grain. The crop can, to an extent which varies with cultivar and husbandry, compensate for variations in plant numbers by adjustments to other components of yield. In particular, tillering (the production of extra side shoots from the base of the plant) can make up for low plant populations. However, the primary tiller or main shoot from each seed is naturally dominant (apical dominance due to internal hormone balance) and tends to be stronger. Thus it is desirable that a high proportion of ears should be from primary tillers and secondary tillers rather than from tertiary and lower-order tillers which will be the case if tillering has to compensate to its fullest extent for low plant density. Excessive tillering could, in fact, reduce final yield because too many short-lived vegetative tillers are produced.

PLATE 5.1 Stages in seed bed formation visible left to right on loam. (Courtesy of Ransomes, Sims and Jefferies Ltd)

This capacity for compensatory growth is a tremendous asset of the cereal plant. There has been debate over the desirability of breeding uniculm varieties, i.e. those which only bear a single shoot carrying an ear. This would adapt the plant for precision drilling and computer modelling but would, in my view, be practically disastrous. We need, and already have, varieties which are more controlled and efficient in their tillering certainly, but tillering ability is a great natural feature to compensate for the vagaries of real farming. Its elimination would be to our disadvantage.

3. The way in which the seed is sown — i.e. depth, spacing, time and method — affects the performance of each seed and seedling considerably.

4. Later variations of considerable magnitude can arise in the life of crops even though they started with the same plant population, viz. disease, pest, weed incidence; weather, fertiliser and growth-regulator effects and the timing of all treatments.

Therefore, in practice, the relationship between seed rate and final yield depends on:

- *Seed factors* — germination, vigour, TGW
- *Varietal factors* — especially capacity for compensatory growth
- *Husbandry factors* — soil management, establishment techniques and timing, subsequent crop management

Soil Fertility

The influence of inherent soil fertility on the seed rate to be chosen has been debated considerably. In reviewing worldwide work on wheat, Hudson (1941) advocated a lower seed rate for poor fertility and low rainfall areas; more seed was needed to explore higher fertility and rainfall. However, Boyd (1952), discussing British trials with all cereals from 1900, found in favour of higher seed rates for lower fertility to compensate! He advocated lower seed rates than in common farm use at the time. Nevertheless, Boyd reported some trials suggesting that extra nitrogen supply raised the optimum seed rate.

More recently (1975), Professor Laloux of Gembloux University, Belgium advocated relatively low seed rates for winter wheat. He based this on extensive trials principally with cv Cama on fairly deep soils in rotation with sugar beet and other crops. The advice was given in the context of a carefully raised nitrogen status and judicious timing of N fertiliser applications. He suggested a target post-winter population of just over 200 plants/m^2 for winter wheat achieved from seed rates of around 100 kg/ha sown in October to give some 220 grains/m^2, whilst later sowings might have increased seed rates to give a maximum of 300 grains/m^2.

In the rich soils of Schleswig-Holstein in Northern Germany, Dr Hermann Effland has advised higher seed rates. This is in the context of high inputs generally with target yields of 10−12 t/ha for winter wheat and a regional average yield some 20 per cent higher than the UK average. Seed rates there for winter wheat are recommended around 220 kg/ha. This should give the order of 400−450 plants/m^2, depending on the factors already discussed.

It has been the general experience from UK trials and farmer practice simulating the two approaches above that higher seed rates and higher resultant plant populations have correlated with better yields. This is good news for seed merchants and cereal seed growers!

Actual Seed Rates

In practice, seed rates actually used range from some 85−315 kg/ha. A narrower band within this range is much more typical at 155−185 kg/ha, though spring wheat and oats typically receive 215−220 kg/ha and spring feed barleys are the most likely to be sown at lower seed rates with less yield penalty. The ADAS suggested seed rates (kg/ha) are as follows:

Crop	Average conditions	Unfavourable conditions
Winter wheat	160	190
Winter barley	125	190
Winter oats	170	200
Spring wheat	200	225
Spring barley	125	160
Spring oats	200	225

Maize seed rates for grain production should give around 90−100,000 plants/ha (9−10/m^2), whereas forage maize should be sown thicker at 125−150,000 plants/ha. Seed rates of between 25 and 40 kg/ha are needed, according to TGW and expected establishment, to achieve these populations.

The criterion guiding seed rates used in practice should be target plant numbers the farmer expects to achieve. A systematic calculation rather than a blind guess at seed weight to choose can be sensibly made according to:

$$\text{Seed rate (kg/ha)} = \frac{\text{Target population (plants/m}^2) \times \text{TGW (g)}}{\text{Expected \% establishment}}$$

The numerator is the 'science' of the exercise, the denominator is the 'art' demanding sensible judgement!

The approximate resulting seed rates for different target populations and thousand grain weights are shown in Table 5.2. It can be seen how great a range is appropriate for typical variations in these vital factors. The original plant population achieved is the yield component most closely under the control of the grower and a straightforward objective to aim for in practice. He would be rated a poor livestock farmer who did not know his stocking rate! The population below which redrilling is sensible is debatable but winter wheat certainly has limited yield potential below $125-150$ plants/m^2 and spring barleys possibly may be left down to 50 plants/m^2. Redrilling is not normally contemplated unless thinner stands than these exist and then it must be decided against the likely success of alternative crops replacing the failed one. Less than 75 plants/m^2 begins to look serious but then poorest patches may be seeded with short-season ryegrass for silage or else redrilled and the better patches left for grain. Cross harrowing is an old-fashioned but still valid method of encouraging tillering in marginal cases.

Raising Seed Rates

The general reasons for raising the seed rate for cereals (within the sensible range 185 kg/ha $\pm$ 65 kg) are:

1. If farming further north, where $10-20$ per cent more seed is needed than in the southern UK for otherwise comparable conditions.

2. If sowing late, e.g. winter wheat may be sown at 160 kg/ha in September, 185 kg/ha in October and 210 kg/ha in November.

3. As an insurance against poor fertility — where some extra seedlings may be lost at the critical stage of transferring from use of inherited seed reserves to a state of independent nutrition relying on their own current photosynthesis and mineral uptake. This risky phase is analogous to the critical period in a young farmer's life when he first has his own tentative current account having previously relied on father's established deposit account!

TABLE 5.2 Seed rates (kg/ha) required for two different target densities (seeds/m^2) at different thousand grain weights (TGW)

		TGW (g)				
		35	40	45	50	55
Seeds/m^2	350 (e.g. WB)	123	140	157	175	193
Targets	400 (e.g. WW)	141	160	180	201	220

4. Crops can afford to thin out as a result of interplant competition and weather damage whereas one cannot fill gaps sensibly.

5. If trash of the previous crop or the weed population is rather high. It is especially the case where pasture has been broken up late prior, usually, to winter wheat. Toxins from slowly decomposing cereal roots and straw can have a similar effect on reducing seedling establishment.

6. If soil conditions are inferior — especially the state of tilth which has been attained and drainage status.

7. If broadcasting the seed. Although broadcasting can be a perfectly satisfactory method of sowing cereals, it is often used by growers as a last resort when weather and soil conditions have prevented drilling. This means that it may not be done as well as planned broadcasting would be. In any case it is more difficult to achieve uniform sowing depth and frequently there is an increased seedling loss, principally from shallow placement.

8. As an insurance against a poorer than expected seed-lot.

It is generally found that suitably dense crops tend to ripen more evenly and somewhat earlier, although late promotion of leaf growth can offset this difference.

Reducing Seed Rates

Arguments in favour of reduced seed rate are:

1. Seed is cheaper to purchase. At 185 kg/ha C2 seed costs about 20 per cent of total variable costs or about 5 per cent of the value of expected output. Thus to reduce to 150 kg/ha would give about 4 per cent reduction of variable costs and reduce seed cost to 4 per cent of the value of expected output — *provided* that the output remained the same. At current relative costs such a saving made over 400 ha of cereals could be equivalent to half the annual wage

for an arable employee. However, a simple grass weed herbicide could cost double this across the same area and may be more necessary in the absence of such dense cereal competition!

2. Handling is easier and there is less seed to dress and prepare if home-saving the seed.

3. Marginally quicker drilling may result owing to the more rapid filling of the drill.

4. Crops may tend to stand better in some circumstances, notably spring barley on exposed downland. This is debatable. When lax crops do lodge they tend to fall in all directions, whereas a thick crop tends to lie one way. Furthermore, the lax crop is likely to have a higher proportion of weaker stems, although these are probably shorter.

5. Less dense crops tend to suffer less from disease. This is probably a microclimate effect. The lusher crop encourages a greater humidity within it which may be more favourable to the onset of many fungal diseases. This is well documented in the case of eyespot. However, systemic fungicides have altered the situation in favour of ample density to secure yield coupled with appropriate crop protection.

6. A less dense crop demands greater attention and therefore one could argue that it promotes better subsequent husbandry because more depends on each individual plant: this requires closer attention to all factors relevant to its health and welfare.

7. Where expensive C1 seed is only obtainable in limited quantities and one wants to maximise seed multiplication, it may be sown at minimal rates.

Whilst in practice cereal crops more usually are found to be less dense than desirable, it is possible for them to be too thick and the detrimental effects of this can be regularly observed on the odd strips that tend to get double-drilled!

Within reason, seed use is not an area in which economy should be sought. It is of paramount importance to achieve a suitable population so that one does not spend more later in cash and anxiety to deal with a thin crop. The trend has been towards rather more generous seed rates to justify the subsequent use of greater yield-promoting inputs as well as to ensure an adequate foundation for yield which correlates with grain numbers produced per square metre, stemming from a thick enough planting. A thick crop of winter barley, provided that it is early sown and thus well rooted, affords the immediate surface soil some frost protection and thus protects its own root system. Whilst older leaves die off dramatically in a hard winter, the root system and developing growing point remain vigorous for an early spring boost. Tables 5.3 and 5.4 indicate the influence of density on winter barley yield, whilst Table 5.5 deals with spring wheat.

Finally, there is some correlation between seed rate and quality of the resulting crop. As seed rate is increased so is the ear population per hectare, but correspondingly grain numbers per ear decline as may the weight per grain and percentage of nitrogen in the dry matter. For example, for spring barley:

Seed rate (kg/ha)	TGW of produce (g)	Percentage of N in DM of produce
125	40	1.71
250	38	1.66

TABLE 5.3 **Winter barley performance (1978/1979/1980) and crop density; crops grouped into high and lower yield categories and compared**

Season	Yield (t/ha) High	Lower	TGW (g) of yield High	Lower	Seed rate (kg/ha) High	Lower	Plants/m² High	Lower	Ears/m² High	Lower
1977/78	7.73	6.65	44.0	50.0	188	162	313	260	920	780
1978/79	7.32	6.32	48.3	47.9	173	171	326	280	868	789
1979/80	7.71	6.30	46.9	46.3	188	163	337	328	930	868
Average	7.50	6.42	46.4	48.1	183	165	326	289	906	812
Contrast (H/L)	+18%		−3%		+11%		+13%		+12%	

Source: Cirencester Cereal Study Group, Royal Agricultural College.

TABLE 5.4   Winter barley: sensitivity of yield to crop components — the influence of adequate density

$$\frac{\text{Yield}}{\text{t/ha}} = \frac{\text{ears/m}^2 \times \text{grains/ear} \times \text{TGW (g)}}{10^5}$$

$$\text{e.g.} \quad \frac{8.1}{\text{t/ha}} = \frac{900 \text{ ears/m}^2 \times 20 \text{ grains/ear} \times 45 \text{ g TGW}}{10^5}$$

To gain an extra 0.4 t/ha (2½ cwt/ac) would require

	(a)	45 *more heads*/m² — *farmer control* by seed rate, sowing date, etc.
or	(b)	(less farmer control) + 1 extra grain filled/ear — breeder chiefly, also season
or	(c)	(limited farmer control) + 2.55 g extra on TGW — weather especially

TABLE 5.5   Spring wheat establishment on 28 Cotswold study fields

	Crop yield level	Sowing date	Seed rate kg/ha	Ears/m²	Days sowing to harvest	Yield (t/ha)
1980	High	25/3	212	670	168	6.40
	Low	5/4	218	504	164	4.73
1981	High	15/3	225	766	174	6.37
	Low	26/3	198	679	162	5.50
1982	High	16/3	224	573	186	6.00
	Low	26/3	201	462	181	4.85

Source: Wibberley, 1984.

Maize, for which a depth of 5 cm is ideal, can be sown at up to 7.5 cm.

If seed is sown too deep the consequences are:

1. It may rot altogether or be taken by soil-borne diseases or pests.

2. The plant runs short of seed reserves. Larger seed may do better owing to greater seed reserves. Evidence of deep sowing is an etiolated plant with a long epicotyl and characteristically yellow-banded leaves known as the rugger-stocking effect and associated with cold conditions; the bands occur at intervals horizontally across the leaf blades (Plate 5.2). This can most commonly be seen on early sown spring

SOWING DEPTH

Incorrect sowing depth is probably responsible for more impediment or failure of plant establishment than any other factor. It is far more important to get this right than to worry unduly over horizontal distribution (spacing) of seeds in practice.

The ideal depth for temperate cereals is 3 cm (2.5–3.75 cm). It is rather more common for seeds to be sown too deep rather than too shallow. Perhaps growers are overanxious to avoid the more obvious fault of shallow sowing. Seedbeds tend to be prepared too deeply: one needs to be able to drill onto a firm base. Spring corn needs to be sown somewhat deeper than winter corn as it is going usually into drying conditions. Yields were cut by 50 per cent if seed was placed at 10 cm compared with 2.5 cm in Norwegian experiments and by 25 per cent at 6.25 cm. When oats were sown at 7.5 cm rather than 2.5 cm, percentage establishment was cut from 80 to 65 per cent.

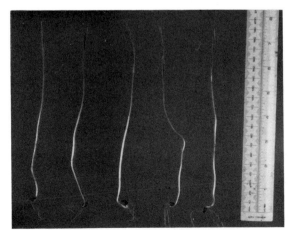

PLATE 5.2   Excessive depth weakens cereal seedlings, which show banded ('rugger-stocking') discolouration.

barley which has gone into soft ground which had been deep cultivated. Weaker seedlings are more liable to loss through wind and other damage even though they have reached the surface. Also the resulting crop will be weaker strawed and so more liable to lodge.

3. Establishment rate is slower and therefore riskier. As a matter of principle, the longer any organism remains in the more vulnerable juvenile phase, the less likely it is to reach strong maturity. Deep sowing tends to give poor tillering.

If seed is sown too shallow the consequences are:

1. It is likely to run short of moisture. Even though it may establish quickly, it will be more vulnerable to subsequent drought.

2. It is accessible for vermin — winter corn in particular needs to be covered against birds, especially if later sown.

3. Frost damage is greater since the roots, the most frost-sensitive part of the plant, are more exposed to ground frost.

The attainment of correct sowing depth of around 3 cm depends on three principal factors:

1. Tilth   This needs to be even and to offer uniform resistance to the penetration of the drill. Uniformity is yet more important if depth is to be even after broadcasting and harrowing in. Level land obviously helps!

2. Drill type   Drills vary as to the degree of depth control available to the operator. In general, direct drills tend to sow at a more even depth than conventional ones. This is due in part perhaps to the more stringent soil conditions needed for their effective operation and in part to the greater consistency of resistance offered by less-disturbed ground.

3. Speed of drilling   This point has been graphically illustrated in the useful film *Cereal Sense* by Nickersons. Miln Marsters carried out trials in 1973 comparing drilling at 4 mph and 9 mph for all the spring cereals. The yield advantage associated with slower drilling was from 5 to 15 per cent. Obviously field conditions and drill type determine the actual speed that is sensible. For instance, Mr Chris Capper's Stanhay precision cereal drill operated very successfully at 7 to 8 mph in good seedbed conditions at the Royal Agricultural College. This drill is unfortunately now in abeyance: it gave better depth control as well as more precise spacing (Figure 5.3).

It is clearly better to drill faster if the weather is about to break on tricky soil in the autumn and speed means the difference between sowing and not sowing that field at all. However, it seems to me that drilling correctly is of such paramount importance that it is desirable to have extra drill capacity (the secondhand value of a corn drill is not high so why not keep and use it alongside a newer acquisition as and when necessary?). Amongst the most successful cereal establishers are farmers who will not drill faster than 4 mph and have the drill capacity to ensure they can keep to that without the disappointment of leaving fields unsown during the optimum period.

Whilst 3 cm can be prescribed as the ideal sowing depth, it may be desirable to go a little deeper, particularly on lighter sandy soils where moisture may not be found until a depth of 4 or 5 cm. Clearly it is important not to deplete surface moisture by over-cultivating such soils, but this can easily occur in reality. In addition rougher tilths necessitate rather deeper sowing in order to ensure that an adequate proportion of seeds are covered.

It is vital that seeds should

- Be covered
- Be in contact with moisture
- Be at a uniform depth
- Emerge as near the same time as possible

There is a correlation between coincidence of emergence and a crop's uniformity throughout its life as well as its final yield. Generally the shorter the time between the first and last emerging seedlings, the more uniform and higher-yielding the crop is likely to be.

In practice it is very difficult to measure sowing depth accurately, though with a fine enough tilth it is amply evident once the plants have emerged!

SPACING

Within the range of spacings commonly used for cereals, differences are less significant as far as final yield is concerned than are differences in depth commonly found. Ever since the development of the drill by Jethro Tull at Crowmarsh Gifford in Oxfordshire, the correct row width for cereals has been debated. Traditional width became accepted as 17.5 cm between rows. The Norfolk Agricultural Station found evidence in favour of closer rows (9 cm) for barley over forty years ago. ADAS (then NAAS, the National Agricultural Advisory Service) conducted

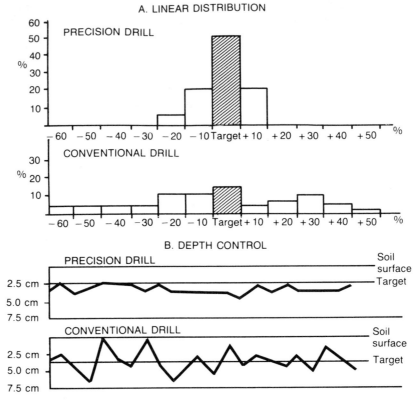

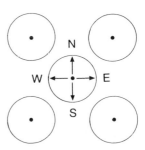

FIGURE 5.3 Drill comparison. (Source: ADAS Bridgets EHF, using spring barley 1977 – 79)

many trials, particularly during the 1960s, on narrower row widths for cereals, concluding marginally in favour (see, for example, Baldwin, 1963). R. Holliday (1963), reviewing work at Leeds University and elsewhere, concluded in favour of rows narrower than 15 cm for both autumn- and spring-sown cereals, recording yield increases of from 2 to 10 per cent. As a result narrow row drills became more widely available. These vary somewhat but typically give around 10 cm between rows or just over.

The advantage in favour of narrow rows is greater with spring cereals which have less time to compensate in growth for variations in spacing than their winter counterparts. Most evidence has been collected for spring barley and winter wheat, which is not surprising in view of the long-term preponderance of these two cereals in the UK. It was believed that winter wheat preferred to be thick in a 17.5 cm row, but farmers using higher populations to secure higher yields are finding benefits from narrow rows, as are recent ADAS trials.

Ideally cereals being sown to a target population of around 350 plants/m^2 should be spaced at 2.5 cm in the row on 10 cm rows. In practice all drills produce a degree of 'dolloping' (grouping in clusters) of seedlings with gaps (Plate 5.3). This effect can be reduced by drilling at a sensible speed (around 4 mph) into a good uniform seedbed with the right drill! Broadcasting may, in theory, give a perfectly uniform distribution of seed so that each seed has an equal distance to explore in all directions before encountering another seed:

PLATE 5.3   Dolloping effect and gaps in adjacent rows of spring barley.

This principle is clearly important in very lax crops where each plant needs equal space, such as many tree crops which benefit from equidistant quincunx planting. Logic suggests it as an ideal for cereals and supports the principle of precision seeding for both vertical and horizontal accuracy in seed placement. In practice we have not yet become so sophisticated! Fortunately our cereals are fairly resilient, adaptable creatures. However, more even spacing has been shown to pay at *higher* cereal populations.

Some farmers have sought to achieve greater precision of spacing by drilling with a conventional (17.5 cm) drill half the seed in one direction and then cross-drilling the remainder either diagonally or at right angles, especially when total seed rate is high as for spring oats and spring wheat. The use of a narrow row drill obviates the need to do this, though the cross-drilling with a conventional drill can achieve the same effect.

The introduction of a narrow-row drill imposes higher standards for land preparation prior to drilling, and these in themselves favour better yields. Such drills will not operate so well in trashy or rough tilths. The conditions under which they can operate properly thus tend to be more restricted. The order of yield improvement attributable to a narrow (10 cm) compared with a conventional (17.5 cm) drill is of the order of 5 per cent. The difference can clearly be masked by other factors quite easily. At high yield levels when other treatments are done correctly the use of a narrow-row drill is going to improve yields, particularly of spring corn, notably barley.

Row widths for maize in major producing countries are often 0.9 to 1 metre. In the UK, grain maize tends to favour narrower rows than forage maize with a range of 50 to 75 cm being suitable.

SOWING DATE

Time of sowing has always been a controversial subject.

Relatively early sowing is again in vogue. It was propounded by Jethro Tull in 1731 and reiterated by William Cobbett in 1822 in relation to wheat. Tull always sowed early on his chalkland in Oxfordshire and the system spread to wheat in stronger land there too. Cobbett recorded the experience of a Herefordshire farmer, Mr Beaman, who had grown six successive above-average wheat crops on wide rows (69 cm) sown 'the earlier the better; should be up before Michaelmas' (29 September); his spring wheat was sown early in March. Cobbett reflected, 'Besides early sowing, care must be taken not to sow too thin. By thick sowing along the drills, you get the plants to stave one another a little, and the wheat ripens at an earlier period.'

John Percival, writing in 1934, considered that sowing wheat from the beginning of September to the first week of October gave the best results in most districts throughout Great Britain.

In 1945 the late Professor Sir Harold Sanders recommended the following optimum sowing dates (common range given in parentheses):

Wheat	winter	20 October	(1 October–10 December)
	spring	15 February	(1 February–10 April)
Barley	winter	15 October	(1 October–10 November)
	spring	1 March	15 January–25 April)
Oats	winter	10 October	(1 October–10 November)
	spring	20 February	(15 January–25 April)
Rye		1 September	(1 August–1 November)

He noted that winter wheat may be sown later on good land and that winter barley was liable to winter killing. He also noted that the common range of dates quoted did not include the extremes found.

Before embarking on a current discussion of sowing dates we ought to note the following:

1. Some farmers have entrenched views about sowing dates, attributing yield differences to sowing date alone when other factors may have been equally or perhaps more important. Certainly a change of sowing date should be accompanied by changes in crop management.

2. The weather is the ultimate arbiter of actual sowing dates achieved. Hence we speak of open and closed autumns, early and late springs.

3. Variations in optimum sowing dates are bound

to occur according to season (a long wet season for instance will result in high levels of disease on unprotected early-sown crops, e.g. *Septoria* on September-sown wheats in 1981).

4. Sowing dates depend on soil types and particularly their drainage status with consequences for cultivating days available. In autumn, drill earlier on heavy land and thin stoney land than on lighter rich land.

5. Experience with sowing dates depends on the cultivars sown. Generally speaking, later-maturing varieties need to be sown *before* early-maturing varieties.

6. It is desirable to have clear objectives in this as in other facets of husbandry. Target sowing dates need to be set and revised according to the season and with experience of new varieties and changing circumstances. The system needs to be geared to these target sowing dates both from an organisation of drilling point of view and that of subsequent crop management.

Timeliness in farming operations is a vital feature. The ability to attain particular target dates for any job, within the obvious constraints of the weather, is a matter of planning, adequate kit and determination. Getting the correct time is not usually a matter of spending any extra money as with the use of an extra input. It does not even involve grappling with metric units! There is an old farming adage, 'The difference between the best farmers and the rest is two weeks'. It is preferable in practice to err on the earlier rather than later side of the possible range. Sowing dates are no exception to these principles. Early sowing makes sense, but not too early (especially not too early in mild districts such as the south-west peninsula and west Wales in autumn).

Reasons for Earlier Sowing

1. The probability is greater of having the required number of good drilling days before the onset of yield penalty through delayed sowing. It is particularly important to start early if you have limited capacity drilling kit, root crops to harvest and/or regularly difficult soil conditions. It is important to complete the target area of the highest value cereal varieties during the optimum sowing period.

2. The crop has the maximum chance to coordinate its growth with its development (Table 5.6). The late-sown crop is busy trying to grow in dry weight whilst hurriedly differentiating in response to changing day length or else, in the case of winter corn, its development is delayed to a less favourable time. Later-sown barleys can tend to accumulate a higher nitrogen content and so may be rejected for malting. This is probably because such crops will receive their N dressing closer to their harvest date as well as producing lower yields than required to dilute the N available.

3. The crop has the longest possible growing season to build up yield and exploit its environment fully. Maximum exploitation of photosynthesis is dependent on early establishment of the ideal leaf area.

4. The probability of having good drilling and pre/post-emergence herbicide application conditions is greater the earlier one starts in the autumn. Soil conditions are the major constraint on the start of drilling in spring.

5. An early-drilled autumn crop establishes more quickly owing to higher soil temperatures and thus seedlings pass more rapidly through the vulnerable juvenile phase. Soil structure also benefits from increased root activity in the autumn.

TABLE 5.6 Effect of sowing date on crop structure and performance of two winter barley varieties on the Cotswolds in 1980

	Variety			
Item	*Igri*		*Sonja*	
Sowing date	14 Sept.	17 Oct.	14 Sept.	17 Oct.
Plants/m²	319	282	322	280
Tillers				
Maximum/plant	5.92	5.72	6.60	5.80
Maximum/m²	1888	1613	2125	1624
Ears/m²	1082	702	1065	707
Tiller efficiency	57%	43%	50%	43%
Grains/ear	18.48	22.25	19.5	20.12
Thousand grain weight 15% mc	48.4g	53.0 g	48.0 g	52.5 g
Yield of grain				
Estimated	9.67	8.27	9.96	7.46
Actual	*8.77*	*7.71*	*8.06*	*6.21*
Actual %	100	87	100	77
Relative %	100	87	91	70
Output @ £85/ton	£745	£655	£685	£527
Variable costs/ha	£200	£195	£200	£195
Gross margin/ha	£545	£430	£485	£332

Source: Data of D.M. Barling.

6. An early-drilled autumn crop is stronger and deeper-rooted before winter temperatures set in. There are circumstances when a crop drilled later may stand frost better than one drilled at an intermediate date which is a delicate seedling just when frost-heave occurs on susceptible lands. However, a crop drilled earlier than either of these is likely to fare best. Extremely early crops of winter barley may become so lush that heavy frosts cause a decaying matt which smothers surviving tillers.

7. An earlier sown crop is deeper-rooted before the onset of any spring drought − this applies to autumn-sown crops as well as to spring varieties.

8. An earlier sown, strongly developed plant may be better equipped to withstand pest (e.g. slug) and disease attack, although generally more prone to pest and disease carry-over from the previous season. Benefit is certainly the case in respect of spring oats which need to be early sown in order to be sufficiently advanced to withstand frit fly attacks around mid-May. With autumn-sown cereals, especially oats, it is important to plough out grassland sufficiently early to allow a clear six weeks before drilling the corn to avoid frit fly. Early sowing has also been found effective in protecting against wheat bulb fly attacks on winter wheat.

9. Earlier-sown crops are generally earlier to harvest, though not by anywhere near so much as the difference in sowing dates. However, in a continuous system of winter cropping there are some extra days after harvest to help offset the need to turn round and drill two or more weeks ahead.

10. An early-sown, well-established crop is reassuring to the farmer and allows better sleep at night! This is a considerable point in real life. However, that reassuring early crop will demand proper management if it is to realise its prodigious promise.

Arguments Against Early Sowing

These largely revolve around the easier transfer of weeds, pests and diseases from previous crops and the probability of greater costs to control them.

1. The 'green-bridge' of volunteer cereals from the previous season means there is little gap between one crop and the next winter cereal if it is to be early-sown. These volunteers can carry diseases and pests.

2. There is less time for trash disposal from the previous crop before winter corn. This trash may carry diseases and pests. It may also release toxins in close proximity to the young seedlings. The trash

problem is particularly troublesome in seasons unfavourable for a good burn of straw when the cultivation system is a reduced one. A commitment to shallow ploughing immediately after harvest may help, when one does not have to wait for good burning weather which may not come at all in some years.

3. Weed control is more difficult since weed germination occurs after the crop is drilled.

4. In the case of winter corn sown very early there is a long period, particularly in mild districts, during which pests and diseases can colonise a crop before winter sets in. Early sowing certainly predisposes winter cereals to attacks by the gout fly, which, though only of sporadic importance, can be serious. However, it is also favoured by very late sowing of winter corn which may have summer damage, and a key strategy against it for spring corn is to sow early! Recent attention has been focused on *Opomyza* shoot fly whose attacks are associated with September-sown wheat crops. This correlation was reviewed in 1957 by D.B. Slope, and Bill Heatherington of the Royal Agricultural College has investigated the pest's recent behaviour in wheat crops. Autumn aphid occurrence can be a threat as some are vectors of virus disease − notably BYDV (barley yellow dwarf virus, which affects other cereal species including wheat and oats). The significance of autumn fungal attacks and the need for autumn control is debated. Protection must generally go alongside the promotion of growth afforded by early sowing but there are strategies to minimise this disease problem apart from just fungicides.

5. Crops may become too advanced too soon and have winter damage, or taller, weaker straw − and they will be at the growth stage for early PGR in difficult early spring conditions.

6. Seed may be unavailable from merchants very early anyhow!

Optimum Dates

Rosemaund E.H.F. in Herefordshire conducted sowing date trials with winter wheat between 1957−61 and found in favour of a relatively early sowing time. They defined optimum sowing dates by the number of days between the sowing dates and 1 December when the 10 cm soil temperature was above 4.5°C at 9 am. Lower yields were correlated with under 30 or over 60 such days, thus suggesting an optimum period related to soil temperatures. This is, of course, the wisdom of hindsight!

In many cases, actual sowing dates occur as soon

as an adequate seedbed can be made! Once made such seedbeds should not be left long exposed to the weather lest they either dry out or turn to slurry and cap in the rain on some soils.

Autumn Older varieties needed more cold to vernalise them, i.e. bring them to the spring condition for full development subsequently. Now many varieties can tolerate later sowing from this physiological point of view.

The suggested order of drilling is rye, barley, oats, wheat (Table 5.7). However, whilst this may be the *technical* order, the *practical* choice depends on which is the most important contributor to farm income and which is on the trickiest land. Few farmers grow all four species anyhow.

Winter rye has a prolific root system if sown in time enabling it to exploit more fully less fertile conditions; it is very winter-hardy.

Winter barley is susceptible to loss overwinter if late sown. Its transition to the reproductive phase, i.e. when the ear is initiated, should take place by early November if it is September sown, putting it well ahead (Barling 1979). Most vital is a well-established root system. Penalties can be very pronounced if sowing is delayed (Figure 5.4). Winter barley crops sown near the ideal date quickly become competitive compared with winter wheats sown near their ideal date (Plate 5.4).

Winter oats are regularly on record as seedling failures or sustaining high winter-kill. Adequate early rooting is crucial because they are our least winter-hardy cereal.

Winter wheat is the most forgiving of late sowing and difficult tilths, although there is inevitably a yield penalty from sowing after November. On many farms it is the principal source of revenue and planted on the strongest land, both of which factors necessitate a sytem geared to achieving optimum planting dates.

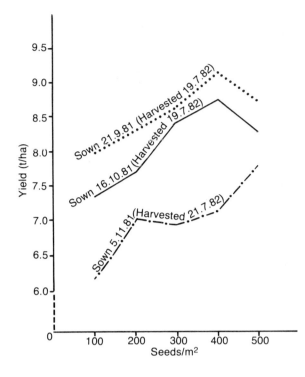

FIGURE 5.4 Winter barley: seed rates and sowing dates. (ADAS at RAC, Cirencester)

Heavy land farmers have a saying born of experience about wheat, 'Drunk or sober, sow in October'!

The latest safe sowing dates for each variety listed by NIAB are a useful guide when one has seed in the barn which could not be planted in autumn. Later-sown winter wheat after cash roots may often give a better financial outcome than waiting for spring corn. Durum wheat may suffer winter damage and disease if sown before mid-October and is sometimes sown in Spring.

TABLE 5.7 Sowing dates for winter cereals

Cereal	Target period	Latest sensible date	Latest safe date (NIAB)
Rye	Mid September	Early November	Mid February
Barley	Mid/late September	End October	End of February for most varieties
Oats	Late September/early October	Mid/late October	End October in south; no later than mid October if north of Yorkshire
Triticale	Late September/early October	Late October	Mid February
Wheat	End September/early October	Mid February	Mid/end February, mid March for some varieties
Durum wheat	October (not earlier)	Early November	March/April if sown as a spring crop

PLATE 5.4   Cereal establishment: crops in adjacent fields in Gloucestershire in mid January after snow thaw (12 cm row width). (Left) − Winter barley cv Video, sown 17 September @ 330 seeds/m²; shown at GS25. (Right) − Winter wheat cv Mercia, sown 30 September @ 400 seeds/m²; shown at GS22 (along tramline).

Spring   Generally, the earlier the better for sowing or as soon as an adequate, warm-enough tilth is achieved and weather is favourable. Early sowing can advance harvest-readiness usefully.

Suggested order of drilling:

Late January	some spring barley
Mid February & March	spring oats, spring wheat, spring barley, durum wheat
April	spring wheat, spring barley
Late April/early May	maize

Spring barley is the most flexible as far as its tolerance of sowing dates is concerned. Like winter wheat, though, there is a yield penalty for sowing too late: after mid-March in most districts its yield potential will be affected.

Spring oats suffer worst from late sowing as a result of both frit fly damage and suceptibility to spring droughts. Oats should be in by the end of March in most districts, except where these two problems are unlikely. The same effect from droughts can occur with spring wheat although it is chiefly confined to stronger, less drought-prone land.

The ideal sowing date for *spring wheat* remains very controversial. Many are in favour of early sowing for some of the reasons already discussed. Some are convinced that the earliest sowing should be followed by no prolonged period to check seedling development often using a late March/early April date. The heavier soils commonly used for spring wheat often do not allow an earlier date anyhow!

Northamptonshire farmer Richard Holbrook has suggested that three consecutive days with a 2.5 cm soil temperature of 7°C, measured at noon, should precede drilling. Interest is increasing in autumn sowing of spring wheat varieties such as Tonic.

Soil temperature is regarded as the arbiter for *maize* drilling. As a species native of warmer climes, maize cannot be sown much before late April/early May in any district of the UK. The 10 cm soil temperature needs to have attained 10°C in order for germination to proceed rapidly and smoothly enough.

SOWING METHOD

The alternatives are broadcasting and drilling (ordinary, plain or 'seed only' drill; combine drill; direct drill; precision drill; pneumatic-feed drill). The choice depends on a number of factors: the farming system, soil types involved, season.

The larger the cereal enterprise, the greater is the likelihood that several soil types will be encountered and the greater is the need for more than one drill, probably of different types. The smaller cereal enterprise may justify only one drill, if any at all, and the drill chosen will need to be flexible in terms of the soil types and conditions under which it can operate effectively. Of course, cereal drills can be used for other crops as well, notably oilseeds and pulses (peas and beans) with relatively minor adjustment.

Broadcasting

It is still possible to find the old seed-fiddle in use

in the UK for small patches of feed corn in more upland districts. This consists of a shallow trough with a slide mechanism which is pushed to and fro as the planter traverses the land. A less accurate system (practised under the author's regular observation thirty years ago in Derbyshire!) consisted of hand-scattering seed from a kidney-shaped container strapped to rest at one side of the waist. Many older farmers have tales of this kind and the student will do well to learn about them, especially if intending to work in a less-mechanised agriculture.

Broadcasting nowadays may be done either from the air or, more usually, from land-operated fertiliser distributors, which may be tractor-mounted or trailed. They may have an accurate, pneumatically controlled feed mechanism or be hopper-fed to one or two spinning discs or to an oscillating spout or spouts. Their calibration should be checked before use!

The advantages of broadcasting are:

1. Greater speed than drilling. The difference depends on the relative spread of machines and on land conditions but is likely to be of the order of 1.5 to 2 times as fast. This can be very useful in a planned system on weak-structured soils which need sowing rapidly once a tilth has been created. Some silty clays and silts which need dealing with all at once at a critical moisture content fall into this category. Unless a harrow is towed behind the broadcaster the total establishment time could be every bit as long as drilling which is less demanding of a following harrow.

2. Broadcasting can be done in late autumn or early spring when drilling would be impossible. The soil damage resulting may be small and can be minimised by wide tyres or cage wheels.

3. Primarily grassland farmers who just grow a little feed corn can be independent of contractors if they use their own fertiliser distributor to sow corn. This is particularly suitable in wetter upland districts of the west and north-west.

The disadvantages of broadcasting are:

1. A less uniform distribution of grains is probable but this can be offset by the use of a skilled operator, markers and a more accurate type of machine.

2. Depth of seed placement is likely to be much more variable than drilling. This is frequently the most significant criticism relevant in practice. If the land can be uniformly C-tined or ploughed before the broadcaster and harrowed over, then most seed can be covered. This may often be preferable to trying to make a full conventional seedbed. In harsh seedbed climates one may expect significant plant loss from broadcast autumn corn overwinter.

3. Seed rates generally need to be higher, typically an extra 25 kg per ha. This is partly to offset the likely disadvantages of seed placement and partly because broadcasting is most often done at a time and in conditions in which each seed stands less chance of establishing itself. To this extent a comparison between broadcasting and drilling is not entirely fair.

Where broadcasting is carried out under good conditions as a planned alternative to drilling, it can give equally good results (Table 5.8). Indeed, on weaker-structured soils it may be superior owing to its speed and therefore timeliness advantage. It is certainly a good backstop strategy to have in mind in a closed autumn when drilling has not been completed. Winter wheat is the most adaptable candidate in these circumstances and it is generally going to be better to broadcast it in November than wait for the *chance* of proper drilling conditions arising up to the end of February.

Drilling

Greater accuracy of seed placement, particularly of depth, and therefore better control of population density are *possible* (Plate 5.5). However, good broadcasting is obviously much preferable to poor drilling! Drills should be calibrated for the particular species to be sown prior to use rather than the guess and mess approach! It is axiomatic that any tool is only as good as its user. There is a danger perhaps in relying on the sophistication of the drill and paying too little attention to the standard of tilth and speed of operation, both of which count for a great deal.

Ordinary, plain or 'seed-only' drill

This is preferred by some farmers if soil nutrient status is ample and time is critical, both of which factors apply to larger, high-yield cereal systems. The time taken to apply fertiliser separately can be saved if the drilling tractor is fitted with a tank to apply liquid fertiliser at the same time; this does not hinder drilling as much as trying to handle solid fertiliser with a combine drill. Many are now wide (4 to 6 m) and have pneumatic (air-seeder fan assisted) feed mechanisms (see below).

Combine drill

This is very popular owing to the completion of two jobs at once, both seed and fertiliser distribution. The

TABLE 5.8 Winter barley establishment in Scotland: broadcasting can equal drilling

	Broadcast and rotavated to 70 mm	Ploughing to 250 mm and conventionally drilled	Direct drilled
Spring nitrogen rate of 120 kg/ha			
Plants/m^2	300	350	280
Yield t/ha	7.1	7.3	7.3
Spring nitrogen rate of 170 kg/ha			
Plants/m^2	320	370	320
Yield t/ha	8.1	7.8	7.9
Estimated total time to prepare the land, sow and cover the crop (hours/hectare)	0.7	2.7	0.8
Estimated total tractor fuel used to prepare the land, sow and cover the crop (litres/hectare)	17	40	13

Notes: (1) Results for the SIAE broadcasting system from a replicated field experiment with winter barley on a sandy clay loam in Midlothian, 1982.
(2) Broadcast by 12 m pneumatic machine.

PLATE 5.5 *Drilling is potentially more accurate than broadcasting. Shown here is the very popular MF30 drill, 2.4 m model, available as 17.5 cm or 13 cm row width, plain or with fertiliser combined (designed for 12 m tramlining and fitted with automatic kit for this).*

fertiliser is released close to the seedling roots. This fact is especially relevant to the sparingly mobile phosphate but only shows benefit on soils of P index 0 or, sometimes, 1 and K at index 0. This method has no advantage on fertile soils and can be some 40 to 25 per cent slower than plain drilling (depending on such factors as whether an extra man is available for filling, size of bags and fertiliser rate). Of course, this refers only to sowing rates because extra time must be added for fertiliser spreading on a separate occasion for the plain drilling. Thus the advantage is in *timing* of sowing for a plain drill, rather than in total time taken for seed and fertiliser distribution. However, there are several timing options for applying the fertiliser in conjunction with a plain drill. At lower nutrient status a plain drill would require more basal fertiliser to perform as well as a combine drill. However, for high-yielding crops at indices of 2, 3 or 4 for P and K, the total input of extra fertiliser may still be high. There is a suggestion of scorch risk if all this is combine drilled in dryish conditions. Another criticism levelled at the combine drill is that it often weighs more than the plain drill when full and is therefore more likely to cause soil compaction. This obviously depends on drill design.

The combine drill is more flexible than the plain drill in that it can be used just to drill seed when time is short and yet is available for closer fertiliser placement on any fields of lower fertility that have to be drilled. However, its mechanism tends to be less efficient and fertiliser corrodes it; it is less suitable for narrow-row work too and costs more per metre width than a plain drill.

Direct drill

These drills are designed to operate in uncultivated or scarcely scratched land. They are of two principal types — those which penetrate the ground by means of discs and those employing tines. Direct drilling is a significant development and is thoroughly reviewed by Dr Harry Allen (1981). To get its current significance in perspective, less than 4 per cent of all cereals are direct drilled (Plate 5.6), but corn accounts for half the direct-drilled crop area (over 95 per cent of this being winter corn).

The advantages of direct drilling over conventional drilling are:

1. Time-saving It is reckoned that by direct drilling one can sow some five or six times the area per man-hour which could have been covered using the whole gamut of conventional cultivations and drilling. Thus

adequate favourable weather is more probable. In fairness, comparison should be made with reduced cultivations systems which are the more common alternative on soils with good direct drilling potential, and the time-saving could then be reduced to half-time. On lighter land in a dry autumn, crops braird (emerge) better. However, direct drilled winter corn is often slower to emerge on denser heavy land (except where moisture loss impairs conventional sowing) and so must be sown sooner. In addition there is a shorter period in the autumn when direct drilling will work well whilst conventional drills can produce good yields from a wider drilling period.

2. Labour-saving Fewer man-hours to pay for per hectare.

3. Fuel-saving Shorter operating time and perhaps as little as 125 t/ha of soil moved by contrast with about 2500 t/ha ploughed.

4. Target winter corn area sown This is a major consideration especially on calcareous clays and clay loams which suit both winter wheat and direct drills but yet often turn too sticky before conventional drilling can complete the desired hectarage.

PLATE 5.6 Direct drilling.

5. Moisture conservation In some circumstances normal cultivations can deplete surface moisture which was already marginal for even germination.

6. Soil conservation The absence of inversion and the little disturbance of soil allow:

- Natural structure to develop
- Stabilising organic matter to accumulate near the surface
- Earthworms to multiply, often threefold
- Less soil erosion on susceptible sites where loosening and drying would render it vulnerable
- Better carriage of machinery for subsequent treatments of the crop
- Fewer weed seeds to rise to the surface for germination
- Less likelihood of soil damage than with cultivations prolonged into a wet period

The disadvantages of direct drilling by contrast with conventional drilling are:

1. Soil types suitable for its effective performance are more restricted. Soils were categorised into three groups for suitability to direct drill by Cannell *et al.* when at Letcombe Laboratory near Wantage. Most suitable soils are calcareous loams and clay loams of good, stable structure on well-drained sites – especially soils overlying chalk or limestone, notably Cotswold brash.

2. Chemical usage is increased. Herbicide use is particularly likely to increase significantly, and more N fertiliser (perhaps 15–20 per cent up) may be needed. Such chemicals greatly increase the energy invested per hectare which can offset the fuel-savings made.

3. Trash disposal is critical for effective direct drill operation. This means burning is more necessary. It is possible to drill into *clean* unburnt stubble after baling and removal of straw but this is not desirable in the UK. Especially in the north of the country where harvest is later and weather often wetter, good burning can be a rare event. True direct drilling does not allow the incorporation of bulky organic manures.

4. Direct drilling allows a higher proportion of winter crops to be planted, thus intensifying the problems of weeds, pests, diseases, trash disposal and soil compaction. Yields from direct drilled spring corn have been less satisfactory than for winter corn and inferior to conventional drilling on more soil types.

5. Biological activity of the immediate surface soil may become harmfully restricted and the rate of turnover reduced. More research is needed on the microbiology of this.

6. Physical restrictions such as smeared drill slots (especially with disc drills) can retard root development in the seedling crop and make it slower to establish, more vulnerable to slugs and perhaps more favoured later by surface grazers such as hares. More subsoil treatment, especially in the top 35 cm or so, seems necessary than with judicious ploughing.

7. Direct drilling has been associated with the spread of sterile brome grass as an increasingly significant weed in some districts. Furthermore, following regular burning and consequent carbon/ash accumulation at the surface, other grass weeds such as blackgrass are not well controlled by soil-acting (residual) herbicides.

8. A direct drill is a more sophisticated, costly and specialised piece of equipment than a conventional drill. A robust, spring-tined type of direct drill is more flexible as to the soils and systems it suits and the range of conditions under which it can operate satisfactorily than a disc-type direct drill. Under good conditions a disc-type direct drill is more likely to achieve controlled depth and distribution of seed; soil penetration can be a more severe problem for this type under dry conditions. It is not necessary to direct drill to achieve single-pass cereal sowing; fewer passes associated with reduced compaction can be very beneficial (Table 5.9; Plates 5.7 and 5.8),

TABLE 5.9 Effect of seedbed tractor wheel traffic on ploughed land on establishment and grain yield of winter barley

Number of wheel passes	0	1r*	1	2	4	6
Air-filled porosity at 3cm (% w/w)	34.5	16.1	14.7	11.4	5.8	4.6
Number of plants m^{-2}	315	189	131	75	36	9
Grain yield (t/ha^{-1})	6.7	5.5	3.9	3.4	2.3	1.3

* Tyre inflation pressures at about half the minimum recommended values.

Source: Campbell *et al.*, 1982.

PLATE 5.7 The Dutzi enables complete soil cultivation in a single pass.

because cultural weed control and natural mineralisation are steadier in less disturbed soils.

Precision drill

A precision drill was developed by Stanhay consisting of individual drilling units incorporating a precise seed delivery mechanism. The drill was expensive and demanded a relatively fine, firm, flat tilth. However, the results under the required conditions were good and it is to be hoped that this sound principle is developed in future.

Pneumatic-feed drill

These drills have conventional drill mechanisms differing only in having large hoppers and pneumatic feed through flexible pipes. They have large capacity in both seed weight and hectares they can cover in a day and can do a good job. They are not particularly expensive compared with a conventional drill of similar width. Their great advantage is that they are light-weight machines. Conventional drills of similar width have to be of much sturdier construction across the whole width and thus result in more soil compaction.

The range of drill coulters is not discussed here and readers are referred to the ADAS mechanisation literature on corn drills or to machinery texts, including the annual 'Green Book' edited by H. Catling which gives an international summary of machines available.

TRAMLINING

Tramlines are regularly spaced tracks established at the beginning of a crop's life and followed for all sub-

sequent treatments applied. They are usually established by periodically blocking coulters of the drill either with manual cable or with hydraulically or electronically operated controls. Alternatively the drill-man may drive wide at intervals to leave a single narrow track undrilled. A single line increases the risk of overlapping subsequent treatments if it is not regularly followed with the same wheel! Furthermore it leaves one track as a persistent wheeling with plants which can remain green and troublesome at harvest.

The tracks may be developed as persistent wheelings but then the above disadvantage applies to them both. A chemical burn out has been used to develop the tracks once the seedling crop has emerged sufficiently, but this is not much used. Both this method and the persistent wheelings waste seed which need not have been drilled in the first place, although this method does allow better travelling conditions in autumn and early spring.

In all events, tramlines must be accurately laid by the straightest-eyed tractor-driver on the place!

Advantages of tramlines:

- Once established, no further marking is needed.
- A fast, straightforward job can be done accurately even at night.
- Access with vehicles for crop inspection is possible over a large part of the season.
- Crop damage from later treatments is reduced.
- The crop protection programme period is extended.
- The crop promotion period is extended, i.e. late fertiliser or growth regulator can be applied with less fear of mechanical crop damage.
- Wheelmarks are concentrated, clearly defined and can be specifically subsoiled later.

PLATE 5.8 Reduced passes with wide harrows; reduced compaction by double rear wheels, but note absence of front wheel attention!

Disadvantages of tramlines:

- Miscalculations in following them are either complete double-doses or total misses. These can be later labelled 'farmer-trials' to save face!
- The cropped area is reduced by some 2−3 per cent depending on tramline frequency. However, the crop rows adjacent to the tramline compensate by an extra yield of as much as 50 per cent, which almost offsets this loss of plant.
- Weeds may colonise the tramlines and interfere with the harvest. However, tines behind tractor wheels can minimise these and frequent crop treatments obviate their development.
- Initial costs of introducing a tramline system put off some farmers. This is not just the modest costs of tramlining kits for drills, if these are the method chosen; it is the cost of fitting all one's kit to a particular module (drill, fertiliser-spreader and sprayer). In some cases there is the nuisance of only having one tractor electrically equipped to take a tramliner.

Overall the adoption of tramlines is associated with higher yield results with cereals. This is partly because more progressive farmers adopt tramlines but also because better treatments, particularly later applications, are given to protect crops.

UNDERSOWING

Cereals may be used as a nurse crop for a grass ley to follow. Spring cereals, especially barley, are preferred to autumn-sown crops because they offer less competition to the grass seeds. Indeed, undersowing is not feasible at all with early-sown, dense winter cereals.

The seed rates of both corn and grass may be cut somewhat compared with sole crops, say 100−150 kg/ha for the corn and 20−25 kg/ha for the grass mixture. If the cereal is much subordinate to the best establishment of the ley, its seed rate may be reduced to as little as 65 kg/ha and it may be taken as a silage cut. The grass mixture may be drilled immediately after spring corn or simultaneously unless it is very fast establishing (e.g. Italian ryegrass/broad red clover), in which case a week or two delay is advisable.

The advantages of undersowing are as follows:

- There is one set of cultivations for two crops.
- The corn acts as a nurse for slower-establishing grasses.
- Some weeds are smothered early on.

- A cash crop is obtained whilst the ley is establishing.
- Full use is made of the land with reduced opportunity for weed colonisation.
- Annual fixed costs are spread over one and a half crops.

And the disadvantages of undersowing:

- There is inevitable competition for resources. For instance, the dry summers of 1975 and 1976 saw many undersown ley failures which consequently cost twice to sow.
- Conditions cannot be made ideal for both crops simultaneously, e.g. fineness of tilth, nitrogen fertiliser policy.
- Harvesting of the cereal grain may be difficult, particularly if the ley was sown too soon after the corn and is already tall.
- Flattened (lodged) cereals may cause great damage to the ley and necessitate patching up later on.
- Later weed control by chemicals is difficult without damage to the sward, especially any legume component, e.g. CMPP kills clover as well as chickweed whereas MCPB kills neither!
- The technique is not recommended for the establishment of longer leys unless cereal yield is sacrificed. The costs of establishing longer leys and their importance over several seasons necessitate the best start possible. The move towards winter barley has favoured a swing away from undersowing.

The early harvest of winter barley allows ample time for land preparation to direct sow leys into good seedbeds to establish well before winter without the complications of undersowing. Nevertheless, undersowing remains quite a frequent practice, especially for shorter-term ley breaks.

OVERSEAS

Paddy rice is commonly scattered in a nursery bed and then transplanted into puddled land as groups of seedlings.

Maize, sorghum and *millets* are still quite commonly established in the tropics either by dibber (stick to make hole for seeds) or by the 'heel-and-toe' method − dropping in several seeds (generally more for poorer-fertility sites and often too many) and covering them as one goes along. A skilled planter can proceed remarkably quickly in this way. Of course, the tropical cereals are also drilled with equipment drawn by draft animals as well as tractors.

Chapter 6

FEEDING CEREALS

Yield and quality of a cereal crop are closely dependent on adequate mineral nutrition. There is a direct relationship between the yield attained and the nutrients demanded and extracted from the soil. Success in feeding crops on a field scale hinges on harnessing and promoting the soil's natural nutrient cycles, supplemented where necessary, rather than using the soil as a mere physical support medium into which all nutrients are added.

PRINCIPLES

Nutrients are taken in as ions (small, charged particles) from soil solution. This uptake uses energy; it is an active process resulting in concentrations often many times higher inside roots by contrast with surrounding soil solution. Roots must respire to release the energy needed for nutrient uptake. Therefore, good soil aeration is vital to supply oxygen for this respiration.

Roots must explore the soil to reach nutrients. This is especially true of the less mobile nutrients, notably phosphate. One cannot have a good crop with a poor, uneven root system. Availability of nutrients depends on the solubility and concentration of sources present.

The correct pH (around 6.5) gives maximum aggregate availabilities of the nutrients present. Whilst some elements become more available at lower pH (e.g. manganese) and others are more available at higher pH (e.g. molybdenum), the whole range of essential elements achieves maximum total availability around pH 6.5 to 7.5 for cereals.

Responses to added nutrients (fertilisers) are best on well-structured soils. Here roots can develop to their full volume for efficient recovery of nutrients.

Best responses occur where fertility has been high previously. Obviously added nutrients display the law of diminishing returns whereby a saturation point is reached beyond which further response will be negative:

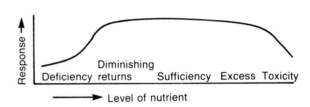

However, extra fertiliser alone cannot transform a poor soil overnight. The healthiest nutrient cycles occur where biological activity is high. This activity by a diverse range of organisms amounting to a high soil population not only releases nutrients more rapidly but tends to suppress pathogenic species. However, if an element is in limited supply, micro-organisms will compete with the crop and can immobilise available supplies in their cells.

Balanced supplies of all essential elements are vital. These include:

Major elements needed in relatively large amounts:
Nitrogen (N)
Phosphorus (P) — expressed as P_2O_5, phosphate
Potassium (K) — expressed as K_2O, potash
Magnesium (Mg)
Calcium (Ca)
Sulphur (S)

TABLE 6.1 Typical nutrient demand in cereals

A. Removal of major nutrients (kg/t @ 15% mc)

	N	P	K	Ca	Mg	S
In grain	17.0	3.4	4.7	0.5	1.3	1.3
In straw	6.0	0.7	6.8	3.0	0.8	0.9

B. Removal of minor nutrients (g/t @ 15% mc)

	Fe	Mn	Zn	Cu	B	Mo
In grain	40	25	25	4	0.8	0.3
In straw	40	60	15	2.5	6	0.3

Note: Species and varieties differ, e.g. durum wheat may
      remove up to 40 kg N/t yield whilst milling varieties of
      common wheat may remove 27 kg N/t.

Source: From J. Archer, 1985.

Minor (trace) elements (needed in relatively small
amounts):
   Copper (Cu)
   Manganese (Mn)
   Iron (Fe)
   Boron (B)
   Sodium (Na)
   Chlorine (Cl)
   Molybdenum (Mo)
   Zinc (Zn)

Other elements are removed by cereals but not con-
clusively shown to be essential. These include
selenium, aluminium and silicon (which, as silicates,
stiffens the straw).

In general a proportionate removal of all essential
elements occurs with an increase in yield (Table 6.1).
According to the law of metabolic equilibrium, the
plant will *tend* to try to maintain a fairly steady ratio
of all elements in its tissues. Obviously a disproportionate supply of one element, notably N, can upset
this to an extent.

Best results occur when nutrient availability is
timed to coincide with crop demand. This applies to
the more soluble sources, especially those needed to
supply nitrogen. The most successful farmers antici-
pate such periods of demand and apply N beforehand
to ensure they are not limiting the crop.

Nutrients exist in the soil in three main *phases*
(Figure 6.1). The reserves are analogous to a grocery
shop (or deposit account at a bank), exchangeable
reserves to a household larder (or current account)
and solution phase to a dining table (or money in
pocket) from which food is directly consumed (or
money directly used). Ions in solution are not only
available for crop uptake but also vulnerable to leaching. Leaching occurs when solution supplies exceed
rate of crop uptake and percolating rainwater is
plentiful.

It is in the long-term interests of the crop that a
balance be maintained in all the above three nutrient
phases. The manuring and fertiliser programme
needs, therefore, to:

● *Provide* all essential elements
● *Start* the crop off well
● *Sustain* supplies of nutrients to harvest
● *Build* up soil nutrient reserves
● *Stimulate* beneficial biological activity

In addition to the above requirements the manures
and fertilisers sought should:

● *Resist* losses by rapid breakdown, leaching and
  volatilisation
● *Store* well
● *Spread* easily and accurately
● *Leave* soil pH unaffected (unless specifically
  chosen to alter it)
● *Benefit* soil structure and moisture-supply
  characteristics

PRACTICE

Three questions have to be considered:

1. What materials to use?
2. How much?
3. When?

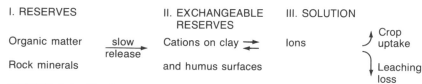

FIGURE 6.1 Nutrient phases in soils.

ORGANIC MANURES

These materials are relatively bulky, slow-release sources of nutrients. Their use is limited or nil on many cereal crops for the following reasons:

- Cartage of large quantities is expensive in fuel, time, equipment and labour.
- Incorporation is difficult or impossible with reduced cultivation systems.
- Supplies are decreasingly coincident with cereal-growing fields.

Nevertheless an ample level of organic matter is essential in the soil, as its quality and a good rate of turnover must be maintained in the long term. Many cereal farmers do use available manures and substantially reduce fertiliser bills.

Light sandy soils will run short of organic matter unless supplies are replenished. In these soils organic manures are a valuable adjunct to crop residues as a source of humus which fulfils the roles of nutrient and moisture-retention carried out by clay in heavier land. Where livestock production and cereal-growing are integrated within the same farm or occur on adjacent farms, organic manures can be beneficially utilised.

Organic manures have four categories of effects:

Chemical Previously living matter has acquired all the essential elements for life and thus, potentially, can return them all to the system. It is possible to calculate ample availability of the total nutrients used as NPK fertilisers in the UK (some 2.25 M tonnes/year of nutrients) within the manures produced on farms and sewage. The problem is balanced composting and distribution which is only feasible in labour-intensive agriculture. Thus fertilisers have become an indispensable input for UK corn yield levels now.

Physical Bulky organic manures have effects on the physical state of the soil. Humus will improve the structure of lighter soils helping to bind sands and increase their plant-available water-holding capacity. On heavy land, bulky manures help to open out the soil.

Biological Organic manures stimulate beneficial organisms such as earthworms and useful microbes. However, spreading these manures may also disperse weed seeds and diseases if decomposition is only partial, which is frequently the case.

Economic Organic manures render the mixed farm more self-sufficient and thus less dependent on bought-in fertilisers. It is possible to supply all the requirements except nitrogen for high-yielding cereals from FYM (farmyard manure) or other manures. This can reduce the costs of nutrients used, depending on the system and distances involved in handling the manure. There are some cereal growers who produce exclusively with organic manures and then sell their grain (especially breadwheat) at a premium via healthfood outlets.

FYM (from cattle yards) may typically contain the following kg per 25 tonnes:

N	P_2O_5	K_2O	Mg
40	50	115	20

In practice, slurry or FYM used as a planned part of the fertiliser policy should be analysed from time to time because its composition varies with:

- class of livestock producing it
- diet of livestock producing it
- type and amount of bedding used
- period and method of storage used
- dilution with rainwater, parlour-washings, etc.

Maize is particularly responsive to slurry or FYM. Stubbles before maize also provide a convenient place to spread muck during the earlier part of the year. There are materials available which are manufactured from organic sources, such as fish offal, coupled with mineral salts. These combine the slower-release properties of organic manures with the ready and tailored availability of most inorganic salts.

Sensible management of organic matter is a necessary base to any sustainable agriculture and a prerequisite of long-term fertiliser policy.

INORGANIC FERTILISERS

Liming

Before any individual elements are considered it is essential to ensure that the correct pH is attained. The over-riding importance of this factor cannot be over-stressed. The British tendency is towards lower pH, i.e. acidity problems ('sour' soils) which are rectified by liming. For cereals the optimum pH is between 6.5 and 7.5. Critical pH level, i.e. below which performance is significantly impaired, is:

5.8 for barley and maize
5.3 for wheat

5.0 for oats
4.8 for rye

To allow for the sampling error it is wise to add 0.2 to each of these to give the lowest acceptable level in practice. It is even wiser to maintain lime status to give the target pH 6.5 to 7.5. This is the setting within which the rest of the fertiliser given will produce the best results.

Statistics indicate an alarming and steady decline in the use of lime in the UK, no doubt encouraged by the removal of the liming subsidy in the late 1970s. As early as 1972 the Strutt report pointed to this decline.

Lime is lost from the system by:

- *Leaching* This is obviously related to rainfall amounts but is never less than 250 kg/ha/year of calcium carbonate equivalent and maybe sevenfold this quantity (Strutt, 1972; Russell, 1987).
- *Crop uptake* Cereals remove about 2 kg of calcium carbonate equivalent per tonne of grain and 6 kg/tonne of straw. Higher yields thus remove more, though straw-burning returns calcium.
- *Use of acid-forming fertilisers* Most nitrogenous fertilisers tend to acidify the soil as pointed out by Strutt (1972). Even ammonium nitrate needs at least the same weight of calcium carbonate (ground chalk or limestone) to neutralise it.

Shortfalls in lime use compared with need probably run at the order of 20 per cent according to the Agricultural Lime Producers' Council. The rule is to lime light soils little and often whilst heavy land needs liming less frequently but with higher doses. As an approximate guide,

Soil type	Lime or calcium carbonate equivalent needed per 0.5 pH rise (kg/ha)
Sandy, light	1250−1850
Medium loams	c. 2500
Heavy, organic	c. 4000

Obviously the cereal grower must:

- Monitor the need or otherwise for lime by regular pH checks. Note that acid soils can even overlie chalk and limestone. Apart from a pH test, acidity is indicated by the prevalence of certain weeds such as corn spurrey and corn marigold, by sickly crops and by slow decomposition of crop residues.
- Have deficient soils analysed to determine how much lime is needed to rectify the low pH. There is a danger of overliming light soils to the detriment of crops. Whilst barley can tolerate pH 8.5, the other cereals are not suited to this level.

Liming is chiefly a contractor's job. Some can supply magnesian limestone which not only provides calcium but also replenishes magnesium. The application of a little lime to cereal straw and stubbles assists decomposition; indeed the use of nitro-chalk at the equivalent of 10 kgN/tonne of straw is a sensible procedure when chopping and ploughing in unless this practice has become the regular straw disposal procedure. This material supplies a little lime as well as nitrogen, both of which promote bacterial activity and hence favour more rapid decomposition of trash.

SOLID FERTILISERS

These consist of granular or prilled materials classified as follows:

- *Straights* supplying chiefly one major element, for example

Ammonium nitrate	(34% N)
Urea	(46% N)
Triple superphosphate	(45% P_2O_5)
Muriate of potash	(60% K_2O)
Kieserite	(16% Mg)

- *Compounds* supplying several major elements together, usually N, P and K, for example,

% N	% P_2O_5	% K_2O	
9	24	24	winter cereals
0	20	20	winter cereals
20	10	10	spring cereals
10	25	15	soils of low P index

Many more exist and manufacturers supply ample literature! In practice, various combinations of straights and compounds are used. Some larger cereal growers find economies in supplying all nutrients as straights. Most growers find convenience in using compounds for basal (seedbed) dressings and then give nitrogen as 'straight' top-dressings. Big packs and pallet systems have eased handling whilst better wide-spreaders (Plate 6.1) and tramlines have improved accuracy of application on a large scale. Significant amounts of solids are top dressed from the air particularly when land conditions remain wet but weather is mild. The cost of doing this is not prohibitive at present: returns from timely application are usually reckoned to offset any extra costs.

PLATE 6.1 *Fertiliser spreader — Accord 12 m turbospreader.*

Fertiliser blends are now being offered which can be convenient and cost-effective providing ingredients are compatible chemically and in terms of particle size and shape to allow homogeneous spreading. Application costs may thus be saved, though this may not compare favourably with rotational fertilising using straights of P and K every third or fourth season to keep up established good soil indices.

LIQUID FERTILISERS

These are aqueous, non-pressurized solutions; only aqueous ammonia is under pressure and must be injected into the soil. They are not the same thing as foliar sprays which are dilute solutions, usually of trace elements, applied as 'first aid' treatments for rapid leaf absorption or in drier weather later in the crop's life, e.g. ear cocktails for winter wheats. Most nitrogen solutions contain ammonium nitrate and urea and can give up to 33 per cent N. Compounds usually contain ammonium polyphosphates for N and P plus muriate of potash for K. It is only possible at present to achieve similar nutrient concentrations to solid compounds by having solid crystals suspended in liquid fertiliser. These require constant agitation to prevent crystal growth and whilst available in the United States do not seem ideal for British farms.

Advantages of liquid fertilisers
1. Ease of *handling* (pump, don't hump!). However, one-tonne bags and pallets for solids help them to compete on this point.
2. *Speed* of application — faster filling time and

large boom widths still giving accurate application have favoured liquids (Plate 6.2). However, accurate 20 m solid spreaders are now available.
3. *Flexibility* of application. Liquids can be applied from trailed sprayers (which can do other spraying jobs when not in use for this, though they should not be the sole spraying tackle available to a serious cereal grower if timeliness of other treatments is to prevail). They can also be applied from mounted tanks — front, rear or saddle. Narrow-row cereal drills which can normally apply seed only can be converted into combine drills with the liquid application kits available. This gives the advantages of closer fertiliser on poorer soils without reducing work rates since liquid can be quickly pumped whilst seed hoppers are refilled.

PLATE 6.2 *Chafer liquid fertiliser equipment; also used for herbicide application.*

Disadvantages of liquids

1. Cost and siting of *storage* tanks. Tanks can be hired, but they are costly. Whilst filling time is rapid, tractors may have to travel some distance to refill unless one is able to justify several strategically placed smaller tanks.

2. *Scorch* of cereal leaves has been a problem. Some dismiss this and crops scorched prior to stem extension certainly grow out of it surprisingly well. However, no farmer likes to see scorch. The use of dribble bars (tubes to trickle liquid down as larger droplets) or stream jets to give similarly big droplets has reduced, though not eliminated, cases of scorch. Liquid top dressings applied earlier in cold weather can still create problems.

3. The *cost* per kg of nutrient is generally over 5 per cent higher than for solid form.

4. The *concentration* of compound liquids cannot be easily made as high as solids. However, owing to the greater bulk density of liquids, the weight of total material to be carried per hectare for an equivalent NPK dose is similar to solids.

The system remains popular with a proportion of chiefly larger-scale cereal growers. In the absence of scorch, liquids have not been shown to give any consistent differences in performance per kg of nutrient compared with solids.

Fertiliser Rates

It is always important to be clear about figures being used in the calculation of fertiliser rates:

- Old unit system

  1 unit = 1 per cent of 1 cwt = 1.12 lbs

  Conversion: Units/acre $\times \frac{10}{8}$ = kg/ha

  e.g. 40 units/ac $\times \frac{10}{8}$ = 50 kg/ha

- Kilogram

  A 50 kg bag of ammonium nitrate (34 per cent N) contains 17 kg N
  Thus 3 bags (150 kg) of fertiliser would give 51 kg N

A 50 kg bag of 20.10.10 contains 20 per cent N, 10 per cent P_2O_5 and 10 per cent K_2O *OR* 10 kg N, 5 kg P_2O_5 and 5 kg K_2O. Thus to give 100 kg N, 50 kg P_2O_5 and 50 kg K_2O would require 10 bags (500 kg) of fertiliser. Very straightforward!

It is all too easy to confuse 'bags of fertiliser' and kg of nutrient required! It is better to think in terms of kg of nutrient you intend to apply and then work out the weights needed of the particular forms of fertiliser in which it is to be given, and the total number of bags per field. For liquid fertiliser, an 11.11.11 compound, for instance, will provide 55 kg each of N, P_2O_5 and K_2O at 500 litres per hectare; 66 kg/ha of each is given by 600 litres, and so on.

Individual nutrients will now be considered in terms of how much and when application should occur.

NITROGEN

Rocks of the earth's crust contain virtually no nitrogen. Soil reserves ('native' N) are thus in the organic matter fraction. Four-fifths of the air is nitrogen gas. Natural addition of this N gas to the soil system occurs through thunderstorms (no more than 2−5 kg/ha/year in the temperate zone) and through nitrogen-fixing organisms. These are both free-living (e.g. *Azotobacter*) and symbiotic in root nodules of legumes (*Rhizobium* spp of bacteria) and in some non-legume species (e.g. alder). It has not yet been possible to confer nodulation and powers of nitrogen-fixation to cereals. The AFRC unit of N fixation in Sussex has worked on this; biotechnology holds promise.

Crop Demand

Cereals are usually more responsive to the addition of this nutrient than any other element. Nitrogen occupies 1.5 to 2.0 per cent of total yield. It is a constituent occupying some 16 per cent of every protein molecule. Proteins are regular components of protoplasm, the ground substance of every living cell. They include enzymes and the additional protein deposits stored in the aleurone layer of the grain.

Applied nitrogen is analogous in the cereal's life to glucose tablets in that of the long-distance runner in so far as it is a readily available source accelerating the current activity of the plant at the time of uptake. It is renowned for increasing the leaf area of cereals: it also prolongs green leaf life, which is known as leaf area duration. It can encourage tiller survival, particularly in adverse conditions, during the normal period of tiller deaths. It promotes stem extension if applied when the crop is so occupied. It can increase the number of grains surviving per ear and late N can boost grain protein content. Overall, N is associated with greater dry matter production and higher grain yields. However, it accelerates water uptake and crops demand more water in order to take up the N.

Excessive N produces surplus vegetative growth including too much dark blue-green leaf of low dry matter percentage which is soft and more susceptible to foliar diseases. It can encourage too many tillers and generally lush, sloppy vegetative growth. In a wet season it further encourages too tall a plant with soft stems liable to lodge. Later applications can delay ripening. In a dry time, on the other hand, later applications can interfere with grain filling and produce a pinched sample.

Whilst grain may contain up to 2 per cent N at a crude protein level of 12.5 per cent, straw remains low in crude protein (CP) — around 1.75 to 2.0 per cent CP for wheat and rye straw; 2.75−3.0 per cent CP for barley and oat straw. This gives a nitrogen uptake of 20 kg/tonne of grain and, allowing for the highest crude protein level, about 5 kg/tonne of straw. On these figures, assuming a harvest index (grain as percentage of total dry matter) of 55 per cent, this gives, for example:

- 5 t/ha grain + 4 t/ha straw, roots, etc. removing 100 kg N + 20 kg N/ha
- 10 t/ha grain + 8 t/ha straw, roots, etc. removing 200 kg N + 40 kg N/ha

As to the timing of this uptake, it roughly parallels rate of dry matter production.

Nitrogen Supply

It is possible to *analyse* quite accurately the total nitrogen present in:

- *Soil*   Chiefly in the organic matter (OM) fraction. This varies depending on soil types, previous cropping and manuring from 0.1 to 0.7 per cent of the topsoil (top 15 cm). This soil can be taken to weigh 2500 tonnes/ha; therefore, from 2500−17500 kg/ha N is present. However, only as much as one-fiftieth of this is likely to be *available* during the life of any one cereal crop, i.e. as little as 50 kg/ha or less.
- *Fertilisers*   For example, ammonium nitrate 34 per cent N; urea 46 per cent N (avoid use on warmer soil from April onwards and on chalky land especially).
- *Plant tissues*   Even simple test papers for the sap are a rough guide to current N status. Laboratory equipment is very discerning.

However, two crucial questions remain:

- What proportion of this total nitrogen is going to become *available* in a form which the crop can extract to coincide with its demand?
- What proportion of the available nitrogen is the root system going to actually *recover* from soil?

Availability depends on solubility and the preferred form is nitrate (NO_3^-) with a limited intake as ammonium (NH_4^+). Thus the pool of available soil N depends firstly on soil microbiological activity, both decomposers and nitrifiers (their activity is related to moisture, warmth, aeration and pH); secondly, it depends on how much of the total is added as available nitrate from the bag.

Recovery depends on the volume and depth of the root system which in turn depends on date and density of planting and quality of soil conditions for root development. In addition recovery may be jeopardised by the premature loss of available N from the rooting zone principally by leaching (also volatilisation losses from concentrated ammonium sources and denitrification losses in poorly drained conditions). Table 6.2 shows N recovery by winter wheats.

Prediction can be seen to be a problem!

Economic Response

At current relative prices 1 kg of N costs 3 kg of grain, approximately, but produces on average 15 kg of grain. Thus the cost/benefit ratio is 1:5. The current level of nitrogen use on cereals in England and Wales is around one-third of a million tonnes. Varieties and sites differ in responses (Figure 6.2).

TABLE 6.2   Fertiliser nitrogen recovery in nine winter wheat crops

	Fertiliser N kg/ha	Yield t/ha	Fertiliser N % harvested	Nitrogen offtake kg/t grain
Wheat after wheat — direct drilled				
Bounty	213	7.39	44.6	27.4
Huntsman	178	5.57	52.6	25.0
Armada	175	7.71	60.5	23.4
Wheat after wheat — cultivated				
Flanders	140	7.11	47.9	22.2
Hobbit	152	8.87	70.4	20.7
Armada	131	6.72	55.5	21.7
First wheat after a break crop				
Hobbit	143	9.84	49.0	17.2
Armada	182	8.60	64.5	26.0
Huntsman	110	6.74	40.4	22.9

Source: Vaidyanathan, ADAS, 1984.

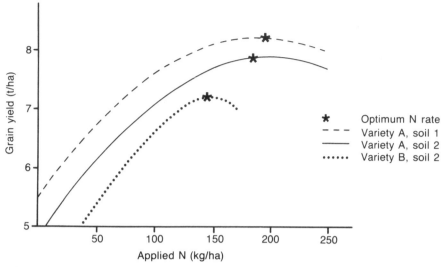

NOTE: Optimum N rate also clearly depends on relative cost/kg of N
and price/kg of grain.

FIGURE 6.2   Nitrogen response curves for winter barleys.

It can be argued that the relative cost of N is too low, causing some farmers to apply unnecessarily high 'insurance' doses. The argument is twofold: (1) the energy cost of producing N fertiliser is about 5 tonnes of oil (or natural gas equivalent) per tonne of N produced, thus it should be used very carefully, and (2) there have been cases where leached nitrates have caused harm, e.g. eutrophication in waterways and nitrate levels in water supplies above the safety limit for babies, as was the case with some East Anglian water supplies during the 1976 drought.

Extra N above the optimum dose *may* depress yields and may increase other costs including extra need for fungicides, growth regulators and greater harvesting and drying costs.

Practical N Fertiliser Policy

The above discussion, I hope, makes clear the difficulties besetting recommendations for N fertiliser. If the *amount* to use is controversial, the *timing* is even more so! In all the multiplicity of trials, theories and pronouncements on the subject the farmer must 'keep his head' and make his own commonsense decisions, taking account of all his knowledge of each particular crop and field. He *must* take account of the total crop management strategy he is using. Because N is such a key factor there is the danger of looking to variations in rates and timing of this

alone to give magic results irrespective of other factors. In reality his policy should be based primarily on *judgement* and *opportunity*: it is likely to vary with experience and with season as well as with new N fertilisers, new varieties or more proven methods of prediction which may come. It seems unlikely that a precise formula will be found — there's still plenty of uncertainty and art to make farming an enjoyable challenge! Real policies are likely to allow a margin of safety to ensure enough N is available to meet demand at all crop growth stages.

The actual N dose will be adjusted in practice by the following:

1. *Expected crop yield* in all the circumstances.
2. *Variety*, e.g. quality wheats may benefit from extra N; stiff-strawed, high-yielding varieties can use more N.
3. *Soil type* — heavier soils are more retentive than light sands.
4. *Previous cropping* — see ADAS soil N index which categorises likely N content of soil (Table 6.3).
5. *Seasonal weather*, e.g. plus or minus 10 per cent if above or below 300–450 mm winter rainfall (1 October–1 April); extra N late *if* moisture is available for uptake without detriment.
6. *Crop appearance*, which includes such observations as colour, density and progress being made in the crop's development. A crop with yellowing leaves

TABLE 6.3 ADAS soil nitrogen index system

	A. Nitrogen index — based on last crop grown	
Nitrogen index 0	*Nitrogen index 1*	*Nitrogen index 2*
Cereals	Any crop receiving farmyard	Any crop receiving large frequent
Forage crops removed	manure or slurry	dressings of farmyard manure
Leys (1–2 year) cut	Beans	or slurry
Leys (1–2 year) grazed,	Forage crops grazed	Long leys, high N†
low N*	Leys (1–2 year) grazed,	Lucerne
Maize	high N†	Permanent pasture — average
Permanent pasture — poor	Long leys, low N*	Permanent pasture — high N†
quality, matted	Oilseed rape	
Sugar beet, tops removed	Peas	
Vegetables receiving less	Potatoes	
than 200 kg/ha N	Sugar beet, tops ploughed in	
	Vegetables receiving more	
	than 200 kg/ha N	

	B. Nitrogen index — based on past cropping with lucerne, long leys and permanent pasture			
		Years since 'ploughing out'		
Crop	*One*	*Two*	*Three*	*Four*
Lucerne	2	1	0	0
Long leys, low N*	1	0	0	0
Long leys, high N†	2	1	0	0
Permanent pasture —				
poor quality, matted	0	0	0	0
Permanent pasture — average	2	1	1	0
Permanent pasture — high N†	2	2	1	1

* Low N = less than 250 kg/ha N per year or low clover content.
† High N = more than 250 kg/ha N per year or high clover content.

Source: After R.J.W. Dight, ADAS.

and with the lower ones dying prematurely from the tips is probably short of N. The significance of this for final yield depends on *when* it arises, how long it persists and how much capacity the crop has left for compensation. Certainly dramatic cosmetic effects can be produced by giving N to yellowed crops! However, not all yellowing is due to N deficiency, though it is often an associated factor.

If a thin crop is struggling, the only obvious thing a farmer can do in most cases is to give a little N (analogous to a little glucose for an ailing child). If it is behind in development and/or at a stage when needed tillers could be lost or grain sites fail to be set, then adequate N must be ensured. Thus crop appearance is more a guide to timing than the amount of N to be used; but a treatment in response to crop appearance could well boost the total N used for the crop beyond the planned level.

7. *Interactions.* Yield responses to N depend on a willingness to support the extra ensuing growth with fungicides and growth regulators if necessary. If irrigation is possible (and it is still rather rarely available for UK cereals), then there is positive interaction.

8. The relative kg *cost* of N and price/kg received for the particular cereal clearly determines the economic optimum N dose. Hitherto the economic optimum dose has almost coincided with the technical optimum needed to maximise yield in most cereal systems.

ADAS recommendations are given in Table 6.4. These are related to soil type and previous cropping. They obviously have to present sensible averages for a wide range of standards of cereal husbandry. Commercial fertiliser companies have sought to incorporate other considerations of crop management in computerised advice procedures such as ICI N-Counter, UKF (now Kemira). These are known as 'balance sheet' approaches to N advice. No doubt computers can play a helpful role in dealing with such an important decision as N policy, especially since it is dependent on so many variables. However, the weather and the farmer's judgement must prevail even over the computer!

In practice, with current varieties in high-yielding systems, up to the following total doses are used to good effect:

	kg/ha	
Winter wheat	200	(+ 50 late on quality varieties if ample moisture)
Spring wheat	175	(+ 35 late if ample moisture)
Winter barley	175	
Spring barley	150−175	(cv Triumph & other high-yielders)
Oats	130	
Rye	130	

TABLE 6.4 ADAS nitrogen recommendations for cereals

A. Winter wheat − spring nitrogen top dressing

Soil type	N index		
	0	1	2
Yield level up to 7 t/ha		kg/ha	
Sandy soils	175	150	75
Shallow soils over chalk or limestone	175	150	75
Deep silty soils	150	50*	Nil
Clays	150	75	Nil
Other mineral soils	150	100*	50
Peaty soils	50	Nil	Nil
Organic soils	90	45*	Nil
Yield level 7−9 t/ha			
Shallow soils over chalk or limestone	225	200	125
Deep silty soils	200	100*	Nil
Clays	200	125	Nil
Other mineral soils except sandy	200	150*	100
Yield level above 9 t/ha			
Shallow soils over chalk or limestone	275	250	175
Deep silty soils	250	150	50
Clays	250	175	50
Other mineral soils except sandy	250	200	150

* Increase by 25 kg/ha where harvesting of the previous crop has damaged soil structure.

B. Winter barley − spring nitrogen top dressing

Soil type	N index		
	0	1	2
		kg/ha	
Sandy soils, shallow soils over chalk or limestone	160	125	75
Other mineral soils	160	100	40
Peaty soils	50	Nil	Nil
Organic soils	90	45	Nil

C. Spring wheat − nitrogen

Soil type	N index		
	0	1	2
		kg/ha	
Sandy soils, shallow soils over chalk or limestone	150	100	50
Deep silty and clay soils	125	50†	Nil
Other mineral soils	125	75	30
Peaty soils	40	Nil	Nil
Organic soils	70	35	Nil

† Increase by 25 kg/ha where harvesting of the previous crop has damaged soil sturcture.

D. Spring barley − nitrogen

Soil type	N index		
	0	1	2
		kg/ha	
Sandy soils	125	100	50
Shallow soils over chalk or limestone	150	125	50
Other mineral soils	150	100	40
Peaty soils	40	Nil	Nil
Organic soils	70	35	Nil

E. Winter and spring oats and winter and spring rye − nitrogen

Soil type	N index		
	0	1	2
		kg/ha	
Sandy soils, shallow soils over chalk or limestone	125	100	50
Other mineral soils	100	60	30
Peaty soils	40	Nil	Nil
Organic soils	70	35	Nil

The rate of N use on winter wheat has increased dramatically in recent years. This is due partly to increased yield expectations and partly to the fact that wheat is much more often grown in cereal runs now. In general the use of N on spring barley is over-cautious, no doubt because of bad experiences in the past with lodging in the absence of any means to control it. Certainly spring barley follows a wider range of crops and manurial residues than any other cereal and some caution is necessary. However, there is fair potential in this crop if high-yielding newer varieties are planted early and densely enough and fed optimally. N treatments should be completed early for malting barleys — by early April.

It cannot be overstressed that N use alone cannot compensate for other gross yield restrictions and surplus N fertiliser will be wasted and even positively harmful in unresponsive situations. However, when higher yields are being consistently harvested owing to ample N as part of good general husbandry, then soil N reserves will also be incidentally built up.

Timing of N

A hotly debated question! Attempts to time N dose to internal development stage are often overridden by seasonal variation. However, N must not be limiting at critical stages of crop development. The quantity available needs to match crop demand at all times. The problem lies in the ready solubility of fertiliser N used.

Progress with controlled-release forms of N is slow. However, urea is available in formulations treated with sulphur or with resins or as urea-formaldehyde or as isobutylidene-di-urea (IBDU). Commercial products such as Didin and Alzone may reduce total N need by up to 10 per cent through slower release and eliminate the need for at least one extra separate top dressing.

The principal form of N used is still ammonium nitrate which is not only readily available to the crop soon after application but also vulnerable to leaching.

Leaching rate depends on:

- rainfall receipts surplus to transpiration plus any existing soil moisture deficit
- soil type: the clay-humus complex can hold ammonium whilst there is some limited retention of nitrate within pores of crumb-type soil aggregates. The maximum percentage of applied N likely to be leached overwinter is 20 per cent

The significance of leaching therefore is greatest:

- overwinter
- on lighter soils, especially sandy land
- where root volume and depth are limiting the capacity of the cereal crop to recover percolating nitrates. It is worth noting that roots will to some extent develop in search of a moving nitrate-rich solution

Owing to the risk of leaching, some authorities have advocated 'spoon feeding' the crop with frequent small doses of N fertiliser. Whilst this may be technically logical, it is neither economically nor practically feasible to carry out many individual dressings. The number of splits of the total N application will depend on the system. When making split dressings some growers like to add a little of other elements, especially K, e.g. using a 25.0.16 or 29.5.5 solid or an N + K liquid.

Autumn Cereals — N Timing

From time of sowing to the middle of February it is not wise to apply more than a total of about 40 kg N/ha; many growers do not exceed 25 kg N during this period. However, winter barley is likely to take in N more voraciously than winter wheat over this time. Reasons for suggesting these limits are leaching and the limited capacity of crops to absorb N until springtime.

The case for more frequent rather than single or no doses of N over the period from sowing to the end of February is greatest when:

- high total N rates are planned owing to yield potential and the crop has been early sown to advance its development
- crop establishment is disappointing, in which case two smaller N doses, say 15 kg, may be given as the sole action possible at that stage which might help

Early springtime is the period for the majority of N to be given.

Seedbed N for autumn corn
Up to 25 kg N/ha may be given in the seedbed. None is needed when:

- richer N residues from previous crops are present
- sowing takes place earlier so that soil temperatures and thus N mineralisation rates are higher
- time is pressing to get sowing done and omission of fertiliser at that stage will accelerate the job

Retaining the N dose until a little later in autumn gives greater management control — some fields may need more encouraging, some none at all.

Some farmers, however, like to use seedbed N, justifying it as follows:

- the job is over and done with, perhaps, for the autumn
- it is an insurance which is beneficial on poorer soils and may show a benefit on better sites in some seasons
- it has never been shown to be detrimental to the crop at around 25 kg/ha

Later autumn N

Especially where no seedbed N has been used, a dose of 30–40 kg/ha is likely to be given at any time from late October to December, according to opportunity (i.e. no other work, mild but not wet weather or on a frosted field which allows traffic). High-yielding crops may be given two autumn doses (e.g. 15 kg seedbed plus 25 kg in November) or similar doses in October and December.

Such double dosing is favoured by:

- the existence of tramlines
- wide-spreaders
- specialised cereal systems where it is desirable to travel and inspect the crops anyhow!

Early new year N

Again, backward and very forward crops are more likely to attract attention at this time than are average crops. Crops may receive 15 kg N/ha in January or 35 kg N/ha between mid and late February. Such dressings are particularly appropriate on shallower soils with early-drilled crops (especially winter barley).

An important argument in favour of such early new year dressings is one of taking advantage of an opportunity when it is offered. Having made these dressings, one is less anxious about the date of achieving the next (often substantial) dose. It is a fact that in many districts and seasons, March is characterised by few travelling days. Thus unless one is prepared to give N from the air in March, one may be unable to apply any at all. The same applies to a lesser extent to April. Therefore, apart from direct (and certainly cosmetic) effects in early new year, dressings at this time are an insurance rendering the date of the next application less critical.

Crops that are earlier-drilled and therefore deeper-rooted into warmer soils are more likely to respond to overwinter N. A thin, backward crop can be aided perhaps in tiller survival by small N dressings overwinter.

It is important to have some early dressing for high yields, but not before February (by GS 25 for winter barley especially).

Main spring N top-dressing

This may be given any time during the period late March to late April/early May, depending on:

- how much N has recently been applied to the crop
- how early the season is: in a wet March a lot of N may leach if a large dose is given early so hold off in such seasons
- what the weather allows!

On forward crops receiving a high total dose, this main dressing may be split one-third/two-thirds between late February/early March and late March/early April (i.e. at GS 31 and GS 32), or even in three lots, perhaps equal or 1:2:1. In general for logistic as well as risk reasons, it is not advisable to apply much more than 100 kg N/ha in a single dose (with current fertiliser materials).

Mid April is a suitable time for winter barley to receive its main dressing (i.e. by GS 31), whilst wheat may be sensibly dosed from then to early May depending on district and its own advancement. Where wheat has been drilled straight after barley, many growers would give N dressings in the same order, quite closely as opportunity allows. Too much N early in this period encourages excessive tiller survival in thick crops and sloppy straw growth, as well as higher disease incidence.

For malting varieties of winter barley it is important not to restrict total N dose, which should relate to yield potential, and to complete all top dressing *early*, ideally by the end of March to avoid prohibitive N levels in the grain (i.e. complete by GS 30).

For low-yielding winter cereal crops or low total requirements owing to high soil index, all the spring N may be given in a single dose in March/April.

Later N

In conditions of ample moisture supply crops may respond to a late dressing of N (late May/early June at GS 37). This can boost grain protein content. The importance of this may be marginal for feed barleys but could be of significant value for quality wheats sold on protein level. There is no evidence to suggest that doses greater than 50 kg N/ha are worthwhile at this stage; 35 kg N may be enough. Response in any event is related to ample moisture for uptake. This late dose should be in addition to the earlier yield-building N doses; it is not a partial substitute for them. Urea is suitable before green leaf is lost and two or three 15 kg N doses have proved useful, though the economics of this treatment may be marginal and only worthwhile if protein level is pushed up into a premium price category.

Spring Cereals — N Timing

Timing of nitrogen is more straightforward with spring than with autumn cereals. This is partly because crops have a compressed period of growth and partly because the risk of leaching is reduced or even eliminated for later sown spring corn. To try to encourage rapid rooting, some farmers broadcast their N ahead of sowing so that it has moved into the soil profile.

At the total levels now beneficial on high-yielding spring corn, many growers like to split the N dose into two, even for later sown crops. This is for two reasons — logistics of application and fear of possible harm if too much N is available in the soil at once. For this latter reason many do not favour combine drilling of high fertiliser rates for spring cereals.

The general recommendations at the bottom of the page are preferred for all spring cereals.

With spring wheat a late bonus for grain protein could be worthwhile on a good crop in a moist season. However, spring wheat varieties are generally superior to winter ones in protein.

Total N for maize should be 125–150 kg/ha for crops expected to yield 5–6 t/ha of grain. A useful part of this can come from previously applied slurry or FYM, sometimes all of it for forage maize.

CAUTION: Use N fertilisers responsibly and avoid nitrate pollution of water caused by excessive or untimely applications, or by leaving fertile land fallow too long.

PHOSPHORUS (P)

This is an important ingredient of such parts of the plant cell as its nucleus and its energy-storage compounds. For historical reasons fertiliser content is expressed as '% P_2O_5', known as 'phosphate' (which is only 43 per cent P). It is absorbed chiefly as $H_2PO_4^-$.

It is the least mobile nutrient, as it is only sparingly soluble. This fact has three practical consequences:

- It does not leach, as release of available forms just matches demand providing soil conditions favour release (pH, structure, microbes).
- Reserves (analogous to a bank deposit account) can be built up in the soil; on the other hand an insidious run-down of deposits can occur over the years.
- Timing of application is unimportant. Indeed, if conditions are difficult either practically or financially, a particular seasons's phosphate dose can be omitted without detriment, provided an adequate soil supply has been previously attained. Varying the dose from year to year will not usually give yield responses.

The available phosphate in a soil can be quite accurately determined from analysis. ADAS have an index from 0 (low) to 4+ (high) availability (Table 6.5).

It is sensible for the serious cereal grower to ensure that the P index is not a limiting factor in securing his yield targets. An index of at least 2 and preferably 3 is desirable. Most soils have substantial P reserves, but removal is proportional to crop yields harvested.

A low supply is associated with:

- acidic soils (or some very alkaline soils)
- wet or compacted soil conditions (restricting rooting); leaves may turn purplish pink in these obvious wet spots
- coming out of permanent pasture into cereals; cultivation systems greatly alter the depth distribution of this and other nutrients (Table 6.6)

Percentage of total N	SOWING DATE				
	Late January/ early February	Late February/ early March	Mid March	Late March	April
In seedbed	20	40	50	80	100*
At three-leaf stage (and by GS 23) (Usually some 4 to 6 weeks later)	80	60	50	20	—

* May be split at high total level of N

TABLE 6.5 Phosphate and potash recommendations for cereals

	P or K index				
	0[a]	1	2	3	Over 3
			kg/ha		
Straw ploughed in or burnt					
Yield level 5.0 t/ha					
Phosphate (P_2O_5)	90	40	40M	40M	Nil
Potash (K_2O)	80	30	30M[b]	Nil	Nil
Yield level 7.5 t/ha					
Phosphate (P_2O_5)	110	60	60M	60M	Nil
Potash (K_2O)	95	45	45M[b]	Nil	Nil
Yield level 10.0 t/ha					
Phosphate (P_2O_5)	130	80	80M	80M	Nil
Potash (K_2O)	110	60	60M[b]	Nil	Nil
Straw removed					
Yield level 5.0 t/ha					
Phosphate (P_2O_5)	90	40	40M	40M	Nil
Potash (K_2O)	110	60	60M[b]	Nil	Nil
Yield level 7.5 t/ha					
Phosphate (P_2O_5)	110	60	60M	60M	Nil
Potash (K_2O)	140	90	90M[c]	Nil	Nil
Yield level 10.0 t/ha					
Phosphate (P_2O_5)	130	80	80M	80M	Nil
Potash (K_2O)	170	120	120M[c]	Nil	Nil

(a) At index 0 large amounts of phosphate and potash are recommended to raise the soil index over a number of years.
(b) Not needed on most clay soils.
(c) A lesser amount may be used on most clay soils.
M This indicates a maintenance dressing intended to prevent depletion of soil reserves rather than to give a yield response.

Source: ADAS.

● situations where previous crops were inadequately supplied, especially on heavy land

Where index 0 to 1 occurs, combine drilling is generally advisable or else an extra 30 kg/ha of phosphate. Indeed, phosphate is generally applied to the seedbed just before, during or soon after sowing. When straw is ploughed in it is sensible to apply the P to this, provided it is going into a soil of index 2^+.

A sensible rule-of-thumb for all cereals is to give phosphate at the rate of 10 kg/tonne of expected grain yield. This will replace P removed and maintain soil reserves. Thus a 7.5 t/ha crop would receive 75 kg phosphate/ha.

Up to 25 kg/ha may be deducted from this calculated requirement if a soil index of 3 is maintained. Burning rather than baling straw makes no significant difference to phosphate supply.

POTASSIUM (K)

This fulfils a role in the salt balance of every plant cell and is important for healthy metabolism. For historical reasons fertiliser content is expressed as '% K_2O', known as 'potash' (which is only 83 per cent K). It is absorbed as potassium ions (K^+). These may be held on the surface of clay and humus in the soil or more tightly between clay plates within soil

TABLE 6.6 Cultivation system and depth distribution of pH, nutrients and OM at Grange Piece, Boxworth (after 8 years)

	Depth cm	pH	mg/l P	mg/l K	% OM
Direct	0–2.5	6.0	66	409	5.4
drilled	2.5–5	6.9	56	390	4.0
	5–7.5	7.3	21	324	3.6
	7.5–10	7.3	15	280	3.6
	10–20	7.3	14	264	3.6
Shallow	0–2.5	6.9	60	428	4.3
cultivation	2.5–5	7.1	57	365	4.3
	5–7.5	7.3	43	309	4.2
	7.5–10	7.4	24	299	3.8
	10–20	7.5	16	280	3.6
Ploughed	0–2.5	7.7	23	336	3.6
	2.5–5	7.7	24	339	3.5
	5–7.5	7.6	24	320	3.5
	7.5–10	7.7	22	295	3.5
	10–20	7.7	26	301	3.6

Note: OM = organic matter.

Source: MAFF.

aggregates. This helps K to resist leaching, though not so much as does phosphate.

Potassium is likely to be short on sandy soils, other light soils and black puffy downland sites. Low status may prevail after cut leys unless slurry is given. Some clays, such as lias and chalky boulder clay, tend to be rich in potassium. Dairy cow slurry is a rich source of K. Soil can be analysed to give a good indication of its potassium-supplying power, known as the ADAS potash index ranging from 0 (low) to 4+ (high) (see Table 6.5).

Crops indulge in luxury uptake of K, i.e. they absorb far more K than is needed to maximise yield. Some of this is returned to the soil during growth. This borrowing of surplus K could be important for the plants' general health and ability to withstand stress (climatic or disease).

Combine drilling of the potash requirement is worthwhile if the soil index is 0, or else give an extra 30 kg of K_2O.

Most growers apply K just before, during or soon after drilling. However, some like to give a little K top dressing in spring. There is a suggestion that in moist conditions which give a late N response there may be a boost to TGW from a little K also.

When straw is burned or ploughed in rather than baled and removed, there is a return of just over half the K taken up by the crop. Where this procedure is repeated, K doses can be reduced accordingly. The cultivation system affects nutrient distribution generally (Table 6.6).

Although the K requirement for cereals is a little higher than for P, many growers use a compound fertiliser supplying equal amounts of each once they have ample indices for both. An index of 2 to 3 for potash is desirable and 3 to 4 preferable for highest yield systems.

A sensible rule-of-thumb to achieve adequate supplies and maintain soil reserves is (like phosphate) 10 kg/tonne expected grain yield, i.e. 100 kg K_2O for a 10 tonne crop! K_2O as muriate is currently about half the price of N and one-third that of P_2O_5 per kilogram (depending on chemical sources and purchasing arrangements).

Overall use of main nutrients on cereals in England and Wales at the time of writing is indicated in Table 6.7.

Table 6.7 Fertiliser use on cereal species in England and Wales in 1987

Crop	Total Dosage								Use of 'Straights' as Sources					
	% area receiving					Actual kg/ha			% area receiving			Actual kg/ha		
	N	P	K	FYM	Lime	N	P_2O_5	K_2O	N	P	K	N	P_2O_5	K_2O
Spring wheat	99	69	68	13	8.5	139	49	54	92	0	0	116	—	—
Winter wheat	100	84	78	13	4.8	193	64	66	99	7	3	185	83	88
Spring barley	98	89	90	25	10.6	101	42	47	57	2	2	84	93	78
Winter barley	100	90	86	19	6.2	152	61	68	98	5	3	143	75	79
Spring oats	90	89	88	45	14.1	79	44	42	35	3	0	72	113	—
Winter oats	95	88	85	20	12.0	109	56	58	83	4	1	109	64	75
Rye	98	71	90	12	0	121	43	78	94	0	20	121	—	114
Maize	88	82	78	82	22.6	100	48	51	33	4	0	116	125	—

Source: ADAS and Fertiliser Manufacturers' Association data, *Survey of Fertiliser Practice.*

MAGNESIUM

It is an ingredient of chlorophyll. Cereals are only likely to respond to Mg fertiliser at soil index 0. However, with increasing yields and rates of potash in particular, available Mg may be depleted. It is wise to monitor the Mg index and maintain an index of 2^+, using either kieserite or, where liming is also needed, magnesian limestone. FYM is a useful source too. As a guide, some 20 kg/ha per year of Mg (given perhaps every third year) should maintain reserves at average cereal yields. Both high potash and high calcium availability are a threat to adequate magnesium uptake by the plant.

SULPHUR

This is a protein ingredient. Incidental addition of sulphur has hitherto occurred in the sulphate form in fertilisers. Most compound fertilisers are now becoming more concentrated and as a result are depleted or devoid of their sulphur. So far this has not shown apparent problems for cereals. Sulphur foliar sprays are showing benefit on some sugar beet crops, with a little incidental contact fungicidal activity also. Cereals may benefit in some circumstances where the air is cleanest.

TRACE ELEMENTS

By definition, the difference between too little and too much of these is very small. Most growers rely on a sound organic matter cycle and natural reserves in the soil to ensure supplies. This usually works except for certain well-known fields where preventive treatment of crops is routine, e.g. *manganese* to some sandy land at about 10kg/ha of manganous sulphate. Manganese can be in temporary deficiency, especially in a 'slow' spring. One of the benefits of a maneb or mancozeb fungicide is no doubt its incidental supply of manganese. It is a matter of experience and field records to determine where such treatment is needed. Manganese analysis of soil is not reliable. Bleaching and speckling of older leaves is a sign of manganese shortage in cereals (white in wheat; dark brown in barley; grey speck with red-brown tints in oats).

Copper is another trace element which may fairly often be deficient for cereals. Here soil analysis is a better guide to predict likelihood of deficiency. However, its actual occurrence is very much a seasonal thing. At high organic matter levels (peaty soils), available copper can be short and cereals will have twisted leaf tips, twisted awns in barley and shrivelled grains. At high pH too, copper deficiency occurs, darkening to olive green the leaves of wheat with poor ears resulting. Barley does not darken but fills ears badly. Copper sulphate may be soil applied for prevention at around 10kg/ha or, better still, for immediate effect, a foliar spray of copper oxychloride at 2.5 kg/ha is given during tillering.

Trace element cocktails are on the market, chiefly for later application to assist ear-filling in the crop. Evidence of the need for these is not consistent. However, at the highest yield levels it is logical that crops endeavouring to balance the much greater major element levels made readily available to them may benefit from an easy supply of trace elements. In addition, copper particularly has fungicidal activity which can prove strategic, as may seaweed extracts as a physical barrier to fungal pathogens.

Chapter 7

CEREAL GROWTH REGULATORS

Growth regulators are basically hormones which are chemicals capable of exerting significant effects on growth and development when present in minute quantities. The effects they have are not always beneficial nor do they necessarily occur consistently! However, they are the subject of considerable investigational work at both laboratory and field level. Commerce foresees an expanding role for them, and farmers' interest in their efficacy and potential is substantial. One of the purposes of this chapter is to discuss the probable factors determining success with plant growth regulators (PGRs) in farm practice.

Plants, like animals, possess internal or endogenous hormones quite normally and these are involved in such phenomena as seed germination, apical dominance and rooting behaviour in the cereals. The external application of hormones − called exogenous hormones − may either exaggerate or counteract the effects of natural, endogenous hormones. Interest in growth regulator treatments to crops really dates from 1932 when A.G. Rodriguez in Puerto Rico demonstrated the promotion of flowering in pineapple by ethene (ethylene) and acetylene.

A group of herbicides introduced in the early 1940s are known as the growth-regulator herbicides because of their mode of action in accelerating growth to the point of extinction: such substances include MCPA and 2,4-D. At very low concentrations these substances can promote the growth of weeds (Hamner and Tukey, 1944), whilst rates close to those recommended commercially will kill weeds. They are only used in this latter role.

Auxins Darwin in 1880 demonstrated coleoptile curvature due to this group of substances, whilst Went further studied their effects using agar blocks. IBA is commercially used in horticultural rooting powder. It is now possible to monitor natural levels in cereal tissue minutely since equipment can detect quantities as small as one thousand-millionth of a gram (i.e. one nanogram)!

Gibberellins (GA) These were first reported in 1938 in significant concentrations in the fungus *Gibberella fujikoroi*, which causes 'foolish seedling' disease of rice. Gibberellin activity can be tested using living plants (bioassay), e.g. dwarf varieties of rice become tall, dock leaves age more slowly.

Cytokinins Discovered in 1955 (e.g. Zeatin), these are especially synthesised in the roots. They appear to have a tillering role and they promote cell division in isolated tissue cultures. They are present in cereals at high levels at the onset of the grain-filling period. Cytokinins also literally seem to keep 'cells moving' and retard senescence (ageing).

Abscisins (ABA) The structure of these was first determined in 1965. Natural concentrations increase when plants are under stress. Sorghum − especially the more drought-tolerant varieties − is notable for its ability to close its stomates rapidly at the onset

110

of dry conditions, and applied ABA also induces closure of stomates. The leaves appear to be the site of ABA manufacture.

WHY USE GROWTH REGULATORS?

Original interest in growth regulators for cereals centred on the prevention of lodging. One of the predisposing factors for lodging is a long, weak straw. Whilst this is a matter of variety, it is also a tendency increased by high dosages of nitrogen fertiliser. The natural hormones increasing internode length are gibberellins; in order to counteract their effect anti-gibberellins are needed. Such a material is chlormequat (2-chloroethyltrimethylammonium chloride, CCC or chloro-choline chloride); it is best known as cycocel commercially or 3C cycocel, and there is a more potent version called 5C cycocel (CCC plus extra choline chloride), e.g. ICI Arotex extra. Even the naming of growth regulators is bewildering!

Chlormequat is designed to shorten and strengthen the straw. Its favourable effect on wheat was reported by N.E. Tolbert in 1960. The visual effect of shortening straw was sought to convince farmers that the treatment 'worked'! However, a strengthening effect is the more important. The treatment has been advocated on the grounds of yield protection, including protection of the previous inputs of fertiliser and fungicides. Now, there is growing awareness of the potential effects on growth which have been research aims hitherto, i.e. the use of growth regulators at various stages in crop development in order to modify growth patterns in favourable ways (Table 7.1). In other words, there is an upsurge of farmer interest in the positive yield-increasing possibilities of growth regulators rather than simply their yield-protecting role. Earlier applications of chlormequat and other materials are being given to crops in order to try to modify growth and secure higher yields. More potent materials have been identified; for instance, paclobutrazol is reckoned 7000 times more active than chlormequat in reducing wheat height.

WHICH ARE THE MAIN GROWTH REGULATORS AVAILABLE FOR CEREALS?

Main groups of endogenous growth regulators include the following:

Chlormequat This remains the most widely used and important cereal growth regulator. The review by E.C. Humphries in 1968 remains a seminal reference. By 1980 around half the UK winter wheat hectarage received chlormequat compared with some 70 per cent in Germany and less than 10 per cent in France.

Ethene (ethylene)-generators such as ethephon These joined chlormequat in the 1970s. Ethene probably acts by inhibiting auxins. Mepiquat chloride (sold in mixture with ethephon as Terpal) also inhibits gibberellins (Figure 7.1). Table 7.2 summarises the currently available products for stem shortening and strengthening in cereal crops. It can be seen that these materials collectively offer an opportunity to treat crops over quite a range of growth stages. However, each product may require a very specific stage for efficacy, e.g. Terpal at GS 32−39 ideally.

WHEN SHOULD GROWTH REGULATORS BE USED?

Timing of growth regulator treatments is likely in general to prove critical (Table 7.3). It is conceivable that growth regulators could be developed to modify growth at all stages of cereal development (Table 7.1). By definition they must be given at the time when the plant is concentrating on a particular phase of development in order to affect that phase, unless

TABLE 7.1 Potential for PGR use during cereal development

Potential PGR use	Phases of growth and development	Zadoks growth stages
Improved establishment. Synchronisation of tillering and improved rooting.	Germination Seedling growth Tiller initiation	0−12
Uniformity of tiller growth, development and improved survival. Stem strengthening, synchronising spikelet development, preventing spikelet abortion.	Tiller growth Spike induction Stem extension Spikelet, floral development	13−39
Uniformity of grain development. Preventing abortion. Control of assimilate partitioning and utilisation.	Grain filling	40−100

Source: Long Ashton Research Station.

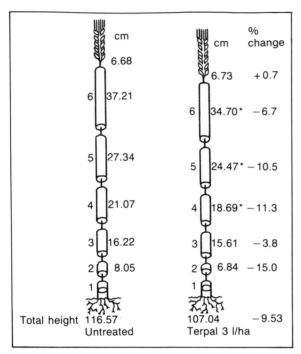

*Significantly different from untreated at 5% level

FIGURE 7.1 Length of winter barley stems after treatment with Terpal (mean value from 14 varieties on 4 locations, 1978, Fed. Rep. Germany). A series of measurements have shown that Terpal shortens all internodes, particularly the internodes 4 – 6. The stem diameter was increased slightly at all internodes. The ear was not shortened. Similar results have been obtained in Great Britain and France. (Source: BASF)

formulated for a slow release over a period when precise effects could not then be expected.

The question of timing must be considered in relation to (1) growth stage of the crop, and (2) circumstances under which the treatment is applied.

1. Growth Stage

Seed and seedlings

Seed priming may incorporate growth regulator to 'pep up' seed prior to sowing. More likely is the incorporation of growth regulators with the seed dressing. Seedlings with thick roots and fatter leaves can be produced by incorporating chlormequat and mixtures with the seed dressing. However, subsequent yield benefit has yet to be proven.

As long ago as 1960 L.G. Paleg reported the use of gibberellin (GA_3) to increase enzyme activity during the barley malting process. One ppm or one gram per tonne is ample to achieve this effect and the treatment is used commercially. GA_3 is also being tried on field cereals with quite dramatic shoot-stimulating effects.

In 1974 a Japanese firm patented a so-called growth regulator seed dressing called Calper, the significant ingredient of which is calcium peroxide, although it also includes calcium sulphate and calcium carbonate. In the moist seedbed calcium peroxide produces three effects:

- *releases* oxygen in the immediate vicinity of the seed, offering promise of successful germination in waterlogged or more marginally anaerobic conditions. This is possibly its greatest bonus. D.A.

TABLE 7.2 **Commercially available chemicals (several products and mixtures) for use as PGRs on UK cereals**

Chemicals	Example products	Normal rate (l/ha)	GS timing preferred	Crops
Chlormequat	Arotex extra New 5C Cycocel	1.75	22−31	Wheat, barley, oats, rye, triticale
Mepiquat chloride and ethephon	Terpal	2.50	32−39	Winter barley and spring barley (Scotland only; not cv Triumph)
Ethephon (2 chloro ethyl-phosphonic acid)	Cerone	1.00	39−57	Winter barley (by GS 45), winter rye and also wheat

Note: Full rates of these currently require yield responses of around 60 kg/ha of grain for chlormequat, 230 kg/ha for Cerone and 300 kg/ha for Terpal plus an extra 55 kg/ha of grain if they are to bear application charges alone rather than as part of a mixed spray. Terpal should not be used on organic or very light soils.

TABLE 7.3 Treatment timing and the expected effects of chlormequat

Timing	Expected effects	
Autumn	Check in vegetative growth Therefore, less winter pride (excess growth)	
GS 14—21	Increase in tiller production	
GS 26—30	Better root growth at the expense of shoot More grain sites set	Best yield response if lodging controlled anyhow
GS 30—31	Stem shortened and thickened Small stimulus to roots	Best lodging control
GS 32—37	Limited shortening of stem Better spikelet retention	

Notes: (1) Most cases of yield depression are associated with late application.
(2) Chlormequat is less mobile in barleys, hence gives less predictable effects than in wheat.

Perry at the Scottish Crops Research Institute has obtained as much as 30 per cent increase in barley emergence from wet seedbeds by this treatment.

• *neutralises* organic acids (notably acetic acid from decayed straw which can impair germination and emergence as shown clearly by Lynch *et al.* at Letcombe Laboratory).

• *counteracts* pathogens of the seed and seedling. Its anti-microbial action is non-specific and covers a wide range of problem species, perhaps notably the *Fusaria.*

Joint work between WPBS and LARS treating winter oat seed with tetcyclasis has reduced mesocotyl and lower internode elongation, thus keeping crowns deeper in soil and so less frost-prone.

Farmer, commercial and research experimentation with seedling growth regulator treatments continues to try to promote early root growth and more synchronous development of tillers by suppressing apical dominance of the primary tiller. Such effects were claimed by M. Sampson of Mandops Ltd. for Halloween on winter wheat and Helestone on winter barley, or indeed early Barleyquat. A cheap treatment is one-fifth of a full dose of chlormequat given at the three-leaf stage. This is advocated by S. D. Bond and field experience suggests it may be beneficial to use higher rates for some varieties, e.g. Vuka. However, the NIAB have cautioned that suppression of apical dominance tends to rob the plant of its normal 'safety valve' of losing weaker secondary tillers under stress conditions. The possibility of altering leaf angle,

making leaves more erect and thus perhaps improving efficiency of light interception has been observed. Leaves may also be kept green and active for a longer period.

Tillering
The control of tillering is sought in practice by:

• choice of variety
• sowing density and date
• timing of nitrogen fertiliser doses

In addition to these, Batch *et al.* (1980) were able to control tiller proliferation using a series of low doses of gibberellin; remaining tillers matured earlier and more synchronously though with variable yield effects. However, the possibility of tiller management has been demonstrated and work continues. There is a trend towards using chlormequat at GS 26—30 to try to strengthen and synchronise tillers, as it can markedly alter crop structure (Tables 7.4 and 7.5).

Stem extension
This is the primary period for growth regulator use on cereals at present. The effects of chlormequat are most obvious when applied at GS 32 as far as stem shortening is concerned because internode elongation at that time is potentially greatest (see Figure 7.1). However, real benefits from chlormequat are correlated with the prevention of *lodging* in higher-input systems, where experience shows that stem strengthening is the significant factor rather than stem

TABLE 7.4   The effects of PGR at two crop densities on yield (t/ha) of winter wheat, cv Aquila, drilled on 7 November, at Wye College, Kent

Plant population		PGR treatments	
		Nil	5C
Lower density	265 pl/m^2	8.03	8.23
Higher density	390 pl/m^2	8.34	8.72

	TGW (g)	Grains /ear	*Effects* Yield /ear	Ears /m^2	Yield t/ha
High density	↑	↓↓↓	↓↓	↑↑↑	↑
5C	↑	↑	↑↑	↓	↑
Both together	↑↑	↑	↑↑↑	—	↑↑↑

Note: PGR treatment was chlormequat (as Arotex 5C at 1.75 1/ha diluted to 400 1/ha volume) at GS 13.21 on 5 March.

Source: Data of P. Hutley-Bull, 1982.

shortening, which is an unreliable guide to lodging resistance.

Lodging is the falling over of ear-bearing stems. Usually stems fall predominantly in one direction, in which case the crop is less difficult to harvest than when stems have fallen over in all directions (sometimes called *straggling*). *Predisposing factors* for lodging incidence are:

- weak-strawed varieties.
- exposed sites — to high winds and/or heavy rainfall.
- excessive nitrogen fertiliser application making growth rank and weak.
- disease — notably eyespot and other stem-base (or 'footrot') diseases. Disease is particularly associated with straggling.

The *consequences* of lodging are:

- obviously greater the earlier it occurs. However, some crops may go flat early under a heavy thunderstorm but still retain sufficient pliability and strength to stand up again.
- direct loss of yield.
- greater variability of grain sample, especially regarding grain size, uniformity of maturity and Hagberg value in milling wheats.
- a more difficult, slower harvesting owing to laid ears close to the ground as well as to secondary tillering and weed growth in the flattened areas.
- greater incidence of late bird damage — notably by rooks, which find flattened areas easy to land on for feeding.
- often higher costs of cleaning and drying samples.
- relative wastage of other treatment costs spent on the crop.
- higher incidence of volunteers growing from shed grains and ears incurring greater disease carry-over as well as extra subsequent control costs.

Clearly the circumstances in which lodging is likely merit serious consideration for growth regulator use. A *delay* in lodging can bring worthwhile yield responses even if the treatment fails to prevent it altogether. Terpal offers a later chance to try to combat lodging at GS 32−39 and Cerone gives a final chance to attempt prevention at GS 39−57; it tends to prevent ear loss by strengthening the neck.

2. Circumstances

The operative factors when a growth regulator is likely to be effective include the following:

TABLE 7.5   Effects of chlormequat (CCC) on yield components of winter wheat, cv Norman (sown 2/10/81)

	Main shoot		Tiller 1	
	Control	+ CCC (at GS 25)	Control	+ CCC (at GS 25)
Grain yield (g)	2.85	3.44	1.44	1.92
Grain number	64.0	69.9	37.6	46.3
Mean grain weight (mg)	44.5	49.2	38.3	41.5
Total biomass (g)	5.44	5.86	2.94	3.68
Harvest index (%)	52.4	58.7	49.0	52.2
Straw length (mm)	696	656	637	592

Source: Data of P.M. Cartwright, University of Reading.

TABLE 7.6   Winter barley and growth regulator use

Growth regulator	Number of crops		N top dressing (kg/ha)		% lodging		Yield (t/ha)	
	1980	1981	1980	1981	1980	1981	1980	1981
With	177	232	138	152	11	17	6.7	5.8
Without	286	396	119	138	12	15	5.9	5.3

Source: ICI 1980 and 1981 crop surveys.

TABLE 7.7   Winter wheat and growth regulator use

Growth regulator	Number of crops		N top dressing (kg/ha)		% lodging		Yield (t/ha)	
	1978	1979	1978	1979	1978	1979	1978	1979
None	259	211	124	134	25	–	6.2	6.2
Once	786	722	143	149	25	–	7.2	7.0
Twice	62	46	152	169	33	–	7.8	7.5

Source: ICI 1978 and 1979 crop surveys.

TABLE 7.8   Winter wheat and growth regulator timing

Mean application date	Total N top dressing (kg/ha)	Yield (t/ha)
4 April	155	7.63
22 April	149	7.36
6 May	145	7.06
18 May	144	6.87

Note: 75 per cent of the 899 crops surveyed in 1980 received growth regulator.
Source: ICI 1980 crop survey.

TABLE 7.9   Yield of winter wheat and CCC use

Year	Variety	N (kg/ha)	With CCC (t/ha)	Without CCC (t/ha)
1971	Widgeon		5.79	5.57
1972	Widgeon	151	4.59	4.71
1973	Widgeon		4.80	5.21
1974	Widgeon		5.32	5.10
1978	Huntsman	120	8.70	8.20

Note: Responses were associated with lodging control which was especially necessary in conjunction with late fungicide use.
Source: Rosemaund E.H.F., Herefordshire.

TABLE 7.10   Winter wheat varietal responses to growth regulator use

Variety	Growth regulator		Associated N top dressing (kg/ha)		% yield increase associated with growth regulator
	Without PGR	With PGR	Without PGR	With PGR	
Armada	27	63	128	136	16
Bounty	14	55	142	150	2
Flanders	38	107	139	160	10
Kador	22	44	134	147	21
Mardler	71	110	150	160	8
Hobbit	18	86	139	154	15
Huntsman	20	89	131	156	17
Hustler	12	35	142	166	13

Source: ICI 1980 crop survey.

- high risk of lodging.
- high input systems. Interactions can be expected with nitrogen and fungicides, both of which promote and prolong green leaf area, thus necessitating extra stem support.
- conditions at the time of application, notably air temperature, e.g. Barleyquat and Bettaquat need >6°C, Cycocel >10°C (ideally >12°C) whilst Terpal and Cerone want >15°C. If temperatures are substantially higher than these approximate thresholds, intake of the applied chemical will be greater, resulting in overactivity. The effect can then be to overcook the crop and achieve yield depression from thin, 'rat-tailed' ears with low bushel weights; thus rates should be reduced at higher temperatures, e.g. three-quarter rate chlormequat at 15−20°C, half-rate above 20°C is an arbitrary guide. If the crop is already under stress following adverse weather or reaction to another recently applied chemical, a growth regulator may be ineffective or detrimental. Stress immediately after application of growth regulator may also produce similar results. Proper surveys of farmers' record-keeping with subsequent computer analysis could help to elucidate the field circumstances under which chemicals do and do not work.
- tank mixes with which they are applied, e.g. there is supposed synergism between chlormequat and carbendazim (for eyespot control) given at GS 31−32. Chlormequat may also be mixed with broad-leaved herbicides such as mecocrop (CMPP) for earlier applications or with wild oat herbicides. Mixing of chlormequat with liquid fertilisers is not recommended since it may induce crop scorch. Conjecture generally supercedes certainty in assessing growth regulator behaviour in the field!

Tables 7.6 to 7.10 indicate crop results, practice, timing and varietal response differences.

HOW HAVE GROWTH REGULATORS PERFORMED IN FARM PRACTICE?

Results of ADAS and other trials have shown inconsistent yield performance (e.g. Table 7.10). However, grain numbers per ear and grain size are often marginally reduced. The cost of commercially used growth regulators at full rates can be equivalent to as little as 0.05 tonnes/ha of grain up to as much as 0.25 tonnes/ha, chlormequat alone being the cheapest. An early one-fifth chlormequat dose thus costs the equivalent of 0.01 t/ha (10 kg).

The ICI surveys of winter wheat and winter barley give correlations between growth regulator use and yield for a wide range of fields. Of course, the use of growth regulator has probably been adopted more widely by innovative farmers who are already using higher inputs − particularly of N fertiliser and fungicides. However, these inputs alone cannot account for the responses associated with growth regulator use (see Tables 7.6−7.9). Nevertheless, the use of growth regulators, particularly in split doses and at earlier timings, is probably associated with closer attention to detail and tighter crop management.

Growth regulators are most widely used on winter wheat. However, the crops of rye, oats, triticale, durum wheat and spring wheat are candidates for chlormequat-containing growth regulators. Robin Child at LARS has obtained a 40 per cent yield increase in triticale (cv Lasko) using sequential chlormequat and ethephon. Spring crops generally require only one-quarter to one-half of the recommended full dose for autumn-sown wheat. Applications are taken up readily in good growing conditions. A practical guide to timing of doses has been when the first node is just detectable at the stem base.

Developments in the use of growth regulators can be anticipated. A better understanding of their effects is needed before many of the currently claimed benefits can be objectively supported.

Part Three

Protection

*Preventing or curing
the negative factors
in cereal growing*

Chapter 8

CROP PROTECTION — OVERVIEW

The best protection a crop can have lies in its own sturdiness through proper timely establishment. Unfortunately, some problems are increased by early sowing and some by late sowing, and all are increased when a crop is weak through quite untimely sowing and poor conditions. Cereal growth is not only impaired by weeds, pests and diseases, as disorders can arise due to weather damage or soil faults. Weak plants are often hosts to a combination of problems compounding diagnosis difficulties (Table 8.1). Table 8.2 indicates key questions to ask on crop protection. Colour plates between pages 148 and 149 aid identification of major pests, diseases and disorders. Ample capacity to treat the likely area of crop is important (Plate 8.1).

The Control of Pesticides Regulations (1986) is a significant piece of UK legislation to control the large amount of agrochemicals used (Figure 8.1):

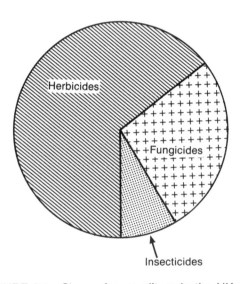

FIGURE 8.1 Share of expenditure in the UK on agrochemicals (total expenditure was almost £300 M in 1985 according to the British Agrochemicals Association).

PLATE 8.1 *Sprayer/spreader.* (Courtesy of Willmot Crop Protection)

- All manufacturers, importers and suppliers of pesticides must obtain an approval for each pesticide product they develop or market for use in the UK.
- From January 1989, certificates of competence, recognised by Ministers, are required by agricultural, horticultural and forestry contractors, unless they are working under the direct and personal supervision of a certificate holder. The certificate is provided by the British Agrochemical

119

Standards Inspection Scheme (BASIS) but other certification may be recognised by the Ministry (MAFF) also.

- All users of pesticides under the age of 25 in 1989 need a certificate, unless directly working under a certificate holder.

- Only adjuvants (additives to sprays) listed by MAFF may be used.

- Tank mixing of pesticides is controlled by the Ministry which only approves specified tank mixes from 1989 onwards.

- Stricter adherence is required to such matters as crop-use restrictions; protective clothing to be worn; maximum application rates; minimum harvest intervals; bee, livestock and human safety.

Spray delivery

Contact herbicides and fungicides have been found biologically most active as large (c. 170 μm in diameter) dilute drops, whilst translocated herbicides work better as small (c. 80 μm) more concentrated drops.

Conventional hydraulic nozzles give most evenly distributed spray deposits from high volumes (c. 200 litres/ha). Hydraulic spinning disc systems deposit heavily on top of crops but with poor penetration and, unless electrostatically charged, drift is a matter of concern.

Closed, electronically controlled systems are being developed to allow addition of extra chemicals during fieldwork for parts of fields requiring special treatment, thus increasing operator safety and economy in chemical use (Landers, 1988).

TABLE 8.1 Diagnosis of problems at different growth stages

Growth stage	Problem	Causes
Germination 00–09	Seed absent or husk residues only	Vermin (i.e. birds and mammal pests)
	Emergence delay/failure	Drought, seed-rotting fungi, waterlogging
	Seeds hollowed out	Slugs, wireworms
Seedling growth 10–13, etc.	Long, weak, colour-banded shoot	Deep sown or struggled through soil cap
	Distorted, stunted shoot	Seed dressing damage, virus
	Yellowed, maybe eaten shoot	Slugs, leatherjackets, wireworms, WBF (wheat bulb fly), frit fly, fungi
Tillering 22, etc.	Plants lifted and dead	Frost, vermin
	Plants cut off near soil level	Leatherjackets, wireworms, cutworms, swift moth, slugs, grass moths
	Plants entire but stunted/ discoloured	Virus or fungus disease. Poor drainage, compaction, wrong pH, mineral deficiency, frost, spray damage, cereal cyst nematode, aphids
	Central shoot or individual tiller yellow	Frit, Opomyza, wheat bulb fly, wireworms, leatherjackets, slugs
	Swollen shoot	Gout fly, stem nematode
	Leaves grazed	Mammal pests
	Leaf tips pecked off (some lying about)	Bird pests
	Shredded, bitten leaves	Slugs, leatherjackets

Stem extension 30—45	Plant stunted, leaves yellow-purple/red	Mineral deficiency, fungus or virus disease, aphids, waterlogging
	Root system much branched, stunted and perhaps showing cysts or galls	Cereal cyst nematode, root-knot nematode, wrong pH, drought
	Tillers dying	Normal intraplant competition, stem-base diseases, some moths, Opomyza, frit fly, wheat bulb fly
	Swollen shoot sections or base	Gout fly, stem nematode, wheat bulb fly
	Leaf strips eaten through	Slugs, leatherjackets
	Leaf strips not usually eaten through	Cereal leaf beetle (larva glistens with surface excreta)
	Leaf mined	Cereal leaf miner (frass inside)
	Leaf edges eaten, some tillers may be cut off near soil level	Moth caterpillars, sawflies
	Various leaf discolourations/lesions	Foliar diseases or spray damage
	Early lodging may be with dark blue-green leaves or later lodging (more prevalent) perhaps with bird attack	Excessive N, weather damage, eyespot
	Upper internodes galled	Saddle gall midge
	Ear distorted	Herbicide damage, weather damage (hail, heat), copper deficiency, frit fly, a feature of certain varieties
	Feeding track down ear and stem	Gout fly
Heading 51—69	Ear entirely black	Smut
	Blind florets	Weather at fertilisation, imperfect nutrient supply to ear, frit fly, thrips
	Grains taken out	Birds, wheat blossom midge (tiny orange-yellow maggot)
Ripening 71—92	Ear discoloured superficially	Fungus diseases, thrips
	Ear white/empty or shrivelled grain, early-ripe	Stem-base diseases, drought, 'loose-ear', sawfly, hessian fly
	Ear sticky, infested	Aphids

TABLE 8.2 An overview of key crop protection questions for a cereal crop

Problem observation			*Husbandry Strategy*
WEEDS	— Spectrum — Severity — BL/grass — Timing of flush	Cultural	— Crop vitality — Weed destruction — Sterile strips
		Herbicides	— Residual — Post-emergence — Pre-harvest — Between-crop — Save costs by treating only parts of fields

(continued)

DISEASES	— Spectrum	— Variety choice
	— Variety susceptibility	— Rotation
	— Affected parts	— Stubble hygiene (green bridge)
	— Timing of damage	— Preventive (prophylactic)
		— Curative fungicides
		— Seed dressing
		— Broad-spectrum/specific
		— Diversification of sprays
		— Earliness with reference to damage
PESTS	— General	— High-risk situation protection
	— Specific	— Autumn — General pests, e.g. slugs
	— Timing	— Aphids
	— Thresholds	— WBF, etc.
		— Spring
		— Late aphids especially
OTHER FACTORS such as	— Lodging	— Variety/site/disease/N/drainage/structure
	— Yellowing	
	— Blind grains	

Chapter 9

CEREAL PESTS

A crop pest here is an animal causing economic loss of the crop, though elsewhere the term may be widened to include diseases and weeds.

Cereal *field* pests are favoured by:

1. Weather conditions suitable for the pest species concerned.

2. A large area of accessible food supply in one place, i.e. large fields of susceptible varieties. It can be argued that a large area masks the effect of some pests whilst small plots can be very vulnerable, e.g. to bird damage.

3. Absence of natural enemies of the pest population which may survive on vegetation around the fields. Clearing of land for agriculture inevitably reduces the population of non-pest competitors of the pests.

4. Easy carry-over. In the case of soil and trash-borne pests, intensive cropping with only a short gap between susceptible crops and inadequate trash disposal aids the pest. Regional predominance of cereals clearly favours build-up of more mobile insect pests too.

Cereal *storage* pests are favoured by:

1. Nooks and crannies in the store which harbour old grain and protect the pests.

2. Uncontrolled microclimate in stored grain; temperature and relative humidity determine ability to breed and survive. Beetles are inherently long-lived insects given the chance (see Figure 9.2).

VULNERABILITY OF CEREAL CROPS

General pests, such as rabbits, birds, slugs, leather-jackets and wireworms, and many specific cereal pests can pose an early threat to cereals. Pests which directly consume crop tissue are clearly potentially most damaging when plants are young and small. A bite taken from a tiny plant represents a bigger percentage loss. The seedling phase is highly vulnerable for this reason and because tissue is tender and attractive at a time when other food supplies are limited; if the single shoot is lost before tillering begins, this means the whole plant is dead. Young plants can be infected with disease, and vector pests, e.g. aphids, can be serious in small numbers during crop youth. Early control forestalls the real danger later if the pest multiplies.

During the tillering phase pests usually appear more threatening than they are since plants can often vigorously compensate for some tissue loss. However, once the ear emerges, damage can again be severe for various reasons. Weather conditions can then allow rapid pest population expansion. Pests on the ear can directly interfere with flowering (e.g. frit fly and wheat blossom midge) and are directly robbing the developing grains (e.g. aphids) or damaging the grain coats (e.g. thrips).

Fortunately, in most seasons pest problems in cereals are much less than disease problems and many crops do not require spraying. Indeed many predators of cereal pests carry out useful biological control provided that they are not eliminated by a spray themselves! However, the potential for rapid multiplication of some pests and risk of very significant occasional damage warrants vigilance: Professor Van Emden of Reading University has noted a potential productivity of 20 tonnes per hectare of aphid biomass within three weeks.

Practical requirements for farmers are:

- allocation of time for regular field walking to inspect crops
- recognition of crop symptoms and of pests in the various stages of their life cycle
- knowledge of the timing and weather conditions favouring attack if any
- understanding of means of spread and of control strategies for the main pests

Advice and annual literature are readily available on the ever-changing range of pesticides, so farmers and students should not be obsessed with learning names and application rates of current products to the exclusion of practical knowledge of relevant pest behaviour and non-chemical pest management methods.

TIMING OF PEST ATTACKS

This is a brief guide to what to expect and when. Summary notes on principal pests are given later.

Obviously timing varies somewhat with each season and the location of the farm. However, the following guidelines to periods of potential damage may help. Some pests can attack any time, e.g. birds and mammals. Effective seed dressing against birds exists for maize seed but so far is available only on imported French seed.

Autumn (September–November)

Aphids can be troublesome especially to early-sown crops in milder southern regions. At this stage some aphids are infected with virus disease and this, rather than direct sucking, is their principal potential damage. The sub-clinical effects as well as obvious symptoms of virus disease can be far-reaching. Rothamsted work indicates aphids' ability to survive in such places as couch patches and migrate onto winter cereals during any mild period in winter. It is some of the bird-cherry aphid population (not more than 10 per cent of these, perhaps) which carry virus most commonly. In high-yielding intensive systems it is wise to control even limited aphid invasions in autumn, with insecticide perhaps included whilst treating diseases anyhow. This can be in the form of barrier-spraying around the first bout or two of the headland whence aphids invade – and where they may be confined if detected early enough. It is not reliable in preventing aerial migration.

Slugs can take seed and seedlings in a wet, mild autumn, especially on potentially fertile sites with abundant trash levels. Routine monitoring with baited traps should be done on susceptible sites where high populations are almost inevitable, e.g. winter wheat after oilseed rape or beans. Slugs are notoriously difficult to monitor and therefore defy accurate prediction. Traps visited weekly indicate activity. Baited tiles are placed every 50 m across the field. Six slugs/tile is used by some farmers as a treatment threshold. Slug damage on a few plants looks alarming and consequently a lot of money may be spent on preventative treatment which with hindsight might have been unnecessary. Hindsight is the commonest wisdom!

Wireworms, the larvae of click beetles, take four or five years to mature. They are active during both autumn and spring, but only attack young plants before nodal rooting. They work their way along cereal rows leaving dead yellow seedlings.

The risk of *frit fly* is greatest following late-ploughed-out grass. An interval of at least 1 month (preferably $1\frac{1}{2}$ months) between sward destruction/burial and drilling of the following cereal is an essential precaution. The central leaf is yellowed and young seedlings may be killed. Other species of cereal and grass flies may produce similar effects; frit is not the sole culprit! The pest has three generations per year so attacks also occur in spring.

Winter (December–February)

All the pests of autumn may continue to effect some damage in a mild winter.

Swift moth larvae cause occasional damage, cutting off plants at soil level in patches. Yellowed and dead tillers are evident quite early in the new year if conditions are mild and crops have been early sown; the causes are *wheat bulb fly* or *Opomyza fly*. Timing of egg hatch, ADAS egg counts and local experience should inform decisions on control. Moist, mild autumns favour attacks by *leatherjackets* during January to May after grass or weedy stubbles. ADAS has operated a prediction service. Spring feeding is more voracious usually and thus more damaging.

Spring (March–May)

All previously mentioned pests may still be active. This is the main period when *wheat bulb fly* and *Opomyza fly* damage is most apparent. Various moths may be active during this period too. In a mild to warm sunny spring *aphid* build-up can begin in earnest. *Cereal cyst nematode* patches show clearly now if present, as does *stem nematode* damage.

Summer (June–August)

Ears are very liable to direct yield loss during flowering and from watery ripe to milky ripe stages. *Thrips* cause brown discolouration and bruising of grains. Thrip damage to grain coats may impair their subsequent germination if for malting or for seed corn. *Frit fly larvae* may damage wheat, barley and oat heads. Grass and cereal *mites* can occasionally cause premature white ears which pull out easily from leaf sheaths, and they may damage leaf margins.

Cereal *leaf miners* and *leaf beetles* both become apparent. They are easily seen and appear more significant than they are. *Wheat blossom midge* may occasionally cause grain loss; *saddle gall midge* and *Hessian fly* may cause ear loss but are rarely considered to warrant control in the UK. *Rook* damage to lodged crops can be considerable.

Undoubtedly the most serious pests likely to occur during this period in economically significant numbers are *aphids*.

CEREAL PEST CONTROL STRATEGY

The aim is always to forestall epidemics in a situation where a susceptible crop inevitably provides a potential banquet for a pest. In prevention or cure the objectives are minimal expense consistent with minimal environmental damage (including least side effects on natural enemies of the pest) and effective pest suppression. Integrated control, using an appropriate combination from the means discussed below, is a commonsense approach.

A three-category strategy is considered below:

1. Prediction
2. Cultural means
3. Chemical means

1. Prediction

Prediction of pest problems is at two levels: (a) official research, surveys, forecasts and imminent warnings and (b) regular inspection of individual crops. There is no substitute for the latter, as early detection is the key.

Research work continues to establish correlations between weather patterns and subsequent pest problems, e.g. autumn rainfall and spring leatherjacket damage. Surveys of farm experience can identify correlations between particular practices, cereal varieties and pest problems. There seems a sensible role for

the computer to process a mass of information from real farm practice rather than spending so much time developing simulated model situations.

ADAS have determined threshold levels for damage worth controlling for various pests, e.g. wireworms and leatherjackets can be counted; egg counts are done in September for wheat bulb fly: a similar scheme can be done later on for *Opomyza* eggs. Late grain aphid spraying is judged worthwhile by ADAS if more than 5 aphids per ear are found on wheat after examining at least 50 ears randomly sampled; the comparable figure for rose-grain aphids is 30 per flag leaf up until milk-ripe stage of the grain. Above average crops can justify treatment before these thresholds are attained. Farmers have received regular bulletins and newsflashes or telephoned the regional ADAS centre for impartial advice on pests and diseases to alert their own local observations.

2. Cultural Means

- *Hygiene* is of first importance. This includes proper treatment of pest-harbouring crop residues and weeds, especially grasses. Furthermore, care and cleaning should be exercised to minimise transfer of soil and debris on implements and boots from fields known to be badly infected.
- *Rotation* can frustrate the build-up of some pests.
- Time of sowing is important. Early sowing in spring favours establishment of a strong plant to withstand frit fly but in autumn allows a long period for aphid colonisation.
- Cultivation can make a contribution. Drainage reduces slug breeding areas, whilst ploughing exposes some larvae to be eaten by birds, e.g. leatherjackets, wireworms.
- It is wise to use resistant varieties, e.g. against cereal cyst and stem nematodes where these are prevalent.
- Crop vigour should be promoted by ensuring ample nutrition of good seed established in a well-structured soil.

3. Chemical Means

- Seed dressings are available for some pests, such as wheat bulb fly and wireworms.
- Baits are appropriate for others, e.g. slugs and leatherjackets.
- Sprays are used for pests such as aphids. Nowadays, systemic materials are generally used and synthetic pyrethroids have been replacing

some of the organophosphorus compounds because of their lower mammalian toxicity. Carbamates are also used.

- Dusts, sprays and fumigants are appropriate for grain storage pest control.

SUMMARIES OF PRINCIPAL CEREAL PESTS

The principal field pests and the major storage pests of cereals are considered briefly below under four headings:

- *Recognition* of the harmful form of the pest and symptoms of crop damage
- *Prevalence* i.e. conditions favouring attack
- *Spread* i.e. summary of life cycle and dispersal
- *Control*

Aphids

Recognition
Grain aphid (Sitobion avenae) = large, about 3 mm; usually dark green; long antennae; long, dark-coloured cornicles on abdomen; can be numerous from June on upper leaves and ears; can be very damaging when landing on the ears of the crop.
Rose-grain aphid (Metopolophium dirhodum) = medium, about 2.5 mm; light green to pale-pinkish; darker green stripe along back; confined to lower leaves; only up to flag leaf, never on ears.

Prevalence
Many other aphid species may be found but not usually in economically threatening numbers.

However, the bird-cherry aphid (*Rhopalosiphum padi*) can spread BYDV, especially in the west of England in mild autumns and springs.

At the ear stage aphid damage is by direct sap sucking, robbing yield by sheer weight of numbers. Honeydew produced by aphids also coats leaves and attracts surface moulds which render photosynthesis less efficient whilst the aphids continue to rob previous photosynthates from the plant.

Spread
Reproduction is rapid, viviparous (live birth without eggs), parthenogenetic (without fertilisation at times) and winged generations arise once colonies become dense. Mild, moist and sunny conditions favour aphids. High-yielding, fast-growing wheat varieties seem more susceptible to late aphid attack.

Control
ADAS recommend spraying up to milk-ripe stage once aphid numbers average five per ear examined or when two-thirds of ears inspected have aphids. High-input farmers often prefer to insurance spray with a product safe for bees (e.g. pirimicarb) whilst passing through the crop with the final ear-protectant fungicide.

Slugs

Recognition
Grey-fawn and some dark slugs feed in the evening leaving slime trails, hollowed seed corn, yellowed shoots and seedling death. Leaves of older plants are shredded and strips eaten out.

PLATE 9.1 *Barley yellow dwarf virus (BYDV) is carried to autumn cereals by flying aphids and then further spread by their wingless progeny.* (Courtesy of IACR Long Ashton Research Station, University of Bristol)

Prevalence
All cereals attacked. Winter wheat favoured (often grown on susceptible land). Infestation early in the life of a crop can produce devastation and necessitate redrilling. Heavier, moisture-retentive, especially calcareous soils favour them; unburnt straw and other crop debris encourage them (Figure 9.1). High risk after oilseed rape, pulses and other crops leaving richer organic matter. Active in wet autumn and especially in spring on late autumn-sown corn, though early spring-sown crops can sustain damage also.

Spread
Round, translucent egg clutches in soil. Hatch in < 1 to >3 months according to conditions and feed directly. Slow mobility but widespread already!

Control
Prediction in dubious cases is based on test-baits (pellets under sack or black polythene, laid in favourable slug weather, ten per field). If four or more slugs arrive per test point within three days, then treat. Correlation between trapping and likely damage is at best crude. Establish crops well; roll loose tilths before and immediately after the drill; above all ensure proper and timely trash disposal from previous crop.

Metaldehyde can be given with seed as an insurance; if high risk, apply pre-drilling and repeat about one week after drilling if necessary: moisture is needed for pellets to work, of course. Methiocarb is more potent in its side effects on other soil fauna,

including earthworms; it is around double the cost of metaldehyde at standard rates but more potent on slugs (and against other soil pests). In trials it has almost doubled BYDV-infected tillers owing to its suppression of aphid predators.

Leatherjackets

These grubs are larvae of the crane fly (daddy-longlegs).

Recognition
Rubbery, fat, grey-brown grubs up to 30 mm or so long. Leave yellowed, dead and eaten plants in patches. Rooks and other birds take special interest in the field!

Prevalence
In a wet, mild winter. Feed most voraciously in spring. Attack all cereals. After grass or weedy stubbles usually. Can do considerable damage.

Spread
Up to 300 eggs per female fly are laid into grassy places in September. Larvae in autumn and until spring. Pupae (June) in soil. Adults (July to September) − very mobile.

Control
Deter egglaying by ploughing out grassland before the end of August (should be done anyhow!). Threshold counts are total of 15 larvae from 10 samples of 30 cm drill lengths. Use chemicals as baits (bran plus HCH applied onto moist soil on a mild evening). A few granules given to any small patches seen very early on can help. Soil can have spray incorporated pre-drilling (HCH or chlorpyrifos) if problems are expected. Curative spraying followed by a little extra N fertiliser should be given to affected crops.

Wireworms

These are larvae of the click beetle (so called because of a spring mechanism which allows it to right itself if it is overturned).

Recognition
Tiny white larvae growing over the four to five years it takes them to mature to be hard, shining, golden-yellow larvae with three pairs of legs just behind the small brown head. Up to 25 mm long or so. Plants yellow, wilt and die, often along the row.

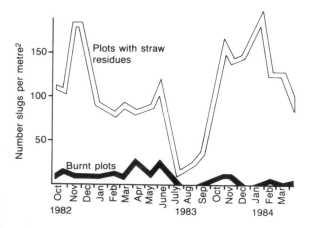

FIGURE 9.1 Numbers of all species of slugs on direct drilled plots at Northfield, Oxon. (Data of D. Glen, Long Ashton)

Prevalence
After old grassland or very weedy stubbles. Heavier soils. Greatest potential damage in spring. All cereals affected but wheat and oats are favourites. Most crops routinely protected by seed dressings now.

Spread
Eggs (late spring) in soil. Larvae (emerge late spring to autumn) have four to five years' life. Pupae (autumn) in soil. Adults (overwinter) – fairly mobile.

Control
Good, firm seedbed. HCH seed dressing is normally adequate. Higher wireworm counts need HCH spray or fonofos granules (at double the cost) to achieve control. Affected crops can be helped to recover by a little extra N fertiliser.

Frit Fly

This small, dark fly is a regular pest of members of the grass family, having a wide host range and distribution.

Recognition
White larvae about 2.5 mm long. Three generations per year. Can cause yellow, dead tillers, leaf distortion and holes, and shrivelling of some grains.

Prevalence
Corn in autumn, especially wheat, but all winter cereals may be attacked. Critical to have $1-1\frac{1}{2}$ month gap between ploughing out grass and sowing autumn corn. Late-sown spring oats can have tillers devastated as can maize. Its larvae may breed in the ears of oats or wheat, causing direct loss of grain in these species.

Spread
Four-stage life cycle. Flies very mobile on wind currents and widespread. Grasses harbour it and are, of course, ubiquitous.

Control
Crops can suffer damage over a long period so complete protection is a problem. Observe ploughing interval in autumn. Sow spring oats early. Preventive chemical treatment is necessary where risk is high or at earliest detection on spring oats, maize or autumn-sown cereals. Curative spraying once damage

is clearly seen is too late. Chlorpyrifos, omethoate or triazophos can be used. Grain maize should be routinely protected.

Wheat Bulb Fly

Recognition
Shiny, tiny (some 10 mm long) white maggots bore into the central shoot just below ground (at the bulb) leaving a brown entry hole. Yellowed shoot or shoots noticeable in late February to April. Larvae may move on to kill or damage other plants nearby.

Prevalence
Winter wheat, especially in eastern counties, including Yorkshire. Also attacks winter barley and rye but not oats. Spring wheat and barley rarely affected if sown after early March. Couch is a major alternative host.

Spread
Female flies (look like small houseflies) lay eggs into rough, bare soil (July–August). Larvae hatch from January and seek host plants but soon die if they fail to find one. Pupate in April/May in soil and adults emerge in June/July.

Control
ADAS count eggs and hence score risk. If risky site, sow winter oats or sow wheat early. Avoid too early sowing of spring corn. Dress seed with HCH or, if no wireworms, use chlorfenvinphos or carbofenothion (up to the end of December only because both these harm birds). Granules of fonofos can be used at sowing or at first signs of damage, or triazophos at egg hatch is cheaper.

Opomyza Shoot Fly (Yellow Cereal Fly)

Recognition
Creamy larvae (about 8 mm) tapered at each end; leave a spiral track down from the second leaf and enter the split sheath. Feed on and kill one tiller (do not spread like wheat bulb fly). Can produce late-maturing secondary tillering problems as crops seek to compensate.

Prevalence
On *early* (September) sown winter wheat, occasionally winter barley. Southern half of England but especially eastern counties.

Spread
Eggs laid October—November. Hatch late January to March and larvae damage during March—April. Pupate in May. Adult flies emerge in June.

Control
Seems seldom worthwhile to treat, especially on free-tillering varieties. Winter wheat sown in October suffers less. Egg counts are a crude guide to risk (ADAS will advise). If spraying at all, do so at egg hatch with triazophos or omethoate.

Cereal Cyst Nematode

Recognition
Stunted yellowed plants in patches. Much-branched roots with white cysts on them.

Prevalence
Especially on light land low in organic matter (e.g. light chalkland) under intensive cereals. Winter oats sustain worst damage but also spring barley.

Spread
Cysts can persist several years. They are spread in soil on boots and wheels of vehicles. Cyst counts can be done on soil samples (separating cysts by flotation).

Control
Do not grow oats more often than one in five years on risky sites. Use resistant varieties such as Rollo.

Thrips

These are also known as thunderflies owing to their abundance during warm, humid (thundery) weather. Their economic significance has been controversial for many years.

Recognition
Thin, blackish imperceptibly winged adults (about 3 mm or so long); juveniles are pale orange-yellow. Silvery white marked leaves of cereals, especially maize (which is young at the right time). Numerous on ears of wheat and barley where they suck and damage grains which may appear bruised, brownish and somewhat shrivelled.

Prevalence
Evident during thundery weather. Provided air conditions do not desiccate their tiny bodies, can become very numerous. Nationwide occurrence. Newer, high-yield varieties with thinner skins and more sugar may be encouraging this pest.

Spread
High fliers (found hundreds of metres up in the air). Very cold-tolerant. Survive winters under pine bark, on grass weeds and in hedgerows. Reproduce prolifically during ear emergence of cereals. Eggs are laid in uppermost leaf sheath; therefore, maximum damage occurs at booting stage and once in ear (especially the sterile florets of two-row barley).

Control
Control seems justified in heavy infestations, especially for seed crops of wheat and barley and for malting barley; this is because thrips can impair the germ of developing grains even though they may not reduce yield substantially. Any damaged grain is also predisposed to later storage problems even though the thrips are killed during grain drying, as mites and fungi in stores can invade via thrip-scars. Chemicals should be sprayed at booting stage for maximum effect and perhaps again at ear-emergence. Fenitrothion, cypermethrin and pirimicarb are alternative chemicals.

PESTS OF STORED GRAIN

Birds and rodents not only consume stored grain but contaminate it with visible excreta. They are controlled by exclusion and poison baits for rats. In addition to these there are three groups of smaller storage pests:

- beetles, two common species being the grain weevil and the saw-toothed grain beetle (Figure 9.2)
- moths, notably the warehouse moth
- mites, especially the flour mite

Saw-toothed Grain Beetle (Oryzaephilus surinamensis)

This is now the most common storage pest. It readily colonises damaged grain, breeding freely between 18°C and 38°C and above only 7 per cent grain moisture content. It generates heat and moisture by its respiratory activity, and thus only a few can soon

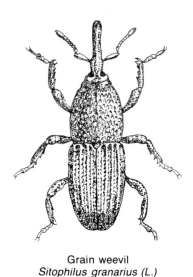

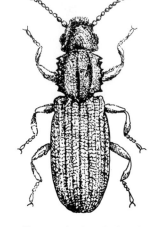

Grain weevil
Sitophilus granarius (L.)

Saw-toothed grain beetle
Oryzaephilus surinamensis (L.)

FIGURE 9.2   Grain storage pests. (Drawings courtesy Dow Chemical Company, makers of Reldan)

start a sizeable colony. Its life cycle takes almost three months at 18°C but is less than three weeks at 33°C. Larvae feed externally on the germ and on broken grains.

Grain Weevil (Sitophilus granarius)

Virgil wrote in Book I of *The Georgics*, 'A mighty heap of corn the weevil wastes,' and this is still possible! It can breed at 13°C and 10 per cent grain moisture. The larvae live entirely inside grains and eat them. It contaminates grain with its white powdery excreta and causes further damage by generating heat and moisture.

Warehouse Moth (Ephestia elutella)

This is common in the UK, especially the north. Its life cycle takes six to nine months, according to temperature. Larvae with a maximum size of over 2 cm do the damage by spinning a web over the grain which seals and cakes it.

Flour Mite (Acarus siro)

Favoured by high humidity and the presence of moulds, the flour mite can live in grain above 12 per

cent moisture. Its life cycle takes only two weeks at 20°C but requires twenty weeks at 5°C. It contaminates grain as an unpleasant brown dust and may destroy milling and baking quality.

Action against stored grain pests
Curing an established infestation is very expensive, difficult and infuriating, often requiring specialist contractors.

Prevention is therefore paramount and can be achieved by:

● Ensuring grain entering the store is dry and cool
● Repairing buildings and clearing crevices and debris which may harbour the pests
● Cleaning conveyors and drier ducts, especially underfloor areas
● Monitoring pest presence in barns, using grain sample baits sited in dark corners or pit traps in bulk grain heaps
● Using insecticides to protect stores and grain

In storage areas containing porous materials, insecticides are preferable as wettable powders and are sprayed at around 5 litres of diluted powder per 100 m^2 surface. Alternatively, smoke or fog formulations can be used to penetrate inaccessible areas. Such protection may last a couple of months only. Chemicals suitable for this are:

- chlorpyrifos-methyl (Reldan)
- etrimfos (Satisfar)
- fenitrothion (Dicofen)
- pirimiphos-methyl (Actellic)

An admixture of *emulsifiable concentrate* dust forms of the above chemicals can be administered as a prophylactic measure when grain is conveyed into store. Resistant strains of the pests already exist, so use full rates.

The treatment of established infestations by various fumigation techniques is a specialist contractor's job.

Chapter 10

CEREAL DISEASES

Several diseases pose threats to the yield and quality of cereal crops. Since the pathogens (disease-producing agents) are micro-organisms, chiefly fungi, the diseases are often recognised in the field only by *symptoms* resulting from earlier activity of the pathogens, rather than by direct observation of the organism. Of course, as the disease advances, the pathogen itself generally becomes more obvious. However, treatment of many diseases at this stage of proceedings is analogous to shutting the stable door after the horse has bolted.

Before embarking on the rather negative and alarming catalogue of possible diseases, it is important to give some positive thought to crop health. A healthy crop is not simply one which is devoid of disease but rather one where growth and development are being positively promoted. Crop health is therefore not simply a list of diseases with a parallel list of potions to kill the pathogen. For one thing many pathogens have so far defied the available fungicides! It is more relevant to consider *first* in practice what measures can be taken to strengthen the crop and, second, in the event of disease, what the underlying *causes* of the outbreak are. Insofar as these causes are within the control of the farmer, it is wise to consider avoidance of the causes before methods of treating the symptoms.

It is relevant to distinguish two facets of crop health: likelihood of infection in the particular farming system, and ability of the crop to withstand attack as far as farm practice can affect it. Some means of encouraging vigorous crop development (e.g. earlier sowing) often predisposes the crop to easier disease colonisation but advantages also accrue, so good farming is inevitably a compromise.

Practical measures for promoting general crop health include:

1. Provision of optimum growing conditions in terms of soil structure, drainage, pH, nutrient supply; date, depth and density of planting.
2. Respect for legislation regarding quarantine, certification and other regulations on transfer of plant material. International travel, especially within the EC, facilitates transfer of disease. Whilst many airborne pathogens, such as rusts, are no respecters of national boundaries, others are only conveyed on live material, soil or debris.
3. Crop rotation. Consider the frequency of cropping of each cereal species. This is particularly relevant for trash- and splash-borne diseases, notably eyespot. Take-all is a highly specialised parasite which normally does not survive more than a season without the host cereal but is subject to decline in long cereal runs. Thus, take-all increases until the fourth or fifth year of a susceptible cereal crop in the same field and yield is proportionately depressed. By the seventh or eighth year in cereals, yield recovers and the take-all level is lowered. Whether by self-stagnation with its own toxins or by build-up of microbes antagonistic to it, the take-all fungus declines (see Figure 14.2). Passing through the worst years is known as 'breaking the take-all barrier'! Nationally take-all is rated by some as one of the most insidious yield-robbers showing its ugly presence clearly in a wet season. Airborne diseases do not respond directly to crop rotation on each individual field (rust spores can travel many hundreds of miles). Nevertheless, the persistent growing of susceptible varieties over a district increases the inoculum present

there. Thus collective crop health in the district is the aggregate of individual farmers' practices.

4. Field hygiene as follows:

- Carry out proper and timely treatment of the previous crop residues whether by soil incorporation, removal or burning.
- Control weed hosts of cereal diseases — notably perennial grass weeds.
- Control insect vectors (transmitters) of disease.
- Use clean, treated seed of appropriate varieties for the disease probabilities of the farm. As with livestock enterprises, avoid introducing new problems on the seed bought in.

THE NATURE OF CEREAL PATHOGENS

Most are *fungi*. These are simple plants lacking chlorophyll and therefore unable to produce their own carbohydrates by photosynthesis. Consequently, some obtain carbohydrates from living hosts, i.e. are parasites, whilst others derive them from dead sources, i.e. are saprophytes. Most parasites are highly specialised to exploit their particular host crop.

Most fungi consist of long, thin threadlike structures called hyphae which absorb the food required and which are known collectively as mycelium. The majority of fungi are aerobic and are sensitive to sulphur, copper and mercury, upon which elements early fungicides were based. Fungi reproduce either by fragmentation of the mycelium or by spore production. Some spores are produced asexually, allowing rapid multiplication, whilst others arise from sexual processes allowing genetic reassortment. Some of these spores are very drought resistant but all require free water to germinate. Spores may be dispersed on seed (e.g. smuts), by wind (e.g. rusts), by rainsplash (e.g. *Rhynchosporium*) or by animals (e.g. ergot, by flies).

As *micro-organisms* they have a short generation interval and prolific spore production. This means that the capacity for the pathogen population to shift in physiology is a constant threat. New races of a fungus can arise far more readily than resistant cereal varieties can be selected or indeed than new fungicides can be developed.

Another group of significant cereal pathogens are *viruses*. These require living host cells to multiply and survive only for a short time outside. They need vectors (usually aphids, sometimes fungi and nematodes) to transmit them from plant to plant.

Climate and, in particular, microclimate within the crop can exert a profound influence on the relative incidence of the various pathogens in any one season. Particular weather patterns can lead to specific pathogen problems.

To perhaps a greater extent in qualitative terms than the weather, the use of *fungicides* alters the spectrum of pathogens liable to be significant. For instance, the use of a specific mildewicide is likely to leave the crop vacant for other pathogens such as rusts if conditions favour their colonisation. Even so-called broad-spectrum fungicides do not cover all diseases, and the absence of competition from those controlled provides greater opportunity for others to establish if conditions favour them. Examples of this principle include the greater prominence of sharp eyespot in some cases when common eyespot has been treated; and the rising in the pop-charts of net blotch of barley in the late seventies which first became apparent in cases where most other common pathogens had been checked. In summer 1982 many wheat crops which had been otherwise protected suffered *Fusarium* damage owing to the season (plus, perhaps, the popularity of particularly susceptible varieties).

Persistent use of the same type of fungicide may induce a resistant strain of a pathogen. This is not an argument against the use of fungicides but rather a comment on some of their inevitable limitations.

By the time that symptoms are readily apparent the pathogen has generally become well established and caused significant harm. In the livestock sphere a veterinary surgeon will frequently speak of sub-clinical disease, e.g. the commonest cases of mastitis in dairy cows: the same behaviour is equally characteristic of cereal pathogens, especially with viruses such as BYDV (barley yellow dwarf virus). Many crops suffered unexpected yield penalties at harvest in 1981, in particular if aphid vectors were untreated in time. Thus it cannot be overstressed that *early detection* of cereal disease is vital. In this connection every pair of eyes on the farm should be mobilised. Means are available to educate and motivate for such observation; for instance, trade literature and advertisements contain colour pictures of symptoms and hand lenses are readily available. The letters CID can take on new meaning if regular short meetings of farm staff are held during critical periods of the season for *crop investigation department* exchanges! Close observation is likely to reveal diseases and disorders new to the farm — so beware over-reacting!

To sum up, practical knowledge of the pathogen for farmers planning and carrying out disease control includes:

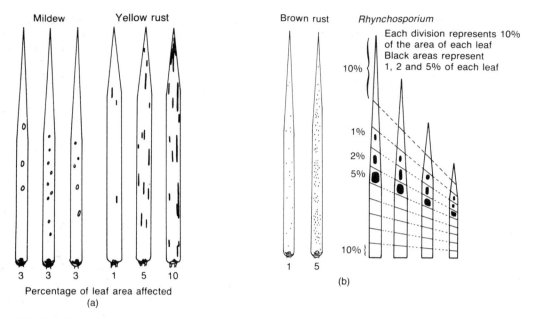

FIGURE 10.1   ADAS disease assessment guide drawings.

- *Recognition* − indispensable! One must be able to identify the cause of disease at an early stage.
- *Severity* − means of quantifying the threat (see Figure 10.1).
- *Spread* − an understanding of the life cycle and means of infection is important.
- *Conditions* favouring the pathogen in terms of climate, crop and site should be known.

CROP DEFENCES AGAINST DISEASE

Crops may avoid colonisation or check secondary infection by pathogens through mechanical means, e.g. they may have a particularly thick waxy cuticle coating the leaves, or closed flowers precluding ergot and loose smut. Some possess chemicals in their sap which are toxic to pathogens in general. Others may preclude disease by hypersensitive reaction, i.e. the initial cell colonised dies simultaneously and thus destroys the pathogen. Some varieties are tolerant of disease in the field, irrespective of race, only allowing its limited development. Such varieties may act as carriers from which pathogens colonise adjacent susceptible varieties.

Breeding for disease resistance is a laborious process. The breeder must combine reliability of yield, desired quality and other agronomic characters such as standing power with resistance to important races of each major disease. Apart from this, breeders must first identify sources of genetic resistance in other plant material and separate this gene or genes from undesirable features which may be linked.

Resistance to disease can be rooted genetically in broadly three mechanisms:

1. Single major gene This generally confers a steadily repeatable resistance from site to site and season to season *until* such time as the pathogen can change to a more virulent race. Security is based on but a single gene, a single padlock with but one key needed to open it.

2. Multiline varieties Here pains are taken to incorporate a range of resistance into a common genetic background. The principle is genetic diversity of disease resistance to reduce the risk of breakdown, consistent with agronomic uniformity. Generally, a multiline variety is a mixture of isolines (pure lines) differing by single major genes for pathogen reaction. The expression of resistance in the field is likely to be more variable than for the first case but risks of breakdown are reduced.

3. Polygene varieties Here two or more genes for resistance act together within a single variety. They can only be overcome by a more complex change of virulence by the pathogen and are thus more durable in resistance. It is probably in this way that Cappelle-Desprez winter wheat retained its performance for so many years against eyespot and yellow rust. Many barley varieties selected purely on visual symptoms probably resist brown rust in a similar way. Such resistance may be incomplete but it is unlikely to break down.

The breeder thus has several bases for procedure in seeking disease resistance. It is in the interests of civilisation that world collections of cereal cultivars and wild relatives are maintained as gene pools from which to select promising material. Mutation has been induced by irradiation to accelerate the natural rate of genetic change in plant material for disease resistance selection and indeed to develop whole new varieties of barley, e.g. Golden Promise.

International cooperation in cereal breeding has obvious benefits and can accelerate the production of improved varieties by exploiting opposite growing seasons of the two hemispheres as well as by awareness of sources for desired traits. However, there is the danger of international cultivation of fewer varieties and thus concentration of risk. Already the occurrence of epidemics on a wider scale has eliminated varieties overnight. In Britain the wheats Rothwell Perdix, 1966, and Joss Cambier, 1972, gave a double lesson for overpopularity and loss from yellow rust. The widespread rapid adoption of a single new variety is always alarming to me. If it is a flexible, forgiving variety like Maris Huntsman, then cause for alarm may be lessened. Different diseases become important and new races of already prominent diseases appear. Thus *Septoria* has been increasing recently on winter wheats whilst net blotch has multiplied in significance on winter barleys.

Therefore, it behoves farmers to minimise exposure of their cereal enterprise to disease, whilst retaining fungicides in a supportive role rather than as the sole line of attack.

CHOOSING RESISTANT VARIETIES

Farmers in the UK are not obliged to choose recommended varieties for disease resistance or for any other reason. However, most non-recommended varieties are excluded on disease susceptibility grounds. The annually published NIAB leaflet of recommended cereal varieties gives increasing information on disease factors:

- Varieties are scored for disease resistance as 0–9 (the higher the better).
- Varieties are divided into diversification groups (DG) according to their specific resistance to particular races. These DG groups are revised annually. Group 1 varieties are almost completely resistant so can be combined with any other variety on a farm: the idea is that otherwise farmers should choose varieties from different diversification groups for a proportion of their fields since those in the same group share a similar limited basis of resistance. The diversification principle is used internationally to some extent, e.g. in North America from Canada to Mexico for oat crown rust. Black stem rust of wheat, which is now a risk to the UK from southern Europe, has been of major world significance, though hitherto only occasionally in the UK. Monoculture of varieties increases the probability of epidemics; therefore diversify. It is not necessary to grow more than, say, three different varieties of a species on any one farm.
- Geographical distribution of major UK cereal diseases is shown in Figure 10.2.

Of factors influencing farmers in choice of variety, disease resistance, not surprisingly, usually ranks after yield, standing power and quality characteristics.

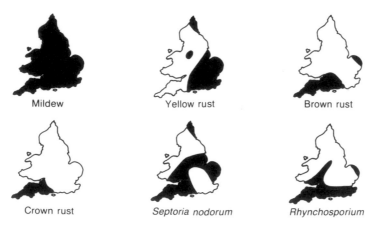

Mildew   Yellow rust   Brown rust

Crown rust   *Septoria nodorum*   *Rhynchosporium*

FIGURE 10.2 Geographical distribution of cereal diseases. (Source: NIAB)

The cost of developing disease resistance has not been borne directly by farmers and it is sensible to make strategic use of it as far as possible so reducing dependence on other measures, in particular the use and cost of fungicides. When varieties share similar yield and quality potential it is especially relevant to take account of differences in disease ratings.

The 0–9 scale for disease resistance is by no means a unique UK system. The idea is to give farmers a rating of the likely extent of yield loss in conditions which favour a particular disease. As conditions vary from season to season there are corresponding variations in disease prevalence. For some diseases there is a clear correlation between the climate of the previous season and the likelihood of problems in the current season, e.g. the chance of yellow rust inoculum is reduced by a previous hot, dry summer. However, NIAB conduct regional trials to assess resistance to natural infection as well as regional tests of resistance to deliberate inoculation of adult plants with pathogens representing known races of significance in the UK. In addition, the NIAB under its director, Dr Graham Milbourn, is now assessing varietal performance with and without fungicides (Figure 10.3). This has become increasingly relevant to the real farm situation as more farmers adopt

fungicide treatment. Varieties differ in their response to fungicides. Indeed most show some response (up to 3 or 4 per cent on yield) in the absence of the target diseases for a particular fungicide. Possible explanations for this include a suggested effect on plant hormones, notably cytokinins, and control of non-target diseases including those not normally considered pathogenic such as ear-blackeners (*Alternaria* and *Cladosporium*).

DIVERSIFYING CEREAL SPECIES AND VARIETIES

As well as some risk-spreading by the use of different cereal varieties it is good to see intensive cereal farms using several species. In particular, the rather less disease-prone species − rye and triticale − are as yet underexploited in the UK.

Mixtures (Blends) of Varieties

An extension of the principle of diversification between fields using different *varieties* is to mix them within the same field. Sinclair-McGill have been in the forefront of this work for some years and they forecast that up to 20 per cent of the UK spring barley hectarage could be planted to mixtures before long. Normally three different varieties are blended. This is enough to maximise the diversity effect whilst retaining a simple enough mixture to be able to match other characteristics such as date of maturity.

Most work has so far been done on spring barley. Varieties yield more in mixtures than as sole crops. The primary explanation for this is that mixtures possess improved disease resistance: this can be both substantial and repeatable from season to season and place to place. An additional explanation for the greater and more stable yield performance of mixtures may be that they are better buffered against variations in the environment such as drought and wetness; different varieties may also exploit the environment in slightly different ways, or indeed their root exudates may in some degree be synergistic. So far evidence for these effects is scarce. However, the disease resistance effect is well established, especially for spring barley mildew. A relatively susceptible variety surrounded by more resistant varieties will tend to have reduced disease. Certainly the spread of disease will be retarded, especially during the later seedling stage. Three varieties blended can show a 50 per cent reduction in mildew compared with the mean for the same three varieties as sole crops. This

Untreated control yield = 100 (5.84 t/ha)
Differences of 5% or less should be treated with reserve.

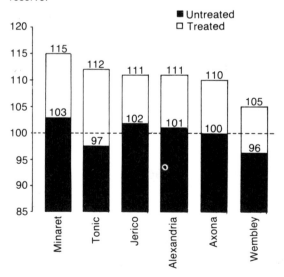

FIGURE 10.3 Yield of spring wheat varieties in fungicide treated trials. (From NIAB)

can mean up to 10 per cent yield increase for the mixture in a severe mildew situation. Effects of mixtures on rusts and *Rhynchosporium* development seem to be less dramatic though still worthwhile.

The case for using variety mixtures therefore hinges on disease resistance, as yield is improved with less recourse to fungicides and thus more cheaply. Fungicidal seed treatment may be given to a susceptible variety whilst the cost of this is saved on the seed of resistant varieties in the mixture. Mixtures are a particularly useful strategy against airborne diseases in areas of intensive cereal cultivation. Quality of mixtures seems to be a straightforward average of the qualities of constituent varieties.

The case against using variety mixtures has included unpopularity with the trade and anticipated difficulties in supplying blends on time for earlier sowing of winter corn. Many farmers already experience quite enough difficulty in obtaining single variety seed on time in the autumn! This particular problem can be sensibly overcome to some extent by multiplying inbreeding cereals for one generation as mixtures when any varietal alteration which might occur would be insubstantial if detectable at all. It can also be argued that a blend is less predictable to manage than a sole variety. Mixtures are not the cure-all and caution should be exercised to avoid any one particular blend occupying too large an area; it would approach the risk of single variety monoculture. This

is perhaps partly why the sale of mixtures has been allowed within states though not between member states of the EC.

Species Mixtures

Mixture of cereal *species* as a blend within one field has been practised for many years. Barley and oats are the usual mixture (possibly including peas or vetches also): in the south-west this is known as dredge whilst mashlum is the term applied in the north of England. The yield benefits from this admixture are partly attributed to induced disease resistance. Certainly the principle is accepted for many tropical cropping systems which often employ complex species mixtures. This is relatively straightforward to manage with a hand-labour system but clearly does not lend itself to mechanisation unless grown in bands (alley cropped).

Overview of the Main Cereal Diseases

These can best be described according to the principal part of the plant affected:

A. Root/stem-base diseases
B. Leaf (foliar) diseases
C. Ear diseases
D. Whole plant diseases

A. ROOT/STEM DISEASES

Recognition

These are caused by various fungi which can infect at an early stage in the crop's life and can be recognised as follows:

Name	Scientific name	Crops affected	Symptoms of attack
Take-all	*Gaeumannomyces graminis*	Wheat, barley. There is a separate oat strain. Also attacks maize and rye	Blackened root and stem base; lost plants in patches — easily pulled up; in early dry summers, white empty ears often eventually turning black with *Cladosporium*
Eyespot (common)	*Pseudo-cercosporella herpotrichoides*	Wheat, barley. Oats less often. Rye strain predominates on winter barley	Diffuse, brown sideways eye-shaped lesions 3 cm or so from stem base with central black dots; lodging in all directions later and white empty ears

(continued)

| Sharp eyespot | *Corticum solani*
(Rhizoctonia) | All cereals | Sharply defined sideways, eye-shaped lesions often spreading in multiples higher up the stem than in common eyespot; thinned crops; lodging later; shrivelled grain |
| Brown foot-rot | *Fusarium*
spp. | Barley, wheat, oats, rye. Also maize. There is a separate oat species | Seedling death; brown staining on stem base and nodes; pinkish spores can appear even up on the ears |

Conditions

Conditions favouring these diseases include aspects of climate, soil and cropping practice as follows:

Disease	Climate	Soil	Cropping
Take-all	Cold, wet spring, with slow growth of crop; persistence of infected material. Dry autumns which deter microbial adversaries of take-all	Light, calcareous, low OM, lower fertility	Spring crop worse affected; winter more *likely* to be attacked. Second to fourth cereals after break are prime targets
Eyespot (common)	Cool (3−14°C), moist (80% RH), rainsplash February−April (but crops colonised October to June)	Heavier types	Dense patches; especially in winter cereals, particularly if early sown
Sharp eyespot	Cool; drier weather suits better than common eyespot	Lighter types, especially if mildly acidic	Encouraged by potatoes and even peas; can appear when other diseases are well controlled. Erratic and unpredictable occurrence
Brown foot-rot	Dry; warm; windy	Acidic; compacted; poorly drained	Intensive cereals; crops under stress

Control

Yield losses from stem-base diseases are typically 10−30 per cent but take-all may exceed this and common eyespot is often at 25−30 per cent if uncontrolled.

Take-all

- This disease has no chemical control yet, though it is claimed that some seed dressings deter it.
- There are no resistant varieties; therefore control is based chiefly on starvation of the fungus, including clean farming, i.e. good early stubble cultivations and destruction of weed hosts (notably couch, fogs and bents). Direct drilling of a second or third winter wheat after oilseed rape may reduce take-all.
- An early, firm but not compacted seedbed with no restrictions to rooting depth and not overlimed will enable early autumn drilling and rapid crop establishment; but avoid seedbed nitrogen which can favour survival of infected residues.
- The disease does decline after the fifth year or so in a continuous cereal run; hence it is advisable *either* to go for a fairly diverse rotation of crops *or* to persist in a long run of continuous, well-managed cereals. A grass break will interfere less with the cycle than crops of other families.

Eyespot (common)

- Tolerant varieties of winter wheat are available and listed by NIAB.
- A two-year break from cereals usually breaks the life cycle adequately as long as it is coupled with good grass weed control.
- Clean farming is vital including good stubble burial and straw burning. Do not use infected straw for bedding livestock since the disease will be later spread with FYM.
- Strategic use of fungicides can be very effective but there is increasing incidence of resistance of the fungus especially to the MBC group and particularly in association with winter barley crops. Use prochloraz (Sportak) where this is the case, and subsequent later season control may switch to another fungicide group to contain mild attacks, e.g. propiconazole (Tilt).
- Timing of eyespot control could be during GS 30−32: inspect crops at GS 30. ADAS threshold is 20 per cent of tillers infected. MBC is still cheap and effective for lower risk situations and where eyespot is not thought to be resistant but 90 per cent of isolates showed resistance after 3 years' carbendazim use at LARS (only laboratory tests can reveal resistant strains of the fungus). For lower eyespot levels a single prochloraz spray at GS 37 has proved cost-effective, saving one spray.

Sharp eyespot

- Many alternate hosts including weeds complicate the exclusion of this disease (said to infect 300 species in 60 families).
- Some fungicide mixtures appear to give some control but not reliably; MBC at GS 30−31 can work.
- Control of common eyespot often seems to increase sharp eyespot.

Brown foot-rot

- Seed dressing controls seed-borne inoculum.
- Soil-borne inoculum is widespread and chiefly prevented from routine invasion by promoting conditions for vigorous crop development.
- Groups 1 and 2 fungicides (see Table 10.2) show some activity against *Fusarium*. Captafol reduces the ear blight stage of the disease.

B. LEAF (FOLIAR) DISEASES

Recognition

See Table 10.1. A guide to the chief features is given below:

Name	Scientific name	Crops affected	Symptoms of attack
Powdery mildew	*Erysiphe graminis*	Barley (especially spring). Wheat, oats, rye and most grasses (there are physiologically specialised strains)	White-grey fluffy pustules on stem-base and lower leaves move up plant; turn browner and later carry black spore cases
Rhynchosporium (Leaf blotch)	*Rhynchosporium secalis*	Barley, rye and grasses	Oval blotches on leaf: pale, green-grey when active, turning fawn in the centre with dark brown, diffuse margin. These lesions often colonise the leaf blade base
Net blotch	*Pyrenophora teres*	Barley (but strain reported on wheat in America)	Tiny brown speckles developing into net-like lesions. Net necrosis is often rectangular

(continued)

Septoria (Speckled leaf blotch)	*Septoria tritici*	Wheat and grasses	Lower leaves have fawn oval lesions developing black dots = pycnidia (spore cases) as they mature

Note: Glume blotch (*Septoria nodorum*) affects upper leaves and ears, has darker brown lesions when held up to light and spore cases are salmon pink/honey-coloured

Brown rust	*Puccinia recondita* *Puccinia hordei*	Wheat, rye Barley	Leaves peppered with brown pustules (gingerish with a yellow halo around them in barley)
Yellow rust	*Puccinia striiformis*	Wheat, barley, rye and grasses	Bright yellow pustules near leaf tip joining later into blocks which stripe the leaves

Note: Other foliar diseases which are far less common include: black stem rust of wheat (*Puccinia graminis*); crown rust of oats (*Puccinia coronata*), so-called because of its crown-shaped teliospores seen under the microscope (it is common in Devon and Cornwall); leaf spot of oats (*Pyrenophora avenae*); halo spot of barley (*Selenophoma donacis*); leaf stripe of barley (*Pyrenophora graminea*); leaf stripe of wheat (*Cephalosporium gramineum*) − these last two increase where straw and stubble remain near the soil surface. (For details of these and others, see Gair, Jenkins and Lester.)

Conditions

The foliar diseases respond especially to conditions of climate and also to cropping practices.

Disease	*Climate*	*Cropping*
Powdery mildew	Warm (12−20°C), humid; light winds; very hot weather (25°C+) checks it	Volunteers. Dense crops luxuriant with applied N. Early-sown in autumn or late-sown in spring
Rhynchosporium (Leaf blotch)	Cool (10−15°C); wet; rainsplash spreads it	Trash, volunteers, grasses; seed-borne
Net blotch	Cool, followed by wet and humid	Trash and infected seed; dense, early-sown crop
Septoria tritici	Wet, cool (5−15°C)	Trash and infected seed
Septoria nodorum	Wet, warm (12−22°C)	
Brown rust	Warm (15−22°C), high RH; windy	Alternative hosts and late N prolonging leaf greenness
Yellow rust	Cool (10−15°C), high RH, windy, overcast. Very seasonal incidence. In some years it quickly reaches epidemic level at temperatures well above 20°C	Nearby hosts; 54 known races 4 common in UK; some varieties have been very susceptible. Late N encouraging green leaf duration favours it

Note: All trash-borne diseases are favoured by reduced cultivation techniques unless accompanied by good burning.

TABLE 10.1 Assessment key for cereal foliar diseases

% infection	Mildew	Yellow rust	Brown rust	Septoria	Rhynchosporium	Net blotch
0	No infection observed					
0.1	3 pustules per tiller	1 stripe per tiller	25 pustules per tiller	1 lesion per 10 tillers	1 lesion per 10 tillers	1 small lesion per 10 tillers
1	5 pustules per leaf	2 stripes per leaf	100 pustules per leaf	2 small lesions per tiller	1 lesion per tiller	1 small lesion per tiller
5	2 lower leaves appear $\frac{1}{4}$ infected	Most tillers infected but some top leaves uninfected	Top leaf — numerous pustules but leaves appear green overall	Small lesions beginning on areas of dead tissue across width of leaf	Discrete lesions on most tillers, about 2 per leaf	2 lower leaves appear $\frac{1}{4}$ infected. Other leaves — few lesions
10	2 lower leaves appear $\frac{1}{2}$ infected	All leaves infected but leaves appear green overall	Top leaf — pustules sufficiently dense to give brown appearance in patches	2 lower leaves; large areas of diseased tissue some covering $\frac{1}{3}$ of leaf	Lesions coalescing but leaves appear green overall	2 lower leaves appear $\frac{1}{2}$ infected. Other leaves — numerous lesions
25	Leaves appear $\frac{1}{2}$ infected, $\frac{1}{2}$ green					
50	Leaves appear more infected than green					
75	Very little green leaf tissue left					
100	Leaves dead — no green tissue left					

Notes:

(1) Examine top 4 leaves. If top leaf has been fully expanded for less than 14 days, refer to 2nd leaf as 'top leaf'.
(2) Ignore all naturally senescent leaf tissue.
(3) Include all chlorosis and necrosis attributable to disease.
(4) Record percentage infection; use interpolated values (e.g. 3 per cent) if necessary.
(5) If foci present, record average over the plot as a whole.

Source: National Institute of Agricultural Botany.

Control

Mildew is undoubtedly the most regular and widespread yield robber, commonly reducing yields by 10–15% unless the top three leaves and ears are kept free. If uncontrolled, it can almost halve yield.

Late drilling increases the risk of mildew, causing significant damage to the smaller plants which result. Yellow rust can be devastating from time to time if uncontrolled. Most foliar diseases can reduce yields by around 10–15 per cent but uncontrolled *Septoria*

nodorum may take 25–30 per cent off yield as may late (June) spread of net blotch.

All foliar diseases demand good field hygiene to reduce inoculum for the next crop if they are to be controlled.

Powdery mildew

● Less susceptible varieties are available for all cereals. Some of these may develop a brown, blotchy rash as a hypersensitive reaction. Use a spread of varieties, guided by NIAB diversification groups.

- Choose different groups of fungicide when spraying more than once in the season. Triazoles are showing reduced efficiency against mildew but trietazine (Calixin) is still an effective specific mildewicide.

- Autumn control may be justified on winter barley where 5 per cent incidence on lower leaves is recorded (see Figure 10.1) on poorer, young crops on light land; 5 per cent incidence on youngest leaves justifies control even on well-established crops on better land. Some crops are protected from the start by broad spectrum seed dressing which may suffice for the whole life of some spring cereals.

- Spring spray on winter crops is best combined with eyespot control where necessary. Winter barley which has suffered uncontrolled mildew through a mild winter may need an earlier spray but this is unlikely. Winter wheat spraying should be withheld until GS 37 or when 5 per cent infection of lower leaves occurs. Late mildew (GS 39−59) warrants treatment at 3 per cent infection of upper leaves, especially in wheats. Fenpropimorph and fenpropidin are effective.

- Likely crop loss if mildew is *untreated after ear emergence* (GS 59) can be calculated at 2.5 times percentage of leaf covered by mildew.

Rhynchosporium leaf blotch

- Only considered of economic significance since 1964.
- Resistant varieties are available.
- ADAS recommend early spring spraying of winter barley having 10 per cent of leaf area infected. Later spraying and spraying of spring barley are justifiable if lesions have spread to any of the top three leaves.
- MBC, triazoles and prochloraz all control it.

Net blotch

- Resistant varieties are available.
- ADAS thresholds are similar to those for *Rhynchosporium*. Late May/June colonisation of upper leaves can be quite rapid in some seasons.

- Both prochloraz and propiconazole control this disease which has only assumed commercial importance since the late 1970s.

Septoria

- Has risen to economic status only since the late 1960s.
- Resistant varieties are available.
- Iprodione (Rovral), prochloraz (Sportak), propiconazole (Tilt, Radar), chlorothalonil (Bravo) plus MBC and captafol all control it. (Captafol should not be used on milling wheats after flag leaf.)
- *S. tritici* is spread by air into the crop in autumn; ADAS thresholds apply only between GS 39−59, when older leaves show 5 per cent infection *or* when at least 1 mm of rain fell on four or more consecutive days during the previous fortnight (this is the typical lag between infection and appearance of symptoms).

Brown rust

- Choose resistant varieties and diversify (a bright yellow halo around early pustules indicates a resistant reaction of the variety).
- Inspect often from GS 37 onwards to full ear emergence. Spray as soon as brown rust is noticed spreading and keep top three leaves green.
- Triadimefon (Bayleton) and propiconazole (Tilt, Radar) are both active, as is fenpropimorph (Corbel, Mistral).
- ADAS thresholds recommend routine sprays for very susceptible varieties (NIAB 3-) at GS 37 and again at GS 59.

Yellow rust

- Avoid susceptible varieties and diversify.
- Always keep top two leaves and ear clean.
- Benodanil (Calirus), triadimefon (Bayleton), fenpropidin (Patrol), fenpropimorph (Corbel, Mistral), and propiconazole (Tilt, Radar) are all active against it.
- Infection can spread dramatically from initial foci. Keep watch! It remains the most aggressive fungal disease of UK cereals under the right conditions and potentially the most damaging.

C. EAR DISEASES

These are most obvious and fairly easily distinguished.

Recognition

Name	Scientific name	Crops affected	Symptoms of attack
Glume blotch	*Septoria nodorum*	Wheat	Purple-brown blotches developing black dots
Sooty mould	*Cladosporium herbarum*	Wheat especially; also barley, oats and rye	Overall superficial, dusty blackening
Black point	*Alternaria* spp.	Wheat especially	Blackened germ end of grain
Grey mould	*Botrytis cinerea*	Especially wheat, barley	Grey spores embedded in glumes; grains gone
Ear blight	*Fusarium* spp.	Especially wheat; also barley, oats and rye	Fawnish-white lesions, sometimes with pink spores
Loose smut	*Ustilago* spp.	All cereals; different smut species for each	Ears just dusty, black and grainless as they emerge
Bunt	*Tilletia caries*	Wheat, rye	Blue-green glumes. Grains filled by fishy-smelling, brown spores inside
Covered smut	*Ustilago hordei*	Barley	Brown spores; not released until combining
Ergot	*Claviceps purpurea*	All cereals, especially rye, triticale, spring wheat; grasses including blackgrass and couch	Purple-black objects replace occasional grains

Conditions

Some of these are indicative of previously weakened crops and the local presence of inoculum on decaying trash as well as on alternative hosts. Humid conditions favour development of these ear diseases. *Septoria* has been discussed above. Smuts are seed-borne whilst ergot is spread in developing ears by flies as well as from sporulating ergots (purple-black curved sclerotia or resting bodies). Ergot is very poisonous, causing localised haemorrhage, diarrhoea, gangrene or even death, although it contains chemicals used medically (e.g. as uterine relaxants to assist childbirth). Its spread is favoured by a cool, moist and thus prolonged flowering period.

Control

Fungicides are available for these ear diseases except for ergot, where the strategy involves using clean seed, ploughing down known ergots and controlling the flowering of grassy areas known to harbour the disease. In the case of smuts, seed must be clean at the outset (OSTS health test), and seed dressings are used (e.g. carboxin mixtures) against loose smut.

D. WHOLE PLANT DISEASES

Recognition

There are two main fungi associated with severe loss of plants overwinter and two main virus diseases which can damage the whole plant quite severely. These can be recognised as follows:

Name	Scientific name	Crops affected	Symptoms of attack
Snow mould	*Fusarium nivale*	Wheat, barley	Distorted seedlings; brown shoot bases covered in mass of pink spores after snow thaws
Snow rot	*Typhula incarnata*	Barley especially; wheat	Dead, shrivelled plants after snow; red-brown sclerotia around bases often about 2 mm in diameter
BYDV	Barley yellow dwarf virus	Barley, wheat, oats, grasses	Barley upper leaves turn yellow from tips; wheat goes pale yellow turning bronze; oats turn purplish-pink. All have stunted patches, very poor ears
BaYMV	Barley yellow mosaic virus (first recorded in the UK in 1980)	Barley	Yellow patches appear overwinter (look like wet spots in the field); leaf colour is streaky yellow. Plants are stunted (may die overwinter); ears poor

Conditions

Both snow rot and snow mould develop, as their names imply, following persistent snow cover, especially in long runs of barley and wheat.

BYDV is favoured by any conditions which prolong aphid vector (carrier) activity for primary colonisation and in-crop spread. September-drilled autumn cereals are at risk especially if a mild, sunny but calm autumn follows. Not all aphids are carriers by any means (see preceding chapter). The southern coastal and estuarine zone is always at risk.

BaYMV is a new disease of increasing incidence in long runs of winter barley. Cultivations will spread the fungus (*Polymyxa graminis*) which carries the virus to the barley roots. The fungus seems capable of retaining the virus for years in the absence of barley cropping. Different strains of the virus complicate the issue.

Control

Snow mould
Some varieties, e.g. winter barley cv Gerbel, show more resistance than others, e.g. Igri; autumn MBC fungicides control it.

Snow rot
Seed dressing with broad-spectrum fungicides helps.

BYDV
Early control of potentially infective aphids in autumn (notably *Rhopalosiphum padi*, the bird-cherry aphid) using cypermethrin. Avoid very early drilling in mild districts, especially of the south-west.

BaYMV
Avoid spreading known patches by boots, cultivations or wheels. Consider the use of a resistant variety, e.g. Torrent, or go out of cereal cropping on that field. Rotations substituting winter wheat are being studied.

DISEASE CONTROL STRATEGY USING FUNGICIDES

The chemical groups and names of major fungicides (including seed dressings) appear in Table 10.2. Typical timing is indicated in Figure 10.4.

Until 1969 only contact fungicides were available based on sulphur, copper or organic molecules

TABLE 10.2   Fungicides grouped according to their mode of inhibitory action and main diseases controlled

Site-specific fungicides	Multi-site fungicides
Group number, name and active ingredients	*Group number, name and active ingredients*

1. MBC or benzimidazoles (Note: E developing resistance)

   benomyl   B, E, M, Sep, Rh
   carbendazim   E, Rh
   fuberidazole   S
   thiabendazole   S
   thiophanate-methyl   E, Rh

2. Dimethylation inhibitors in ergosterol biosynthesis

   Imidazoles   imazalil   S
                   prochloraz   E, M, Rh, N, Sep, R
   Piperazine   triforine   S, M
   Pyrimidines   nuarimol   S, M
   Triazoles   propiconazole ⎤
                   triadimefon   ⎥
                   triadimenol   ⎬ B
                   flutriafol   ⎦

3. Ergosterol biosynthesis inhibitors

   Morpholines   fenpropimorph   M, Rh, R
                   tridemorph   M, R
   fenpropidin (a piperidine)   M, Rh, R

4. Hydroxypyrimidines

   ethirimol   M

5. Carboxyamides

   benodanil   R

   carboxin   S(L) − only in mixtures

6. Guanidines

   guazatine   S

7. Organophosphates

   pyrazophos   M, N, Rh

8. Dicarboxyamides

   iprodione   Sep

9. Phthalimides

   captafol   Rh, Sep

10. Phthalonitriles

    chlorothalonil   Sep, Rh

11. Dithiocarbamates   B

    mancozeb

    maneb
    manganese/zinc dithiocarbamate complex
    Polyram
    propineb
    thiram
    zineb

12. Mercurials

    organomercury seed treatments (e.g. Phenyl mercury acetate, PMA)

13. Sulphur

    Sulphur   M esp. on barley

KEY: B = broad spectrum; E = eyespot; M = mildew; Sep = *Septoria*; Rh = *Rhynchosporium*; N = net blotch; R = rusts; L = loose smut; S = seed dressing.

Source: Based on MAFF classification.

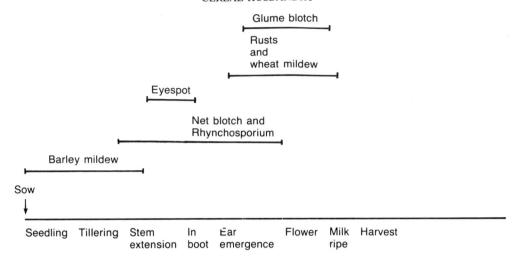

FIGURE 10.4   Periods of maximum need to consider fungicidal control of the main diseases of cereals.

including organomercurial seed dressings. The introduction of systemic materials, adequately persistent and non-phytotoxic, has transformed the scene. Systemic materials are readily translated through the host cereal's vascular system, protecting it wherever the pathogen chooses to invade. Their widespread adoption has been based on their ability to combat deeper-seated infections, protect new growth and off-set inaccuracies in spray application.

In addition to specific materials, broad spectrum compounds and mixtures have enabled control of several important diseases simultaneously. The pesticide legislation of 1986 has limited the former use of farmer-concocted tank mixes of great miscellany.

The efficient use of fungicides must be based on:

- regular crop inspection
- timely application
- costing of treatments against the potential value of crop saved

Application costs can often be reduced by mixing compatible chemicals, e.g. eyespot control in winter wheat at GS 31 together with chlormequat growth regulator and a spring herbicide dose. In practice, a typical strategy would be to provide fungicidal seed dressing initially for autumn-sown cereals followed by an eyespot spray in early spring (GS 30–32), which may involve a fungicide which is active against other diseases present or likely to be soon present. Between GS 39–55, control late foliar diseases and potential ear diseases. The concomitant development of tramlines has facilitated the adoption of these later fungicide treatments to protect the flag leaf and ears

which are so crucial to the actual material to be stored in the grains.

The great debate is whether to rely on a fire-engine approach to established diseases *or* to go for a prophylactic (preventive) spraying policy in anticipation of potentially troublesome diseases.

There are several considerations to bear in mind about the fire-engine approach:

- Its successful execution demands regular vigilance in order to catch damaging diseases very early (Figure 10.4). Otherwise not only may crops suffer higher losses but one may end up giving a higher dose of chemical later on in attempting to control an established severe attack.
- Costs can be saved when threatening diseases do not materialise: it is easy to be scared into insurance spraying by the tales of woe associated with possible disease effects!
- Fungus resistance has already developed to a number of systemic fungicides. One way of reducing this happening is to cut down on frequency of their application.

Against the fire-engine approach are the following:

- The tendency of disease symptoms to appear only after some damage has already been inflicted subclinically.
- The ability of diseases to spread rapidly under favourable conditions — sometimes ahead of the farm's capacity to administer curative sprays.
- The increasing research and development on disease forecasting which is improving the prediction of disease incidence based on climatic

and agronomic data as well as the more accurate estimate of cost/benefit ratio which is now possible, enabling a more precise judgement of the need for prophylactic treatment to be given.

CEREAL DISEASE FORECASTING

The possibilities for more accurate forecasting stem from advances in micro-computing which facilitate data collection including climatic information and effective communication of disease intelligence into crop management 'packages'.

Effective forecasting is most justified for high-cost/high-value control situations, especially where disease incidence is irregular but its damage has been previously measured accurately. Forecasting can cover seasonal shifts in disease patterns, likelihood of damage in a particular area and the probable course of disease development. At Long Ashton Research Station, University of Bristol, Dr David Royle and team have studied the epidemiology of several diseases, enabling them to quantify likely responses to weather variations. For instance, in the case of brown rust on spring barley, at 8°C a wet period of fifteen hours was needed to induce the same amount of infection as occurred in only six hours at 12°C; for *Septoria* on winter wheat, a lax crop allowed splashed spores to leap 50 cm up the crop canopy following a single June downpour, whereas dense crops did not permit such access for raindrops. The team have developed a splashometer to measure spore movement by rainsplash up the canopy!

In the Netherlands, the EPIPRE system of disease forecasting has been operating since the mid 1970s (Figure 10.5). The title stands for epidemiology, prediction and prevention. It has been extended to other European countries including the UK. Linked to the Dutch National Agricultural University computer at Wageningen, the scheme covers *Septoria*, mildew, yellow rust, brown rust and aphids. Farmers send in observations made at critical stages and receive advice by return on whether to spray; they also have telephone access. Most Dutch farmers participating claim a saving equivalent to the cost of one fungicide application — scope for some confidence in it! Another reason for greater confidence in the future of disease forecasting is the scope for improved bio-assay techniques for diseases. BYDV could previously be confirmed only after long laboratory tests but the severity of the particular strain of virus present can now be assessed instantly by a colorimetric test developed at Long Ashton Research Station. This offers the possibility of rapid testing of

the virulence of virus present in adjacent grasses and volunteers *before* the cereal crop is even at risk.

Commercial firms are also offering predictive advice based on ADAS thresholds, weather conditions and historical damage data, e.g. *Cereal Disease Alert* (Schering).

THE PROBLEM OF FUNGICIDE RESISTANCE

Resistance to antibiotic sprays by the rice blast fungus (*Pyricularia oryzae*) was reported as long ago as 1962. Mildew resistance to ethirimol was detected in 1971, only two years after the introduction of this systemic fungicide.

Farmers first encounter resistance as a failure of an approved fungicide to achieve control even though properly applied. The mechanism seems usually to involve a population effect whereby susceptible strains of the fungus are removed, leaving an already present resistant minority scope to develop unhindered by their competition. The problem is especially associated with chemicals having a very specific mode of action. Development of resistance in barley mildew can be a result of effective fungicide action on the majority of the fungus population which

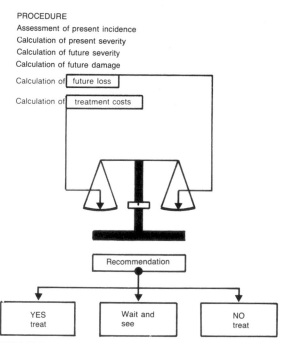

PROCEDURE
Assessment of present incidence
Calculation of present severity
Calculation of future severity
Calculation of future damage
Calculation of future loss
Calculation of treatment costs

Recommendation

YES treat

Wait and see

NO treat

FIGURE 10.5 Steps in the procedure of EPIPRE. Every time new information from a particular field is entered into the data bank, the programme goes through all these steps for that particular field.

was susceptible. Triadimefon (Bayleton) became less effective than tridemorph (Calixin) by 1983, illustrating the dynamic nature of the situation (Figure 10.6).

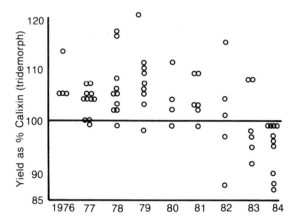

FIGURE 10.6   Development of barley mildew resistance to the broad spectrum fungicide triadimefon (Bayleton). (ADAS data of J. Jenkins, Leeds)

The response to the problem must be to:

- minimise fungicide use where other control measures can be used.
- avoid repeat treatments with fungicides from the same groups (see Table 10.2) on the same crop. Ring the changes.
- apply fungicide mixtures and mixed formulations where possible.
- monitor and report any apparent failure of fungicide performance.
- revise fungicide use as choice of chemical and parasite resistance changes.

INTERACTIONS BETWEEN FUNGICIDES AND OTHER TREATMENTS

The use of a fungicide to reduce disease will alter the nature of the crop either by promoting greater duration of green leaf area or otherwise altering plant morphology and physiology. In turn, the use of other inputs may influence the crop's reaction to disease and hence the level of response to fungicides. Data

of Jordan and Stinchcombe (1986) illustrate an example of this for winter barley. This work developed out of observations during nitrogen rate and timing trials at one place (Figure 10.7).

At another level than the physiological or biochemical ones mentioned above, there is a straightforward interaction between disease control response and agronomic practice. The example of the interaction between BYDV infectivity and sowing date of winter wheat illustrates this (Table 10.3).

Timing is critical to achieve cost-effective fungicide use. The cost/benefit ratio has to date been very favourable in the UK cereal crop, commonly around 6:1 as far as estimates allow. Responses are commonly 10−15 per cent extra yield, sometimes more, according to variety. Fungicides can also affect quality, sometimes beneficially, sometimes not, e.g. reducing Hagberg falling number in milling wheat as shown by Paul Davies and others (1982).

The duration of fungicide effect (days of antisporulant action) varies somewhat. Triazole is effective against mildew for 21 days, and the addition of morpholines extends this to 31 days; imidazole against net blotch and *Rhynchosporium* lasts 34−38 days (LARS data). Triazole treatment also seems to increase frost-hardiness of plants.

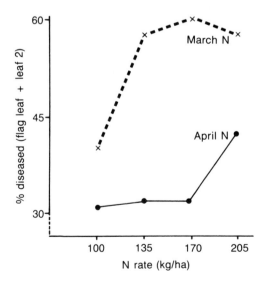

FIGURE 10.7   Effect of timing of the main N dose on disease in winter barley leaves 1 and 2, Eastleach, June 1981.

Cereal pests

Aphid

Wireworm

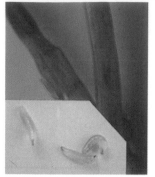

Wheat bulb fly larvae

Slug

Cereal mineral deficiencies

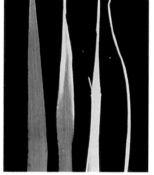

Copper deficiency

Calcium deficiency

Potassium deficiency

Magnesium deficiency

Phosphorus deficiency

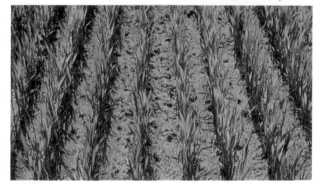

Nitrogen deficiency

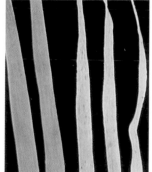

Manganese deficiency

Cereal stem diseases

Fusarium W/B

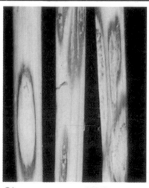

Sharp eyespot W/B

Sharp eyespot, lodging W/B

Eyespot W/B

Cereal foliar diseases

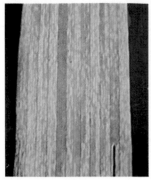

Yellow rust W/B

Brown rust W/B

Septoria nodorum W

Septoria tritici W/B

Mildew W/B

Rhynchosporium B

*Barley yellow dwarf
virus W*

Net blotch B

Cereal ear diseases

Wheat mildew W

Whiteheads (Foot diseases)

Glume blotch W

Cladosporium W

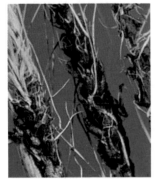

Barley, loose smut B

Stinking smut W

Ergot W/B

Fusarium W/B

TABLE 10.3 Cumulative weekly infectivity index for barley yellow dwarf virus; Rothamsted, autumn 1982

Crop sown in week beginning	Infectivity index on									
	5/9	12/9	19/9	26/9	3/10	10/10	17/10	24/10	31/10	4/11
1/9	2	2	86	93	102	117	129	134	145	145
6/9		0	84	91	100	115	127	132	143	143
13/9			84	91	100	115	127	132	143	143
20/9				7	16	31	43	48	59	59
27/9					9	24	36	41	52	52
4/10						15	27	32	43	43
11/10							12	17	28	28
18/10								5	16	16
25/10									11	11
1/11										0

Note: Infectivy index = number of flying aphids potentially capable of infecting cereals with BYDV multiplied by the proportion that actually do so.

Source: Rothamsted Experimental Station.

Chapter 11

WEEDS AND THEIR CONTROL IN CEREALS

WEEDS OF CEREALS

Weeds are classically defined as plants in the wrong place. The most troublesome in cereals, as in other crops, are members of the same family. Grass weeds are not only suited to similar growing conditions as for cereals but are also more difficult to control within cereal rotations. Weeds are any plants successful in the struggle for existence with cereal crops. In short, weeds interfere with cereal production. They do so as follows:

- compete directly for resources of the environment and inputs, reducing yield (e.g. as much as 1 per cent yield of wheat/wild oat/m^2, Figure 11.1)
- harbour diseases and pests, e.g. couch with take-all; various grass weeds with frit fly; sterile brome with BYDV
- may be poisonous to livestock (more of a problem with some poor grassland than in feed corn crops) or may produce root exudates toxic to the cereal crop
- slow the harvest by delaying maturing and hampering combining, e.g. bindweeds, cleavers
- increase costs of cleaning and drying grain, e.g. mayweeds
- reduce grain sample quality, e.g. wild oats
- promote crop lodging, e.g. cleavers, sterile brome
- lower land values or increase the dilapidations payable by a tenant on the termination of the tenancy, e.g. persistent perennials, wild oats
- may be parasitic, e.g. witchweed (*Striga* spp) on cereals, notably sorghum and millet in the tropics
- increase the weed seed bank in soils

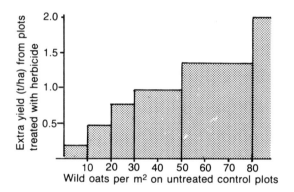

FIGURE 11.1   Yield response of winter wheat to removal of various populations of wild oats. (From Baldwin, J.H., 1979, *ADAS Quart. Rev. No. 33*)

Table 11.1 illustrates an effect of weed control on cereal yield and sample quality.

WEED CHARACTERISTICS

The characteristics of plants which make them successful as cereal weeds include:

1. They are usually self-fertilised and produce many seeds in relation to their size, e.g. chickweeds up to 2500 per plant, corn poppy 17,000, mayweed 300,000. These seeds may be viable even when immature.

2. They have efficient fruit and seed dispersal. This includes ability to pass incognito down the spout of

150

TABLE 11.1   Effect of weed control on yield and proportion of matter other than grain (MOG) in an Oxfordshire barley crop combined on 24 July 1980

	No weed control	Blackgrass and BL weeds controlled
Grain yield (t/ha)	4.36	8.00
MOG (t/ha)	11.10	11.24
Grain: MOG	0.33	0.71
Length of straw (cm)	54	67

Source: After Elliott, 1980.

corn drills! They may also spread via straw, boots, implements (especially combines) and by cultivations.

3. They establish rapidly, and this couples with the ability to adjust their individual size to population density like goldfish.

4. They are an effective means of vegetative spread, notably couch and perennial bindweed.

5. They are tolerant of a wide range of environmental conditions such as temperature, rainfall, soil type, i.e. successful weeds are not highly physiologically specialised.

6. The seeds are capable of long-term survival over periods of unfavourable conditions, e.g. chickweed, charlock and mayweed may remain viable after 60 years; wild oat after 9 years plus.

WEED LIFE CYCLES

Annuals

These complete their life cycle within one year. They are sometimes very short-lived ephemerals with perhaps a six-week life cycle, e.g. annual meadow grass, annual nettle, which can germinate at any time of year except when the weather is very severe. Ephemerals flower at all seasons and are prolific seeders. Indeed all annuals tend to be prolific seeders, quick to establish and shallow-rooted. Many annuals germinate in two 'flushes' at spring and autumn, e.g. chickweed, charlock, corn poppy, spurrey, blackgrass and wild oat. Others germinate only in spring and early summer, e.g. fat hen and redshank, whilst some germinate to any extent only in autumn, e.g. hemp nettle, parsley piert (Figure 11.2). Cleavers (goosegrass) germinates during milder parts of winter.

Perennials

These plants produce seed but also continue from year to year vegetatively. Means of vegetative propaga-

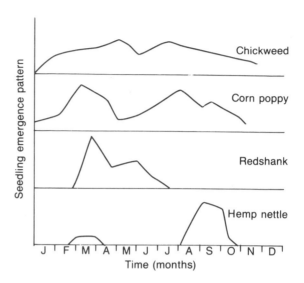

FIGURE 11.2   Periods of emergence of some annual BL weeds.

tion include rhizomes (couch), stolons (creeping bent), corms (onion couch), persistent roots (perennial bindweed).

Within each of these two life-cycle categories it is useful to distinguish *grass weeds* and *broad-leaved (BL)* or *dicotyledonous weeds* .

It must be realised that weeds visible as seedlings or mature plants at any one time are but the 'tip of the iceberg' of the total weed population present as seeds and other propagating parts in the soil.

THE WEED FLORA IN CEREALS

Whilst successful weeds are adaptable, differences in the composition of the weed flora occur from place to place according to variations in the following:

- *climate*, e.g. hemp nettles are much more prevalent in the north of the UK
- *soil conditions*, viz. type, structure, pH, nutrient status, moisture status. The weed spectrum is a useful indicator of soil conditions (Table 11.2)
- *husbandry practices*
  Frequency and depth of soil disturbance: although ploughing can achieve good burial of existing weed plants, it also resurrects more weed seeds to the soil surface
  Crop rotation: in a predominantly winter cereal

system, grass weeds tend to increase, as do cleavers and dead nettles
Use of herbicides: these remove the competition of susceptible weed species, thus allowing non-susceptible ones to expand (Figure 11.3)

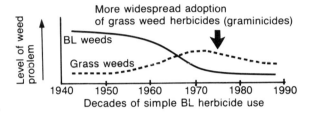

FIGURE 11.3   Changes in the problem weed flora.

WEED MONITORING

Individual field records should contain an assessment of the weed spectrum present. This should obviously include the *range* of predominant species and the *level* of infestation of any troublesome ones. Any species present, even in small numbers, which may pose a rapidly expanding threat should be recorded, e.g. blackgrass.

A rough sketch map of the field is useful to indicate the distribution of weed problems — headlands generally have concentrations of weeds, some species occur in patches, e.g. onion couch. Rogue wild oats and blackgrass where possible and mark where they were found on the field map to check again.

The ability to *identify* weed seedlings is a most important skill for a competent cereal grower to develop. Distinguishing features are the number, size, shape, hairiness or otherwise, and the arrangement of cotyledons (seed leaves) and true leaves. Details of currently recommended control measures and chemical products can be looked up in booklets but there is no substitute for field recognition of the problem. However, students are advised to make up their own scrapbook of pictures and specimens of common weed seedlings — pictures are to be found in trade leaflets and farming magazine advertisements. In addition there are several books which contain pictures to aid recognition. If new to the subject, it is better to aim to recognise one or two species each time one visits a field than to try to assimilate names of a whole host at once. It is important to distinguish

TABLE 11.2   Soil conditions and weed types

Predominant weeds	Soil conditions
Spurrey, corn marigold, hemp nettle	Acidic
Corn mint, horsetails	Wet
Ox-eye daisy, brome-grasses	Low fertility
Chickweed, fat hen, cleavers	High fertility
Corn buttercup, couch	Heavier
Couch, onion couch, charlock	Calcareous
Mayweeds	Lighter, sandier

serious weeds from those which are common but relatively innocuous. In any case the total list of regularly troublesome cereal weeds is not too formidable!

Broad-leaved (BL) Weeds

Chancellor and Froud-Williams (1984) present the results of surveys in central southern England which indicate the prevalent broad-leaved weeds of cereals (Tables 11.3 and 11.4). See also colour plates.

It is clear that cleavers remain troublesome, as do field bindweed and knotgrass. Field pansy, forget-me-not (FMN), cranesbills and red dead nettle (RDN) are all increasingly encountered (see Table 11.5). ADAS thresholds based on this work give plants/m^2 worth controlling as 1 for cleavers, 2 for poppy, 3 for chickweed and mayweeds, 10 for speedwells, FMN and RDN, 20 for field pansy.

Grass Weeds

Many species can qualify for inclusion here for cereals somewhere in the world (see Table 11.6 and Behrendt and Hanf, 1979). They often complete their life cycles faster than the hosting cereals.

In the UK, there are three major species (couch, wild oats and blackgrass) and other occasionally or potentially damaging species (meadow grasses, sterile brome, onion couch and a type of awned canary grass called *Phalaris paradoxa*, newly established in Britain in 1978 though first recorded here in 1847). See Figure 11.4 for identification of a number of grass weeds found in cereal crops. ADAS thresholds for control (plants/m^2) are 1 for wild oat, 5 for blackgrass and 25 for RSMG. These do not take account of the seed bank factor or of the value of keeping land clear for seed production.

TABLE 11.3   The percentage of fields in which the thirty-one most frequent BL weeds were recorded

Weeds	Winter wheat	Winter barley	Spring barley	All crops combined
Galium aparine cleavers	12	3	0.8	9
Viola arvensis field pansy	11	12	4	10
Convolvulus arvensis field bindweed	7	2	0.8	5
Myosotis arvensis forget-me-not	7	4	2	6
Stellaria media chickweed	6	7	8	7
Polygonum aviculare knotgrass	6	5	4	6
Bilderdykia convolvulus black bindweed	5	0.5	5	4
Lamium purpureum red dead nettle	4	2	7	4
Rumex obtusifolius broad-leaved dock	4	2	2	3
Cirsium arvense creeping thistle	4	2	2	3
Papaver rhoeas poppy	3	1	2	2
Veronica persica common field speedwell	3	2	4	3
Aethusa cynapium fool's parsley	2	2	2	2
Matricaria perforata scentless mayweed	2	2	2	2
Heracleum sphondylium ssp. *sphondylium* hogweed	2	3	0.8	2
Polygonum persicaria redshank	2	0.5	3	2
Polygonum lapathifolium ssp. *pallidum* persicaria	2	1	2	2
Veronica arvensis wall speedwell	2	2	0	2
Chenopodium album fat hen	2	0	3	1
Chamomilla recutita scented mayweed	1	0.5	0.8	1
Sonchus arvensis perennial sowthistle	1	0	2	1
Lamium amplexicaule henbit dead nettle	1	2	2	1
Anagallis arvensis scarlet pimpernel	1	0.5	2	1
Ranunculus spp. buttercups	1	0	0	0.7
Epilobium spp. willowherbs	0.9	0	0	0.6
Equisetum arvense horsetail	0.9	0	0	0.6
Aphanes arvensis parsley piert	0.7	0	0	0.5
Buglossoides arvensis field gromwell	0.7	0	0	0.5
Chamomilla suaveolens pineapple weed	0.7	0	0	0.5
Atriplex patula common orache	0.6	0	0	0.4
Cerastium fontanum common mouse-ear	0.6	0	0	0.4

Source: From R.J. Chancellor and R.J. Froud-Williams, 1984

TABLE 11.4   The percentage of fields of three cereals in which no weeds were observed

Crop	Clean of grass weeds	Clean of BL weeds	Clean of all weeds
All winter wheats	11	47	7
Short-strawed winter wheats	8	44	4
Long-strawed winter wheats	17	52	12
Winter barley	42	65	32
Spring barley	28	67	23

Source: From R.J. Chancellor and R.J. Froud-Williams, 1984.

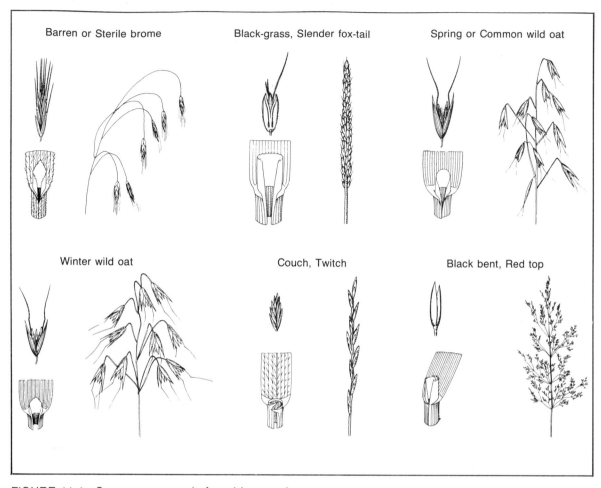

FIGURE 11.4   Some grass weeds found in cereal crops. (Courtesy of CIBA-GEIGY Ltd)

TABLE 11.5   The crop equivalent concept to express weed competitiveness (dry weight per weed/dry weight per cereal plant ratio). A value of 0.5 means that two weed plants are equivalent to one wheat plant. Ranking is from most to least competitive

Weed	Crop equivalent
Cleavers	7.2
Wild oats	2.5
Poppy, mayweeds	0.6
Blackgrass, chickweed, field speedwell	0.5
Ivy-leaved speedwell, red dead nettle	0.3
Forget-me-not	0.2
Field pansy, parsley piert	0.1

Source: From work by George Cussans and Bernard Wilson of Long Ashton Research Station.

Couch (*Elymus repens*, formerly *Agropyron repens*); sometimes includes black bent (*Agrostis gigantea*) and other species. Many local names indicate its widespread distribution in the UK: twitch, squitch, wix, wickens; it is known as quackgrass in the United States and as scutch in Eire.

It is a perennial spreading by rhizomes (underground whitish creeping stems) which can produce a new shoot from each node (joint) which occurs every 2 cm or so along the rhizome. Any cultivators, particularly powered ones, which split the rhizome into pieces thus assist its multiplication. Indeed, prior to the advent of glyphosate, the standard policy to control bad couch infestation was to alternate rotavation with periods for shoot development prior to paraquat spraying, repeating the process to starve out the

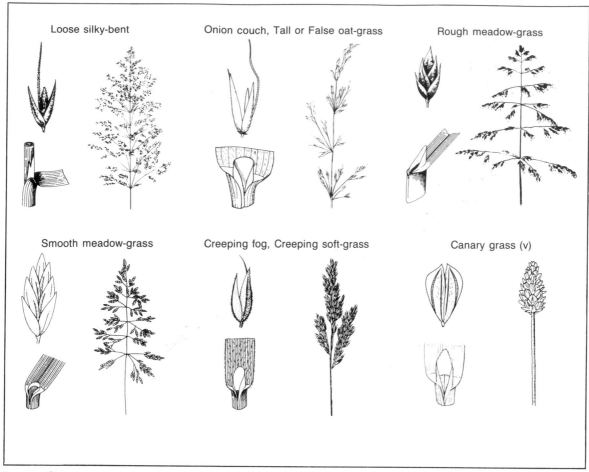

Loose silky-bent   Onion couch, Tall or False oat-grass   Rough meadow-grass

Smooth meadow-grass   Creeping fog, Creeping soft-grass   Canary grass (v)

(continued)

rhizome reserves as necessary. Couch can also spread by seeds.

Wild oats *Avena fatua* is the spring type; in the UK only since 1917 is *A. ludoviciana*, the winter type with clustered seeds.

This weed is a worldwide problem (see Price-Jones, 1981), and it is reckoned to be a major problem in many areas of the UK, particularly on heavier land (Figure 11.1). In 1974, National Wild Oat Year was launched as the start of a major campaign. A 1977 MAFF survey of 2250 UK fields reckoned it to be still expanding onto some 60,000 hectares each year and it was present on 97 per cent of English farms surveyed. The problem of wild oats arises from the seed's capacity to lie dormant in the soil for nine years, perhaps more. Hot, dry summers reduce this dormancy capacity; dormancy is broken by cultivations which aerate and scarify the seeds. Straw and stubble burning reduces dormancy and also destroys some seeds. Up to 90 per cent of seeds left on undisturbed stubble disappear.

Critical population for yield depression is reckoned at 12 plants/m^2, which may mean some 30 panicles/ m^2, allowing for tillering. However, once the problem has been brought down to manageable proportions, there is much to be said for keeping it there. Apart from the satisfaction of a tidy farm, there is then the option to grow seed crops and crops for quality markets such as malting and breadmaking, plus the avoidance of dilapidation charges at the end of a tenancy.

TABLE 11.6   Grass weeds in cereals: percentage of 2626 surveyed fields infested in 1981 in central southern England

	Winter wheat	Winter barley	Spring barley
Wild oats (*Avena* spp)	32	31	52
Rough-stalked meadow grass (*Poa trivialis*)	29	6	0.5
Couch grass (*Elymus repens*)	26	20	53
Blackgrass (*Alopecurus myosuroides*)	23	9	11
Barren brome (*Bromus sterilis*)	12	4	–
Italian ryegrass (*Lolium multiflorum*)	10	6	5
Bent grasses (*Agrostis* spp)	9	5	4
Timothy (*Phleum pratense*)	6	2	4
Onion couch (*Arrhenatherum elatius* var. bulbosum)	4	1	2
Meadow brome (*Bromus commutatus*)	3	0.3	–
Yorkshire fog (*Holcus lanatus*)	1	0.7	–
Perennial ryegrass (*Lolium perenne*)	1	0.3	–
Soft brome (*Bromus mollis*)	0.4	0.4	–
Creeping soft grass (*Holcus mollis*)	0.5	0.1	–
Loose silky-bent (*Apera spica-venti*)	0.3	0.1	–
Marsh foxtail (*Alopecurus geniculatus*)	0.2	–	–
Smooth-stalked meadow grass (*Poa pratensis*)	0.1	0.1	–
Wall barley (*Hordeum murinum*)	0.1	–	–
Tall fescue (*Festuca arundinacea*)	0.1	–	–

Source: After R.J. Chancellor.

Blackgrass   Alopecurus myosuroides, also known as slender foxtail.

This weed germinates mainly in autumn but also in spring, tillers densely, flowers early and is phenomenally prolific, producing 100 seeds per head on average (Figure 11.5). It can adjust its size incredibly to growing conditions and develop a few seedheads in the most frugal circumstances. At least the seeds only have very little dormancy. It burgeons in winter wheat on heavier land, and it is now a widespread problem, the east, East Midlands and southern areas of Britain being most affected.

PREVENTION OF SPREAD OF WEEDS IN CEREALS

Sow clean seed   The use of seed cleaning equipment has virtually eliminated the formerly prevalent weeds of darnel and corn cockle; the latter has a phenomenal growth rate and once posed great competition.

Use clean equipment, especially drills.

Headland hygiene
Prevent seeding and spread of weeds on waste areas, ditchsides and field headlands. Generally the wayside species which colonise field headlands but do not invade the crop are more susceptible to herbicides than weeds. This being so, the total elimination of headland flora by spraying right up to a wall or other field boundary usually results in recolonisation by a flora dominated by weed species. For instance, pure stands of sterile brome and/or cleavers can be found along some headlands in this way. Furthermore, hedge roots can be damaged and wall bases exposed to frost-collapse after spraying too close.

Weed problems are often greatest extending into fields from the headlands. This is chiefly because the headland can be a source of weeds (e.g. sterile brome, Table 11.7), even though these have to compete with wayside species in unsprayed field edges. Another factor accounting for the greater abundance of weeds around the perimeter of fields may be surface soil compaction due to turning tractors which allows the weeds to root relatively more effectively than the corn.

There are a number of strategies to deal with the headland problem:

1. Carry out subsoiling/paraplowing treatments around the first 20 metres or so of the field perimeter where compaction has been evident and favouring grass weeds whilst rendering the cereal stand weaker.

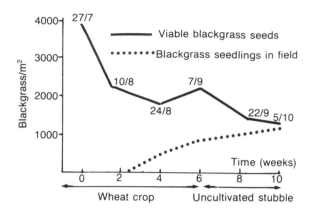

(A) *Blackgrass seed bank and seedling production in wheat stubble. (After Moss, 1980)*

(B) *Heavy loam soil changes with contrasting cultivation systems over 8 years.*

Topsoil	Direct drilling	Ploughing
Organic matter %	5.4	3.6
Adsorption (Kd)	10.8	2.5
pH	6.5	7.7
P (ppm)	66	23
K (ppm)	409	336

(C) *Effect of cultivations and burning on the percentage of blackgrass control needed from a chemical for containment of the problem.*

	Straw burned	Straw not burned
Ploughed	49	66
Direct drilled	88	92

FIGURE 11.5   Blackgrass and its control. (A – after Moss, 1980; B & C – reported by G.W. Cussans, Long Ashton Research Station)

2. Ploughing the headlands to 15 or 20 cm depth achieves burial of existing weeds and their seeds and is appropriate for suppression of sterile brome. In addition, providing it is done in favourable soil conditions immediately after harvest, it can overcome surface compaction and bury straw and stubble residues which would otherwise harbour pests and diseases more easily. For reasons of caution and legislation, burning straw up to the boundary is not correct and immediate ploughing is thus appropriate. 'Plough once or thrice but never twice' or you may bury sterile brome seeds one year and resurface them the next!

3. Planting of spring corn around the perimeter 20 metres or so of a winter cereal field allows more time for control of headland weeds by perhaps cheaper means than selective herbicides (and overwinter stubble suits game!).

4. The establishment of *sterile strips* between the immediate headland flora and the crop is becoming popular. Although some land remains unplanted with this technique, the strip need not be wide − one metre is ample (this would represent less than 1 per cent of a 20 hectare field). Methods of establishing sterile strips are by persistent cultivations, e.g. rotavation or ploughing (turning the slices outwards) perhaps three times per season, or by the use of total chemical control once per year, e.g. high-rate atrazine.

Chemicals may be applied from a knapsack plus hand-held devices or via the end nozzle or two of the spray boom. Various gadgets can be seen doing such work! Wide-tyred tricycles are used. Separation of crop from immediate headland reconciles the potentially conflicting requirements of high-standard cereal

TABLE 11.7   The occurrence of *Bromus sterilis* and headland management

	Headland management				
Location of B. sterilis	*Untreated*	*Sprayed**	*Cultivated†*	*Totals*	*% of fields*
Absent	1734	42	63	1839	84
In hedge only	124	10	13	147	7
In headland, but not in rest of field	109	3	12	124	5
In the whole field	75	2	5	82	4
Totals	2042	57	93	2192	100
% of fields	93	3	4	100	

*   Narrow strip at edge of crop sprayed shortly pre-harvest with a non-selective herbicide.
†   Narrow strip of cultivated ground around the edge of the crop.

Source: After R.J. Chancellor.

production and conservation of the broadest spectrum headland flora. Indeed, when headlands immediately adjacent to a boundary are left in as near to an undisturbed state as possible, the flora tends to be more diverse and therefore useful to a wider spectrum of wild creatures including beneficial insects and nesting birds. At the same time weed species are kept in check by the rest of the population there. In addition, the sterile strip makes an easy highway for larger creatures (including humans!) which use headlands for access from one covert to another, e.g. from wood to wood. Furthermore, the sterile strip provides a dusting and sunning area for gamebirds, notably partridges. It is also a start in the establishment of a firebreak for straw burning and can act as a limited firebreak in its own right in cases of accidental fires.

The crop can be more cleanly combined at the edges without the combine picking up headland weed seeds and dispersing these into the field as it turns. Indeed, for cereal seed crops the perimeter can develop such an encroaching weed problem from the headland that as many as six passes round of the combine may have to be discarded for seed purposes.

The cereals and gamebirds project of the Game Conservancy at Fordingbridge, Hampshire is encouraging farmers to leave the outer boom-width round fields unsprayed to leave weeds for partridge food. This means reduced crop yield plus weed spread possibilities out into the field, perhaps necessitating overall spraying, whereas previously only the outside of the field may have needed treatment.

Roguing troublesome weeds within the crop should be done whilst they remain at very low populations. The surest method of roguing is to hand pull, collect together and burn. It is most appropriate for grass weeds which mature ahead of the crop and, with ever shorter cereal crops, project above crop height. Appropriate candidates for this approach are the wild oat and occasionally blackgrass. Crops have to be inspected anyhow and the existence of tramlines facilitates access for roguing. Cereal volunteers may also need roguing from seed crops of another species. The maximum rogue population considered to justify this method of control is officially put between 250 and 350 per hectare.

The Weed Research Organization's chemical glove can accelerate roguing. Here a translocated herbicide plus dye is applied to each plant by squeezing it with a glove which is supplied with a trickle of glyphosate (Roundup). This method is not as sure for seed crops as is hand pulling. The cost of judicious roguing

should not be allocated purely to the crop involved: instead the rogues should be viewed as a threat to the system and costs of their exclusion allocated to the whole threatened hectarage.

METHODS OF WEED CONTROL

There are several considerations in choosing a method of weed control:

1. Accurately identify the weed(s).
2. Assess the population density and severity of the problem.
3. Investigate the reason(s) for the problem, e.g. if related to soil conditions.
4. Consider cultural control options.
5. Select safe chemicals where these are needed.
6. Select effective chemicals to control the problem(s) (see Appendix 6).
7. Compare alternative chemicals on the basis of

   - costs (assess per tonne of expected grain yield, not just per hectare!)
   - ease of application
   - timing of application required
   - impact on farming system

8. Consider the ability to forestall the future build-up of a weed problem as well as to cope with the existing one: is the policy containment at tolerable levels or eradication?

Government legislation against weeds includes the Seeds Act (1920) and subsequent amendments which require the seller to declare the number of injurious weed seeds (wild oats, dodder, docks and sorrels, blackgrass and couch) as well as the percentage by weight of all weed seeds. The Weeds Act (1959) requires farmers to control five named species (spear thistle, creeping thistle, broad-leaved dock, curled dock and ragwort), which are largely pasture weeds.

CULTURAL MEASURES AGAINST WEEDS

These include:

1. Increasing the *crop's competitive ability* by providing good drainage, structure, pH level and nutrient supply; sowing on time and densely enough.
2. *Rotations* to exploit the different crops, their different competitive abilities and weed floras (they also allow the use of different selective herbicides). Undersowing of smother legumes such as trefoil

might be considered in some systems. Spring cropping helps.

3. *Cultivations* to destroy existing weed growth or to create a false seedbed to encourage near-surface weed seeds to germinate for subsequent destruction. A fallow allows cultivations to destroy weeds during hot, dry spells. However, apart from the cost of cultivations there is the associated problem of moisture depletion within the cultivated zone. On the other hand, minimal cultivation, particularly when associated with paraquat use, can promote annual grass weeds, notably sterile brome. Ploughing, when reinstated, controls this because the weed has virtually no dormancy capability. Judicious use of weeders is needed to keep organically grown cereal crops clean (see Plate 14.2).

4. *Hand operations*, which, whilst costly and laborious, are useful to clear weeds from seed crops and high-value crops and to maintain a clean farm that may have been won at considerable graminicide cost during more prosperous cereal-growing times.

5. *Straw burning* is a valuable means of weed control (Figure 11.5), but it breaks the dormancy of remaining seeds which can infest early-sown autumn cereals more spectacularly and thus necessitate autumn spraying.

CHEMICAL MEASURES (HERBICIDES)

Herbicides typically now represent about 80 per cent of crop protection costs and 20 per cent of total variable costs for UK winter cereals grown intensively. The expense of controlling grass weeds to sustain long runs of winter cereals represents a leading single cost item (often over 10 per cent of gross output) in the total cost per tonne to produce such grain. Single applications of some graminicides cost the equivalent of 0.5−0.6 t/ha of grain.

Herbicides are no substitute for good, positive crop husbandry but they have proved a great aid to it in the face of a declining UK agricultural labour force.

History of Cereal Herbicides

In France in 1886, copper sulphate was found to kill broad-leaved weeds without affecting the cereal crop. Sulphuric acid was subsequently used in the UK early this century but, apart from anything else, proved corrosive to the machinery applying it. In 1933, a more effective, less corrosive broad-leaved weedkiller for cereals was introduced − DNOC (dinitro-

orthocresol). However, this was still a contact herbicide and so needed high-volume application but was of little use on perennial weeds because it was not translocated. Therefore the introduction in 1942 of MCPA (methylchlorophenoxyacetic acid) and its close relative 2,4-D (2,4-dichlorophenoxyacetic acid) was a breakthrough allowing lower volumes because translocated and proving more effective against perennials. In addition, they proved less toxic to use and disappeared within six weeks of application without residues. However, as hormones they could distort the cereal crop if applied at too late a growth stage. They controlled charlock, fat hen and poppies rather well but removal of their competition allowed chickweed and cleavers to dominate until mecoprop (CMPP) was introduced to curtail them in 1956. This allowed the *Polygonum* species (black bindweed, knotgrass and redshank) to take over until dichlorprop appeared in 1961 to at least partly control them. The past quarter century has seen a plethora of new products and mixtures totalling 100 chemicals and around 500 brand names. Recent developments have been against the grass weeds which remain the most persistent and sometimes seemingly intransigent problems of cereal crops.

The introduction of paraquat (Gramoxone) in 1960 facilitated ground clearing prior to cereal drilling and made way for the direct drilling technique to develop. However, toxicity remains a problem and the introduction of glyphosate (Roundup) in 1974 has proved a great boon, being more slowly translocated and therefore capable of killing deeper-seated weed tissue as well as being much safer to the user and in the environment. Its more recent pre-harvest use (grain moisture content must be below 30 per cent) often at low rates with wetters has revolutionised control of remaining field bindweed, thistles and grass weeds − notably couch (probably best controlled through winter wheat) and onion couch (best controlled through winter barley since it matures earlier than true couch) − see O'Keefe (1981).

A recent development has been the use of crop safeners (such as Naphthalic anhydride, NA), which are applied to cereal seeds to allow more cost-effective or potent herbicides to be used such as EPTC in maize or the development of wild oat control within crops of cultivated oats.

Great strides have also been made in reducing spray volume and better targeting onto the weeds. Chemicals having a more persistent weed control effect from autumn into spring have appeared such as pendimethalin (Stomp, active against annual BL and grass weeds) and isoxaben (Flexidor, active only against BL annuals).

Herbicide Classification

Herbicides may be classified according to the following:

- Total effect (e.g. diuron)
- Selective (e.g. MCPA)
- Application and mode of action
  Soil-applied, residual, e.g. pendimethalin
  Foliage-applied — contact (i.e. kills more or less what it hits), e.g. paraquat or translocated (i.e. moves around the plant killing parts beyond its point of entry), e.g. glyphosate
- Treatment method
  Overall
  Spot — on troublesome patches of the field
  Headland
- Timing of treatment
  Pre-sowing, e.g. aminotriazole
  Pre-emergence, e.g. tri-allate
  Post-emergence, e.g. mecoprop (CMPP)
  Preharvest, e.g. glyphosate
- Chemical group (Table 11.8)

Factors Affecting Herbicide Performance in Cereals

Herbicide performance can be assessed simply by the percentage kill of target weeds that is achieved. The need for this can be quantified against the potential multiplication of a known starting population — modelling of this relationship has been done for wild oat and blackgrass by George Cussans and his team (now at Long Ashton). In addition, the absence of crop check by herbicide damage is sought (see Tottmann, 1982).

The effectiveness of a herbicide depends on:

1. Correct formulation, concentration and spray-droplet-size distribution.
2. Compatibility with other chemicals simultaneously applied.
3. Avoiding particular chemicals on sensitive varieties. Check manufacturers' warnings because damage can be spectacular, especially from such chemicals as chlortoluron (Dicurane) and metoxuron (Dosaflo).
4. Correct stage of crop growth.
5. Suitable weather for application in respect of minimal drift, minimal volatilisation, maximum weed uptake, minimal crop damage.

6. Application at the most vulnerable stage of weed growth possible, i.e. when the weed is most actively growing and assuming low-resistance in the weed population to the chosen herbicide. Resistant populations of cleavers and chickweed exist, for instance, to mecoprop. Sterile brome has become tolerant of herbicides on field margins. Localised blackgrass resistance has also been recorded by Stephen Moss (LARS, 1986).

7. Soil conditions. Natural soil components such as high surface organic matter following direct drilling, as well as persistently incorporated straw or ash residues plus charcoal from regular straw burning, can all adsorb herbicides, thus rendering them inactive against weeds. It is possible to measure the Kd value of fields to obtain a direct indication of this adsorption capacity. Kd can be quadrupled by direct drilling over eight years rather than ploughing to dilute this effect. A very useful chemical like pendimethalin (Stomp) will not work if soil OM exceeds 6 per cent.

8. Associated cultivations. The type of cultivation system and whether or not straw is burnt have a profound effect on weed control. Work at Letcombe Laboratory and now at Long Ashton Research Station has illustrated this (Figure 11.5). Under direct drilling for eight years on a heavy loam soil, chlortoluron gave only 50 per cent blackgrass control as against 100 per cent under ploughing. Burning is reckoned to destroy some 30 per cent of wild oats and reduce dormancy in a further 10–15 per cent, whilst some 60 per cent of sterile brome is destroyed.

Strategies for Herbicide Use on Cereals

It is not clever to try to memorise application rates: always check label recommendations and reduce rates only if appropriate.

Autumn-sown crops
Early weed control is vital to remove their competition and to allow crop competition (especially of winter barley) to overtake future weeds which may germinate. The adoption of herbicide mixtures has provided wide-spectrum control of most of the troublesome weeds.

Autumn herbicides are applied just pre-emergence of the crop to achieve control of expected weed problems *or* increasingly just post-emergence once the weed threat, both quantitative and qualitative, has actually revealed itself.

TABLE 11.8   Chemical types of herbicide and main uses

Group name	Chemical name	Well-known brands*	Mode of action†	Note on uses‡ (BL = broad-leaved)
Aliphatic acids	Dalapon TCA	various various	foliage; wax reduced soil; can last 5 months	Couch; volunteers
Amides	Benzoylprop-ethyl Diclofop Flamprop	Suffix Hoegrass Commando	foliage; stunt, inhibit cell division	Wild oats Wild oats, blackgrass, awned canary grass Wild oats
	Aminotriazole	Weedazol	soil; bleaches, lasts 1 month	Broad spectrum, pre-sowing
Anilines	Pendimethalin	Stomp	soil; stunts, lasts 3 months	AMG, blackgrass, BL annuals pre-emergence
	Bentazone	Basagran	foliage	In mixtures, BL weeds
Bipyridils	Diquat	Reglone	foliage; desiccant	Desiccating lodged crop, BL weeds
	Paraquat	Gramoxone	foliage; bleaches	Total, pre-sowing
	Difenzoquat	Avenge	foliage; stunts, gradual bleach	Wild oats (plus some mildew control in spring barley)
—	Glyphosate	Roundup	foliage; stunts, red/gold bleach	Total, pre-sowing and just pre-harvest
Hormones or alkanoic acids	MCPA 2,4-D and 2,4-DB MCPB MCPA+2,4-DB + Benazolin Mecoprop Dicamba Clopyralid Fluroxypyr	 various Tropotox Legumex extra CMPP various mixtures various mixtures Starane	foliage chiefly; distorted, excess growth before death	BL, especially poppy and charlock BL, especially polygonums (redshank, etc.) In undersown cereals with leys containing legumes As above plus chickweed and cleavers Many BL post-emergence Many BL post-emergence Mayweeds, thistles, corn marigold Cleavers, hemp nettle, forget-me-not
Nitriles (HBN)	Ioxynil Bromoxynil	Deloxil	contact; foliage	Many BL, especially speedwells and mayweeds
Substituted ureas	Chlortoluron Isoproturon (IPU) Linuron Methabenzthiazuron	Dicurane Hytane, Arelon Mixtures usual Tribunil	 soil; inhibit photosynthesis	Blackgrass and BL pre- and post-emergence in autumn cereals Broad spectrum, especially corn marigold Moderate blackgrass and meadow grasses

(continued)

	Metoxuron	Dosaflo		Blackgrass and BL post-emergence
	Chlorsulfuron+	Finesse, Glean C		Annual BL and grasses (low rates)
Thiocarbamates	Triallate	Avadex BW	Soil; lasts 2−3 months	Wild oats and meadow grasses (MG)
Triazines	Atrazine	various	soil; inhibit photosynthesis,	Total, or selective in maize, for couch control
	Cyanazine	Fortrol	kill leaf tips	BL annuals and AMG
	Terbutryne	Prebane	and margins first	Pre-emergence blackgrass and BL

* Approved *mixtures* are also often used.
† Soil ≡ residual.
‡ Always follow label advice.

Autumn herbicide use in winter barley usually obviates the use of any spring treatment, and in other winter cereals it may at least limit the scale of the spring problem to a few surviving weeds, which nevertheless must be prevented from seeding and from competing with the crop in spring. If spring spraying can be omitted this saves not only cost but also avoids the greater risk of crop check or later damage. This may occur unless crops are sprayed with hormone herbicides between a 5 and 10 cm length of the leaf sheath of the largest leaf, and this stage may only last up to a fortnight during which good spraying days may be few. For winter wheats which often need spring herbicide, it is wise to choose materials that can be tank-mixed with eyespot fungicide and PGR given around GS 30−31.

The use of sequential doses of low rates of herbicides can reduce the need for full spring rates since earlier control weakens weeds and favours the crop. It does not, however, deal with the need to prevent remaining weeds from seeding. Earlier sowing has increased the problem of grass weeds exacerbated by continuous cereals under reduced cultivation regimes. Tilths are now made finer for such crops partly to ensure the efficacy of residual herbicides used early on.

Summary charts of the currently available products for autumn and spring use are found in Appendix 6.

Spring-sown crops
These are usually sprayed between the three-leaf stage of the cereal, when the weeds may be more vulnerable, and the five-leaf stage, which is a safer time for the crop. Half the recommended rate of herbicide is often used, and the treatment is mainly for BL weeds and wild oats.

MCPA deals with many broad-leaved weeds including poppy, charlock, fat hen, shepherd's purse and thistles and also volunteer oilseed rape and weed beet. Mecoprop (CMPP) also deals with the above plus chickweed and cleavers (only to some extent; large, persistent cleavers succumb to fluroxypyr, i.e. Starane). Dichlorprop (formerly 2,4-DP) adds to the above coverage of *Polygonums* (except knotgrass, which is poorly controlled) and spurrey.

For spring cereals undersown with ley mixtures containing legumes, MCPB should be used or alternatives such as benazolin plus 2,4-DB (Legumex extra), which also combats cleavers and chickweed (which is little affected by MCPB). Annual grass weeds in spring cereals may be dealt with using flamprop (Commando), which is particularly useful against wild oats, as is triallate (Avadex), though this has low action against blackgrass.

NB: It is vital to ensure that only currently approved products are used, as legislation changes when either environmental risk or ineffectiveness develops.

Part Four   Performance

Assessing progress, preservation, potential and profitability of cereals

Chapter 12

CEREAL CROP MONITORING

Monitoring here includes crop walking and inspection, recording, judging, field comparisons and trials to assess the potential of new varieties and new husbandry techniques.

CROP WALKING

A balanced approach to cereal husbandry requires the perspectives gained only by regular walking (measuring stick and lens to hand!) of crops and working with them. This might be known as lens and stick farming to future generations! Additional perspectives come from:

- discussions with other farmers and students, researchers and extensionists
- study of the literature
- performance records kept in the farm office

But these are no substitute for field study with others, and practical work. Besides, the challenges and uncertainties of the real field situation are rather more enjoyable than indoors! What is advocated then is management by detailed, regular field observation — methodical monitoring management (MMM).

The cereal grower needs clear objectives (expected outcomes) of his farming, subject to constant review, perhaps by having a scheduled 'thinking time' out in the field or Land-Rover (Figure 12.1). First, set realisable targets and then adapt to practical constraints as they arise. If one starts thinking negatively it is a bar to progress; 'Aim for the stars and you might reach the top of a straw-stack' goes the (modified) old farming saying! Gone are the days of wheat, wheat, wheat, West Indies when some people drilled in autumn, shut the gate (if there was one!) and disappeared to winter in the sun only to

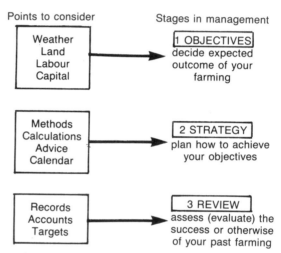

Note: Deciding to do nothing to a crop at a particular stage can be harder than deciding to apply some treatment!

FIGURE 12.1   Components of cereal management.

reappear in time to see the combines greased and supervise the harvest! To be successful, cereal crops require fortnightly inspection overwinter, weekly at most other times but daily or every other day during periods of rapid development coinciding with critical input timing requirements.

CROP RECORDING

Good records need to be:

- simple, or they will not be used again!

- tidy, for ease of reference
- relevant for practical use
- regularly kept, for complete history
- both physical and financial for comprehensiveness

A system with the above qualities has the following advantages:

- It effectively programmes regular crop inspection, so encouraging timely, more thorough crop care.
- It can promote greater observation and crop understanding by all contributors.
- It can save money by encapsulating past experience for future reference.

Many systems can be devised for cereal crop recording according to an individual farmer's requirements. However, it is all too easy to generate a paper-chase! Figure 12.2 suggests a system for crop appraisal.

The NAC Cereal Unit (now the NAC Arable Unit) devised a useful field recording system and also operator sheets for instructions when treating cereal crops. They have recently published a cereal crop manual. An increasing number of UK farms have computerised their records and financial accounts though far fewer use computers for budgeting crop alternatives. A cereal field record summaries sheet which I have used for some years is shown at Figure 12.3.

CROP JUDGING

Many district agricultural societies operate competitions to promote higher standards of husbandry. This means that some farmers and other agriculturists find themselves with the interesting but formidable task of trying to judge other people's crops fairly!

I first make use of the sheets normally used for more routine purposes and already mentioned (Figures 12.2 and 12.3). I reckon to do some head counts, remove samples taken at random in at least ten places in the field and later inspect and photograph these. I then score each crop using my form shown at Figure 12.4. When judging cereal crops, I try to take account of:

- How easy or difficult the land is by contrast with other sites in the class (this is especially necessary when judging cereals at vastly different elevations or on diverse soil types) — though I do not give a large weighting to this. All the weightings used are open to debate! However, the form is not

---

FIELD NAME: DATE OF REPORT:
SIZE (Ha_____; Ac_____) Grade _____

SOIL
  Type
  Tilth state
  Other (depth, slope, aspect, altitude, wetness, etc.)

FIELD HEADLAND
  Weeds
  Wayside species
  Separation (crop/headland)
  Condition of boundary

CROP CONDITION
  Establishment (Date? Rate? Method?)
  Population (Suitability? Uniformity? Depth of planting?)
  Growth stage
  General appearance (Colour, etc.?)
  Likely treatments to date

CROP PROBLEMS
(Identify and assess severity & distribution)
  Grass weeds
  Broad-leaved weeds
  Diseases
  Pests
  Any other disorders

Likely future treatments
Expected yield
Other details/comments
Overall standard of management
Key limitations
Use back of sheet to record map of field and distribution of any relevant features, e.g. soil variation, weed incidence

FIGURE 12.2 Cereal crop appraisal form. The following is a guide of points to consider/record when inspecting and evaluating a field.

---

intended to be slavishly followed but rather used as a general guide for more objective assessment.

- Expected crop performance, taking account of likely yield and quality-determining factors.
- Cost effectiveness of the management given (maximum net margin, not high yield at any price, should be the criterion of profitable cereal production).
- Standard of hygiene and soil state, because this has longer-term cost implications. In the case of

Field name	Variety	Drill date	Seeds/ m²	Plants/ m²	Autumn weed control	NPK kg/ha	1st N date (GS)kg	2nd N	3rd N	4th N	Total N	PGR	Spring weed control	Aphid control
1														
2														
3														
4														

(cont) Field name	Diseases present	Diseases not controlled well	Fungicides I (GS)	II (GS)	III (GS)	IV (GS)	Lodging %	Ears /m²	GS when wet/ drought	TGW (& HLW)	Harvest date	Yield (t/ha)	Special remarks
1													
2													
3													
4													

FIGURE 12.3 Cereal field record summaries sheet.

low levels of wild oats, for instance, it also has to do with future cropping opportunities such as fitness to grow seed corn crops.

Friendly crop competitions are a worthwhile stimulus to both the progress and the enjoyment of cereal husbandry.

FIELD COMPARISONS

It has been my privilege to coordinate, for the past twelve years, the Cirencester Cereal Study Group. The structure of this group is indicated in Figure 12.5.

Foundation

The Cirencester Cereal Study Group was set up in 1977 linked to the NAC Cereal Unit, founded in the same year. The base for the group's work is the Royal Agricultural College. Some eleven meetings are held each year, omitting only the harvest month of August.

Administrative costs of running the group are minimal.

Farmer Dominance

It was felt important that such a group should be farmer-dominated though including representatives of research, advisory, educational and commercial branches of agriculture – i.e. AFRC (the IACR part now), ADAS, RAC, seed, fertiliser and agro-chemical personnel.

Integrated Realism

The overall aim of the group was to adopt an approach of integrated realism towards the improvement of cereal crop performance on thin Cotswold soils of the Sherborne and the Elmton series especially. The focus has been on specific inputs and aspects of crop behaviour but all the time the group

CLASS OF CROP	Year:					
Number/Name of field or farm						
POTENTIAL OF SITE /10 (Lower score for higher inherent fertility)						
EXPECTED YIELD /30						
HEADLANDS/BOUNDARIES/ TRAMLINES /10						
UNIFORMITY /10						
CLEANNESS OF LEAF/EAR /10						
WEEDINESS /10						
COST EFFECTIVENESS /20						
TOTAL /100						

FIGURE 12.4   Cereal crop judging: summary form.

has had the overall crop management strategy firmly in view through attention to field visits above all.

Field Focus

Every season we have monitored and compared ten study fields, each on a different member's farm. These have provided a focal point for visits, discussions and correlation of experience.

Initially, for three seasons, winter barley management provided the most urgent need for enquiry. Subsequently spring wheat and winter wheat occupied

three and four seasons respectively. Over the whole time since its inception, the group has been linked with a replicated trial on the nitrogen requirements of winter barley. In the 1984—85 season, our focal crops were six fields each of six-row winter barley and of triticale; these were continued into 1986, but with fifteen study crops of six-row barley since it yielded 7.84 t/ha (8.9—6.7) in 1985, and triticale ranged too widely (from 7.5 to 4.5 t/ha) to encourage more than five study fields for 1986. All the winter barley met specific weight quality standards.

Our monitoring involves plant counts (20 × 0.5 m row-length samples taken randomly per field) and ear counts (20 × 0.1 m^2 ring counts taken randomly per field) as basic plus additional tillering and development studies in some cases. Members have also pursued various on-farm studies of their own and currently several have straw incorporation trial plots imposed in commercial crops. Weather data and the full sequence of treatments and performance have been recorded and compared at the end of each season. Much of this work has been published through the NAC already.

The emphasis on study fields has not prevented us from straying onto other crops, farms or places relevant to cereal management as well.

Specialist Stimulus

It has been our policy to spice the programme with contributions on specialist topics. These are often made over lunch prior to the usual two field visits per meeting as well as out in the field. Topics are chosen which are relevant to the immediate management of the crop as well as to wider changes affecting cereal production. Our aim has been to stimulate close observation, to widen awareness and to foster understanding of cereal crop behaviour.

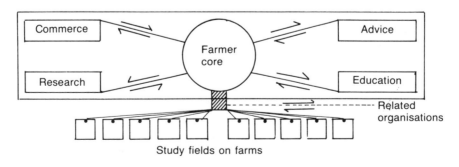

FIGURE 12.5   Structure of the Cirencester Cereal Study Group.

The successful farmer

The bankrupt farmer

Happy farmer going to market

Miserable farmer coming home

Aristocratic farmer

Tenant farmer

FIGURE 12.6   To see ourselves as others see us? Level-headed and even-tempered: 'Groups exploit the diversity of farmer types'! (Drawings by Brian Walker, courtesy of The Countryman, Burford, Oxford)

The group has been fortunate to have Dai Barling of the RAC conducting his studies on internal development of primordia in both winter barley and winter wheat. His work has been published by the NAC. Other special topics have included PGR work, disease prediction, pest control, subsoil management, straw decomposition and product quality assessment.

Unified Understanding

It is most important to make the link between research findings and farm practice. Feedback of farming realities to research, advisory, educational and commercial members is equally valuable. Communication between all members is fostered in a group of this size which has some 35 on the mailing list and usually 20−25 attending each meeting.

Farmers in the group were instrumental in the staging of Barley '79 at the Royal Agricultural College and the subsequent founding of the Cotswold Cereal Centre based there and now directed by Dr Mike Carver. This organisation is much larger than the original study group but performs a different role —

that of replicated trialling on behalf of its members, most of whom are farmers. As well as this, group members are active in the Cotswold Direct Drilling Club whose founder and both subsequent chairmen are study group members. As a result of the initiative of one farmer member the group also sponsored work towards a crash programme on net blotch control at Long Ashton Research Station. Conservation and public image aspects are also within the group's programme.

In spite of these wider involvements, the group continues by popular request as an enjoyable but educational team effort seeking to pioneer and compare improved cereal crop management. Groups exploit the diversity of farmer types and experience (Figure 12.6).

Many forms of farmer group exist in the UK and overseas. I believe there is tremendous benefit and scope for such farmer study group work as an efficient (psychologically-sound, cost-effective and developmentally-beneficial) means of improving crop performance. This comment is perhaps particularly relevant in the UK at present when the advisory

PLATE 12.1   ICI arable trials site Ropsley, Grantham, Lincolnshire.

services are undergoing major upheaval and review (see Wibberley, 1988, JRASE Vol. 149).

CEREAL TRIALS

Properly replicated, randomised and representative trials are vital to cereal crop progress in order to evaluate the new products of research and plant breeding. These require experience and back-up services to conduct adequately (Plates 12.1 and 12.2). There is a demand for them from cereal growers.

Imprecise trials which are later quoted as if accurate can be most misleading. It is generally too expensive and undesirable for individual farmers to attempt to conduct more than simple observational/yield measurement tests on their own farms such as leaving a tramline unsprayed to provide a comparison to indicate the effect of a particular spray. For one thing, many cereal farms still lack adequate yield measurement equipment! A lot of grain yields are still not known accurately until heaps are sold. These may be difficult to allocate accurately to fields or even to varieties grown unless storage facilities allow clear separation of lots — never mind trying to distinguish plot results.

Therefore, in 1979 the Cotswold Cereal Centre was set up by farmers for farmers and funded by their subscriptions (originally £200 or £250 per annum,

depending on whether less or more than 200 hectares of cereals were grown). This agronomy centre is based at the Royal Agricultural College, Cirencester, though financially independent of it. Its total trial sites normally occupy 20–25 hectares per season near Cirencester. All the results are confidential to the membership since it is their subscriptions which finance the trials. Through a farmer-dominated technical committee, the director and his colleagues are guided in their plans and provided with many of the innovative ideas to pursue. Controlled, accurate experimentation at the Cereal Centre removes the need for an individual farmer to conduct his/her own trials. Results are provided within a month of harvesting and give the farmer the key source of independent trials information on all inputs that are involved in the farming of cereals and other combinable crops.

Through fortnightly field days when all the members can visit the trials plus monthly newsletters advising on trials progress and the farming activities relevant to that particular period, the members are kept fully informed. Since the Cereal Centre has charitable status, its intention is clearly to turn members' subscriptions into trial plots in the most cost-effective way.

In 1985, Arable Research Centres Ltd, based at the Royal Agricultural College, was registered as the umbrella charity organisation for the Cotswold Cereal

PLATE 12.2 Long Ashton Research Station, University of Bristol. A general view of the experimental plots, with the main laboratory buildings in the background.

Centre, Cambridge Cereal Centre (started in 1984), Lincolnshire Cereal Centre and Hampshire Cereal Centre (both started in 1985); other Cereal Centres have been added since in other areas. Membership of these Cereal Centres is open to any farmers who feel they would benefit from independent trial results and advice on topics that they have indicated are of concern to them. Farmers who subscribe to their local centre also receive results and are welcome at field days of the other centres. At the Cotswold Cereal Centre cooperation with ADAS has been mutually beneficial from the start. Such centres seem to me to provide a possible future network to which ADAS extension officers may be attached if constructive dialogue continues between the private sector and the diminishing government-funded UK development work on cereals.

Development trials are not a substitute for fundamental research work on cereals. This continues for England and Wales under Long Ashton Research Station, University of Bristol and at Rothamsted Experimental Station, Hertfordshire. Both are funded by the recently restructured Agricultural and Food Research Council (AFRC), part of which has been constituted as the Institute of Arable Crops Research (IACR). Long Ashton has its own flourishing members' association which any cereal farmer can join, receiving research reports and invitations to open-days and cereal workshops. Rothamsted too has a 'friends' association. These are linking their activities. Extension (i.e. communication of information) is the vital bridge between these various activities, giving some priority to the opinions, experience and needs of farmers!

Chapter 13

HARVESTING, PRESERVATION AND STORAGE OF CEREALS

HARVESTING

Harvest is 'attended with the most anxious solicitude; for it is a business which cannot be for a moment neglected, and the man who wishes to get it rightly managed must superintend it without intermission from the dawn of day until its final close. He should previously get rid of all other work, and make every preparation for the due performance of this; the barns should be thoroughly swept out ... and every tool should be in complete condition' — so the graphic language of *British Husbandry*, Volume II, 1837!

Harvest is also the most potentially exciting or disappointing season according to results, representing as it does the culmination of every effort and event of the past season. The keynotes from the above quote are relevant, namely *preparation* — have the kit and storage ready — and *timeliness* — do not miss the chance of suitable harvesting conditions.

The importance of thorough harvesting has three main facets:

- recovering a maximum amount of usable dry matter produced — this coincides with about 35 per cent moisture content of the grain though, with respiration causing losses as the ear ripens!
- securing optimum quality for subsequent use, i.e. minimal damage to germination capacity when relevant (e.g. malting barley) or to any other features
- leaving as little as possible behind of any material which may hinder the following crop, e.g. diseased or excessive amounts of straw, numerous volunteers

HARVESTING TECHNIQUES

A high proportion of the world's cereal harvest is still gathered by sickle (Stalin's combine!). Various mechanical devices exist. The reaper and binder of Appleby (1875) is still in use in parts of the UK and provides a sample of long, unbroken straw for the specific market of thatching. This machine also has the advantage of dealing with less mature crops, so obviating excessive shedding during cutting of the crop. The stalks are cut near the base and bundled with ears still attached into sheaves. These are grouped butt downwards in the field to form stooks (shocks); in these stooks the grain is exposed to dry well off the ground and they withstand rain. Whilst oats often stood over ten days in stooks to dry — this period being called 'field room' — wheat generally required only a week or so, whilst barley was often fit to cart after only two days. Carting or 'leading' the sheaves into the barns was a triumphal and celebrated village occasion.

Threshing (thrashing) teams would then deal with the sheaves, separating the grain for subsequent winnowing or 'dressing' of the grain to remove chaff. The simplest thrashing device is the flail — apart from rubbing in the hand (as done to obtain a sample for rapid assessment of harvest readiness). A simple flail consists of two sticks attached by the tips with a flexible short leather strap so that one stick is held and swung, making the free-moving stick attached strike the heads of corn releasing the grain. Such devices are still widespread tools in the world's cereal harvest. However, the vast majority of grain in Britain is now harvested by combine harvesters!

These have developed via tractor-drawn models and those which delivered grain into sacks for discharge by chute to the ground for subsequent carting. Now most are large, self-propelled machines capable of high output and equipped with bulk tanks for grain which is subsequently discharged to trailers for carting.

Recent advances have included:

● the attachment of electronic monitors to detect grain losses and ensure efficient operation
● the development of multiple-drum thrashing mechanisms, which enable gentler, more complete separation
● a new thrashing mechanism in the axial flow principle (Plate 13.1), which aims to thrash grain in one continuous swirling action. This batters the straw but works well
● special hillside combines for effective work on steep slopes
● whole-crop harvesting by a self-propelled forager prior to static separation at the store later
● redevelopment of the Roman head-stripper principle into new combines in order to reduce ear losses, front end losses and amount of straw passed through the machine

PREPARING FOR HARVESTING
Machinery

It is obvious that any necessary overhaul of field and barn machinery should be complete well ahead of harvest time. Equally, it is important to maintain a stock of all items subject to wear and tear during harvest. There is little more frustrating or ridiculous than queuing for minor spares during fine harvesting weather at a distant machinery depot!

Storage

Cleaning and chemical treatment of grain stores needs to be done in plenty of time. This is more vital the warmer the climate of the region. Contaminated tropical grain stores can lead to losses of the majority of the new grain put into them and make a singularly depressing sight.

Staff

This is a time of year when the need for co-operation and a happy team atmosphere is paramount. It is a time when heat, long hours and perhaps difficult crops unite to impose an extra strain. In addition, casual and student labour very often has to be integrated into the team. The importance of this 'team' factor above technical slickness cannot be overstressed; indeed it is a pre-requisite of it. Thus proper rewards, incentives and above all appreciation of effort and roles should be established long before harvest begins.

Crops

The benefits of a uniform crop and of avoidance of

PLATE 13.1   Interior view of axial-flow combine harvester. (Courtesy of International)

lodging become very clear now! Desiccation of some crops can help to speed the combine, as indeed can the use of pre-harvest glyphosate which reduces the hindrance from any green matter such as couch grass that it reaches.

It is particularly important that the earliest harvested cereals should be relatively easy to get, notably winter barley. Difficulty at this stage can set the scene for poor morale throughout harvest. Lodging can be devastating, especially when bird damage (from rooks, etc.) ensues.

HARVEST READINESS OF CROPS

Whilst the great majority of cereals in the UK are harvested to be stored as dry grain, there are alternative states and ways in which crops can be harvested:

- As grazing crops or green silages (but *not* some sorghum cultivars which contain HCN, i.e. prussic acid). Only rye and triticale are grown specifically for this purpose, and then special forage varieties are used. However, forward crops of winter cereals can be grazed (wheat up to the end of March in England), using sheep which like the clean grazing and help to tiller a lax crop.

- As whole-crop cereal silage, which is cut when the grains are cheese ripe. Barley may be so treated. This used to be a way of removing embarrassingly diseased crops early in the days before systemic fungicides! This silage tends to be of high dry matter and thus produces no effluent (by contrast with much grass silage), but it is low in protein.

The majority of the maize grown in the UK is of forage varieties for ensilage. It is bulky (10–12 t/ha DM), palatable, energy-rich but low in protein. Being cut at over 25 per cent DM, it usually produces no effluent. It needs to be precision-chopped by a forage harvester with a special header. If it is not fed quickly enough after exposure, it is very prone to support mould growth which causes deterioration. The snag is the lateness of harvest, though newer French varieties like Bastille bring it forward three or four weeks from the end of October/November towards the end of September/early October.

The whole wheat or barley crop can be cut near maturity without separating straw from grain and the whole above-ground material delivered for processing at a refinery (Figure 1.2). This concept is being developed partly as a solution to cereal surpluses. The cereal crop offers a wide range of raw materials (Figure 1.3).

The readiness of grain for normal combining or dry harvesting is indicated by straw shrinkage at the nodes, which become greyish in wheat, and by the grain, which rubs easily from the ears and is hard when bitten.

Durum wheat and rye should be cut at about 20 per cent or above moisture content (mc) to preserve quality. Barley plants tend to brackle over and tangle, their heads nodding right over and sometimes snapping off altogether at the neck. The grain develops a wrinkled husk and is flinty hard. Maize, sorghum and millets are ready in three or four days once a black layer is visible at the point of attachment of the grain (the hilum).

Shedding of whole heads means around 20–22 grains/ear average for two-row barleys and 35–40 for six-row varieties and for wheats. Oats are prone to shedding. It is common for 40 kg/ha of grain to be shed but this figure can surprisingly often climb towards 125 kg/ha or more (sufficient to seed the next crop if it was shed uniformly!). Perhaps the renewed penchant for free-range poultry will solve this problem as it did for us formerly when we folded mobile arks across our stubbles for pullets to pick up this shed grain. Grain loss points on a conventional combine harvester are shown in Figure 13.1.

Delayed harvest in a wet year may lead to sprouting grains whilst still in the ear (especially in rye and some wheat cultivars) with corresponding loss of quality (Figure 13.2).

The order of grain harvesting in the UK is typically as follows:

1. Winter barley — two-row is easier than six-row (end of June in an early season in the far south-west; during July in most places). Rye is also often early.

2. Winter oats   (Incidentally, this commonly topped all cereal yields when grown as a break from long runs of spring barley on chalkland in the 1960s. I remember harvesting 6.5 t/ha then. It is very awkward in a wet harvest and when lodged unevenly.)

3. Durum/triticale/winter wheat/spring barley/spring oats — together. This is the main harvest period of August/September but in the north of Britain may extend into October and beyond in difficult years. The actual sequence is dependent on relative sowing dates and varieties; moral: think twice before sowing too many of these species on the same farm.

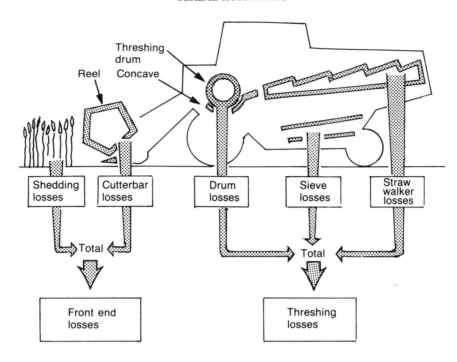

FIGURE 13.1   A conventional combine harvester and its grain-loss points.
(After ADAS mechanisation dept., 1980)

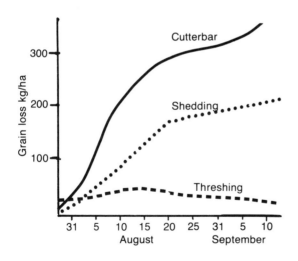

Harvest dates

Note: The contrast between threshing losses
and front end losses clearly confirms
the results of earlier work.

FIGURE 13.2   Spring barley grain losses at
harvest 1972–73. (Source: ADAS)

4. Spring wheat — in practice modern varieties may
follow on directly after winter wheat and can be ready
in August in southern England.

5. Grain maize — late! Second half of September and
of October for maximum yield, though sweetcorn
varieties grown on a horticultural scale are harvested
earlier but less ripe, of course. Maize for grain is cut
at 35 per cent mc often and up to 45 per cent mc.
Early frosts advance its maturity and reduce the
mechanical hindrance of leaves and stems (stover).
Maize is still commonly gleaned and fed into a static
combine in southern Africa and elsewhere. Cob-
picker machines can also be used.

GOOD COMBINING

Combining is best done when grain moisture content
lies between 14 and 24 per cent. Below this, grain
is drier than needed, can break in threshing and is
likely to shed badly; if above 24 per cent mc, drying
costs are high and grain damage by crushing resulting
in reduced germination capacity is more likely during
combining. In practice, people do harvest at up to
30 per cent mc and occasionally more; this can be

wise in a very wet season like the 1985 UK harvest when it may be the only way to catch a milling wheat crop before it loses its Hagberg value. Milling wheat wants harvesting relatively moist whilst malting barley and seed corn want to be as perfectly ripe as possible. Various moisture testers are available (see McLean, 1980).

The following should be noted in combining:

- Settings should be correct according to crop conditions and grain size, e.g. reel bars about 15 cm below ears. When adjustment is needed, alter one setting at a time systematically.
- Forward speed, concave clearance, drumspeed, sieve settings and fan speed will all need adjustment to the crop.
- A smooth *flow* to the machine improves the efficiency of threshing with less damage to grains.
- A sign that the operation is proceeding well (i.e. drum speed and concave clearance are appropriate) is that only the occasional grain can be found left in the ear amongst the straw.

Combine types vary, of course, but there are general indications from the crop and trash of malfunctioning:

- If trash is green and damp, then there may be an over-low cutterbar setting. Also the sieve may be overwide.
- Excessive levels of cracked or skinned grains indicate overthreshing — too high a drum speed coupled with too small a concave clearance resulting in too much rethreshing; periodic sample examination can tell the driver how well the threshing and separating mechanisms are performing.

Combines frequently cause serious soil compaction, especially in wet harvesting conditions, so they should be equipped with larger section tyres or have tyres which can be operated at lower inflation pressures.

Grain trailers and particularly lorries can cause undue soil damage if they run alongside moving combines, and they are now often left on the headland to be filled. It is useful to have a spare trailer parked on headlands from early morning in case of traffic or other delays to shuttling grain trailers. Old lorries are very useful for fast road-carting and in that capacity they avoid tying up another tractor with a grain cart, but they can often carry heavy loads and are seldom equipped with a means of reducing soil compaction.

Minimise losses. Even if a slightly dirtier sample results from cutting lower, it can be cleaned in the barn but not (yet) vacuum-cleaned from the field! If there are many hectares ahead to be cut, do not wait too long either. It is not foolish to begin when 23–24 per cent mc level is reached if you would otherwise lag behind with later-harvested cereals. Earlier harvesting also means the combine works in longer days. Normal crops are not combined until the dew is off in the mornings (after 10 am or so), but dead-ripe and over-ripe crops may be better cut at night when the extra humidity may reduce shedding liability. Timely harvesting preserves germination capacity for seed and malting crops.

Adequate combining capacity for the area of cereals grown is clearly vital. In practice this varies from about 30 to 40 hectares per square metre of straw-walker area. However, this is a crude guide because some larger combines devour grain more efficiently and crop size varies too.

Laid crops need lifters on the front of the combine set as close to the ground as possible without picking up stones and soil. Modern machines can 'hug' the ground undulations.

Badly laid crops and those with well-grown undersown leys can be swathed, i.e. cut and left in windrows to dry off more uniformly (weather permitting) and picked up using the reel on the combine a week or so later. Swathing is still routine for North American prairie cereals to limit shedding and encourage steady drying out in order to preserve milling quality of wheat.

GRAIN PRESERVATION AND STORAGE

The reader is referred to McLean (1980) and Nash (1985) for details of equipment and conditions; only a brief treatment is possible here.

PRESERVATION

Basically, grain is a living thing and its germination capacity needs to be preserved for many purposes. Its moisture content and storage temperature greatly affect the duration of safe storage to preserve germination, which for barley is:

15% mc at 15°C	40 weeks
15% mc at 20°C	19 weeks
16% mc at 15°C	20 weeks
16% mc at 20°C	10 weeks

In order to control spoilage, grain must be maintained in a state of suspended animation or killed deliberately.

The requirements for activity of both grain and spoilage organisms include four conditions with four consequent means of preservation:

For life	*To preserve*
1. Water	Dry (often at a high energy cost)
2. Warmth	Cool (usually to avoid the need for excessive drying) Refrigerate (too costly and impractical except for research)
3. Oxygen	Seal (too unreliable alone, probably; dangerous gases too)
4. Right pH	Pickle (treat with acid) or add alkali

Options 3 and 4 may suit livestock feed grain only.

STORAGE

Storage is necessary because:

● There is an aseasonal demand for grain.
● Orderly marketing needs to be achieved.
● It is difficult to organise the sale of large quantities off the back of the combine for many farms.

Table 13.1 indicates the storage capacities needed for the common grains. Clean and treat the store for pests. Conditions for long-term safe storage of grains are represented in Figure 13.3 with reference to probable pest damage.

TABLE 13.1   Grain storage capacities

	Space needed (m³/t)	*Weight stored (kg/m³)*
Wheat	1.27	790
Barley	1.41	713
Oats	1.94	519

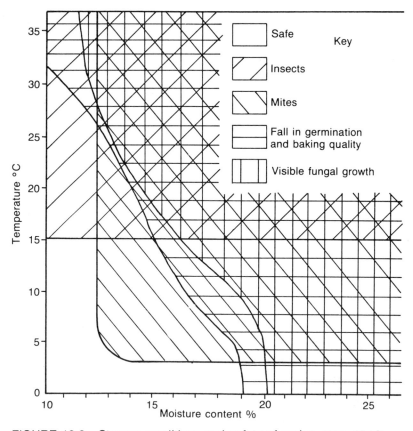

FIGURE 13.3   Storage conditions and safety of grains. (After ADAS)

DRYING

This is the most common method of preservation. Grain is often still at over 40 per cent mc a fortnight before combining at around 20 per cent mc, and the target for marketing is often 14—15 per cent mc. For grain to reach 14 per cent mc, water must be driven off according to the moisture content when combined:

25% mc	130 kg water/tonne
20% mc	70 kg water/tonne
15% mc	12 kg water/tonne

Once dried, the grain must be watched, i.e. kept cool and dry and checked for pests. Always avoid zoning of moisture which can easily accumulate in top layers if the bulk of the grain is too large for the capacity of the fans and blowing has occurred at too high a temperature for too short a period.

Choosing a System for Drying/Storage

Whatever the system chosen, it needs to:

- provide reliable holding conditions for grain prior to marketing at optimum quality.
- cope with grain at the rate of arrival at harvest time.
- have capacity for all the expected grain.
- utilise adaptable buildings where possible.
- cost as little per tonne to install as possible. In practice, costs have varied from as little as one-quarter to one-third of grain value per tonne to over double its full value.
- cost as little per tonne to run as possible (often 3—5 per cent of grain value).
- allow adequate separation according to intended markets of different grain lots produced or which may be produced in future.
- allow convenient handling of grain.
- allow rapid enough despatch into collecting vehicles.

In practice, compromises have to be reached between these ideals.

Drying rate depends on ventilating air properties: temperature, relative humidity (65 per cent RH is in equilibrium with 14 per cent mc, 88 per cent RH with 20 per cent mc) and rate of flow. Uniformity of the grain presented — level depth, absence of dusty dense zones — is also crucial to achieve uniform drying.

It must be remembered that viability is impaired when grain temperature exceeds around 50°C, whilst nutritional value is lessened above 90 to 100°C. Grain can also crack under stress if it is either dried or cooled too quickly, thus rendering it vulnerable to future deterioration. It should be cooled to below 18°C.

Cleaning Facility

Consideration should be given to the provision of a cleaning/grading unit (Figure 13.4). This may consist simply of a three-sieve mechanism or be equipped with indented cylinders and wild oat separators for specialist seed crops. With less sophisticated grain storage systems a mobile seed cleaner is appropriate, either purchased or from a contractor for smaller farms. Grain sample presentation is increasingly important (Plate 13.2).

Pre-cleaning is sensible since there is no point in drying rubbish and post-cleaning is increasingly essential for many markets to satisfy tightening sample standards.

PLATE 13.2   Milling wheat samples (cv Avalon): 1=before cleaning; 2=after cleaning.

Drier Types

There are a number of types of driers (Figure 13.5):

Batch driers
Portable models suit smaller farms (say with 500 tonnes of grain or less). These hold grain in 25—50 cm columns around a central air distribution plenum. They can normally dry 5—15 tonnes of grain at around 90°C and cool it, driving off around 5 per cent moisture within three hours. Airflow is rapid (Plate 13.3).

Larger units are often designed to dry more slowly and need to be equipped with augers to stir the grain or else differential drying will occur. The augers will

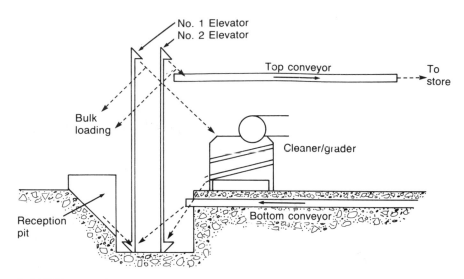

FIGURE 13.4   Cleaner/grader for pre-cleaning prior to drying and storage and cleaning/grading prior to dispatch. (From McLean, 1980)

eventually damage the grain if it needs a long drying period.

Storage driers

Bins can be designed for low temperature drying of 100–200 tonne lots of grain at low rates of aeration with or without some extra 10°C of supplementary heat. Such drying may take up to a month and the bin then becomes a permanent store which should be equipped with a temperature gauge for regular checking (Plate 13.4).

Bins can also be designed for higher temperature (45–60°C), more rapid airflow drying and cooling of shallower (1–2m) grain lots for a working day prior to storage elsewhere. Systems can be automated to replace dried grain layers with incoming wet grain.

Sack driers in which hessian-bagged grain is placed over holes in a platform through which hot air is blown have now been superseded on most farms.

On-floor storage of grain in bulk is now very widespread (Plate 13.5). Low rate aeration of ambient air, coupled with monitoring by temperature probes, cools and maintains these grain stores. Air is supplied through 1m-spaced lateral ducts or via a large central duct or through loose brick floors (which are cheap but hard to clean) or through grills in false floors and/or walls. Some are designed so that air can be sucked out from the grain, which is a good technique for keeping it cool and pest-free. Grain up to 18 or even 19 per cent mc can be stored with this cooling facility. On-floor systems can be installed in adapted or multi-purpose buildings as well as purpose-built portal frame stores (Figure 13.6). It is relatively cheap per tonne of grain held and can be easily designed to accommodate all grain so that there is no harvesting delay.

Continuous-flow driers

These are high-cost installations which operate at 40–100°C air temperatures and therefore are costly to run. Large models can extract 5 per cent moisture from 20–25 tonnes per hour. Continuous-flow driers are usually equipped with exhaust-air recycling to improve fuel efficiency by some 15 per cent.

Cross-flow driers are common (Plate 13.6). Air flows across the grain as it passes down through columns, being cooled when it reaches the lower sections of these columns.

In *counter-flow* driers the air flows against the grain flow.

Concurrent-flow driers have the air flowing with the grain to keep the hottest air in contact with the wettest grain, thus improving efficiency and reducing the risk of overheating.

Mixed-flow driers incorporate all three air-flow systems and are increasingly popular.

Other designs include the horizontal belt drier which can accommodate a range of crop seed sizes

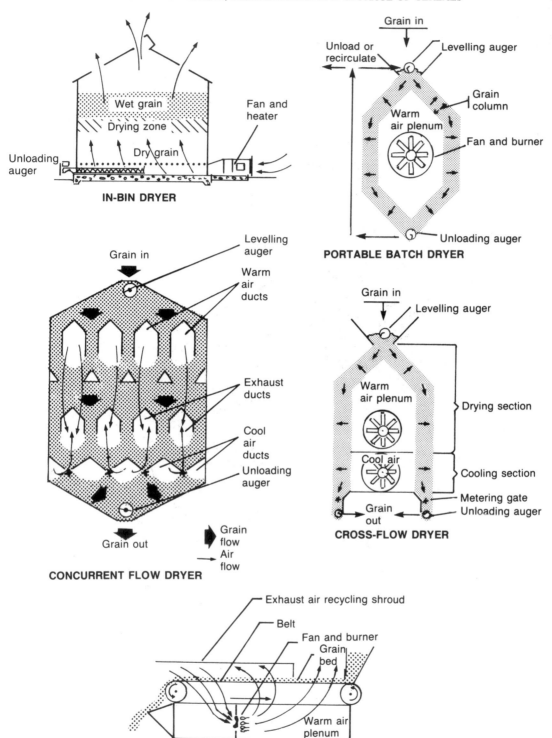

FIGURE 13.5   Types of grain drier. (After OMAF Canada)

PLATE 13.3   Portable batch-drier.

PLATE 13.4   In-bin drier/storage for grain
(Brice-Baker).

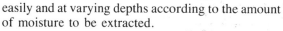

PLATE 13.5   On-floor drying/storage system.

PLATE 13.6   Cross-flow, low-energy drier.

easily and at varying depths according to the amount of moisture to be extracted.

Figure 13.7 summarises comparative energy requirements of alternative grain drying systems.

MOIST GRAIN STORAGE

This eliminates drying costs and allows earlier combining when grain is less liable to shed. It suits grain of 18–20 per cent moisture content intended for livestock feed which will be dust-free and more palatable.

However, moist grain is hazardous (generating lethal gases in sealed stores) and more difficult to handle than dry grain, which flows easily through augers and conveyors.

Moist grain must be either sealed in airtight silos or chemically treated to prevent deterioration. Sealed silos are expensive and prone to leaks which cause caking and bridging of the grain coupled with heating and deterioration. The gases they contain are lethal and have caused several fatalities in workmen entering to remedy faults. Grain must be removed regu-

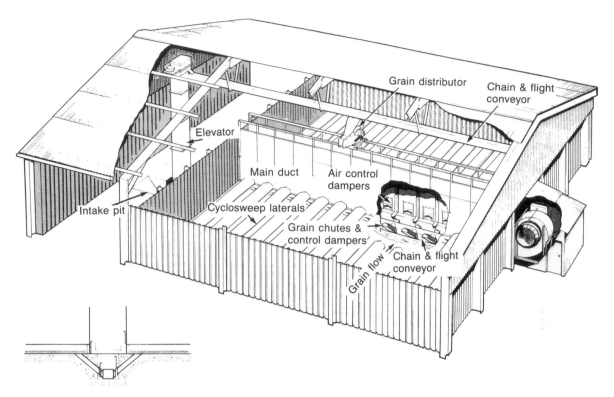

FIGURE 13.6   A typical grain store installed with the Cyclosweep Single Conveyor System.

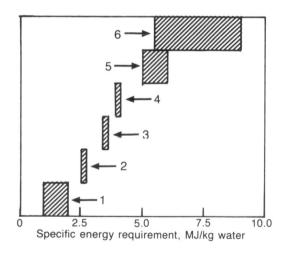

FIGURE 13.7   Specific energy requirement of grain drying systems: 1. Bulk drying with ambient air only. 2. Latent heat of vaporisation of water. 3. Radial bin drying system. 4. Mixed-flow high temperature drier. 5. Cross-flow high temperature driers with air recirculation. 6. Cross-flow high temperature driers without air recirculation. (From McLean, 1980)

larly and silos resealed immediately. Once removed, grain must be fed immediately since it will soon heat and spoil. In hot weather, moist grain silos must be emptied since pressure build-up can be dangerous. Therefore, sealed moist grain storage is normally only for one to eight or nine months' duration at most.

Chemical treatment is an easier, more reliable means of preventing moist grain spoilage (low rates are sometimes used as an adjunct to semi-airtight silos). Propionic acid was first introduced as BP Propcorn. It has a pungent clinging smell, is an irritant and is corrosive but appears to improve palatability. The acid is applied when grain enters the store as harvested. Rate of application is proportional to grain moisture content, e.g. it varies from around six litres/tonne at 17 per cent grain mc to ten litres/tonne at 24 per cent mc.

Caustic soda is an alternative material applied at 2.5—4 per cent by weight to whole grains; it raises their pH and improves voluntary food intake, digestibility and feed conversion ratios. Grain thus treated has proved very suitable for the last two or three months' feeding of beef animals before slaughter. Once treated, grain can remain in open storage without specialised buildings and is safe to handle. How-

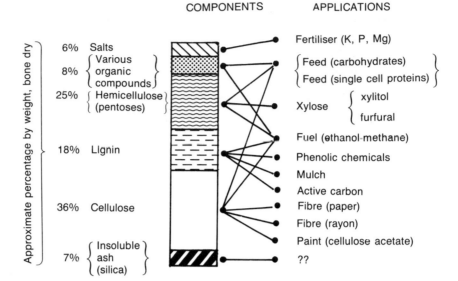

NOTE: For barley, chemical composition deteriorates from top to bottom of straw – e.g., crude protein declines from 4% to 3%, crude fibre increases from <40% to >50%.

FIGURE 13.8   Wheat straw: chemical composition and possible utilisation for each main component. (After Rijkens, 1977)

ever, caustic soda is extremely corrosive and users must wear full protective clothing during applications through a special unit.

STRAW

Whether straw is regarded as a useful by-product of the cereal crop or as a nuisance depends on its abundance relative to local feed and industrial demand. When cereals are widely cultivated, as in parts of East Anglia where the cattle population has fallen by 25 per cent since 1955, straw can be a surplus embarrassment. If straw burning is restricted and shallow or heavy soils make incorporation difficult, excess straw is unwelcome. The problems in handling and using straw are its low value and its low density which makes it bulky to transport. Straw composition and structure are indicated in Figures 13.8 and 13.9.

In 1980–85 the UK straw crop amounted to some 13 M tonnes/year compared with some 24 M tonnes of grain. Straw (including stubble) to grain ratios are now reckoned to average 0.6 for spring barley, 0.65 for oats, 0.7 for winter barley and 0.75 for winter wheat.

There are three options for straw disposal (Table 13.2):

• *Bale* and remove it (use of a forage harvester and loose handling are generally regarded as uneconomic).
• *Burn* it, usually after spreading as it left the combine.

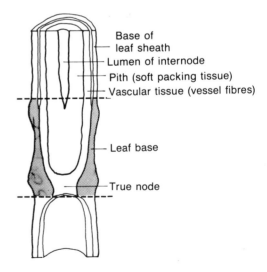

FIGURE 13.9   Structure of straw. (Adapted from Lynch, 1983)

TABLE 13.2 Methods of straw disposal in England and Wales

| | | | Percentage | | | |
| | | | 1983 harvest | | | 1984 harvest |
	Wheat	Winter barley	Spring barley	Oats	Mean	Mean
Baled and removed	39.6	80.4	80.2	84.7	60.5	55.9
Ploughed/cultivated	2.1	0.7	2.2	0.5	1.8	7.1
Burned (excluding stubble-alone burns after baling)	58.3	18.9	17.6	14.8	37.7	37.0

Source: MAFF.

PLATE 13.7 Straw-chopper in action on a Massey Ferguson combine.

- *Incorporate* it into the soil, usually after chopping, preferably as it left the combine (Plate 13.7).

These options are briefly reviewed below but considered in detail by Staniforth (1982), White (1984) and Butterworth (1985).

TABLE 13.3 Estimates of baled straw utilisation in England and Wales, 1982 (thousand tonnes)

On-farm straw uses	kt
Fuel	166
Chemically treated livestock feed	125
Potato storage	70
Sugar beet storage	50
Horticultural uses	71
Livestock bedding and feed (plus wastage)	4266
Total	4748
Off-farm straw uses	
Stables and mushroom production	300
Livestock feed (processed off-farm)	150
Thatching	15
Building board	16
Briquettes for fuel	1
Total	482
Total (all uses)	5230

Source: White, 1984 (MAFF).

Baling

The advantages of baling are as follows:

- It makes available for use the around 35 per cent of total harvestable dry matter in cereal crops which is straw and so makes a potential contribution to profits (Table 13.3).
- Straw removal from the field is possible immediately after combining. Livestock-keeping neighbours can be invited to bale and take it or contractors can be used (Table 13.4). Straw once fed through the animal or mixed as bedding with faeces and urine is more valuable to the land as FYM (Plate 13.8).
- Straw removal avoids the problems associated with the other options.

PLATE 13.8 Straw as feed and bedding for sheep housed prior to lambing.

TABLE 13.4   **Range of feeding values of cereal straws**

Type of straw	Metabolisable energy* MJ/kg DM	Crude protein per cent of DM	D value†
Spring barley	5.8−6.7	2.5−4.5	40−46
Winter barley	4.9−6.4	2.5−6.5	34−44
Spring oats	6.4−7.1	2.5−6.0	44−49
Winter wheat	4.6−6.2	3.5−4.0	32−43

* Calculated from *in-vitro* D value.

† Digestible organic matter in the dry matter (*in-vitro*).

Source: ADAS.

The disadvantages of baling:

- It can take a limited labour force from the more important tasks of harvesting the next field of grain or from stubble cultivations prior to the next crop.
- It incurs direct costs and often needs storage space too (straw of wheat 12.5 m³/t, barley 11.2 m³/t, oats 10.5 m³/t).
- Getting a good burn of stubble is more difficult if desired once straw has been removed.

PLATE 13.9   Round baler from John Deere, offering easy hook-up and a low power requirement. It is a high-performance economy machine for bales up to 1.2m in diameter.

Methods of baling

Conventional small bales produce bale weights of 20−25 kg. Large square or, more often now, large round bales produce around 400 kg bales of straw (Plate 13.9). High-density bales (Hesston) can reduce

PLATE 13.10   Old rick.

normal bale volume to one-third and the Holden equipment can virtually double this compression again. The high-density principle is good but expensive, suitable for contractors and when straw is in worthwhile demand at a distance. (Stand back when opening bales later!)

There is now a multiplicity of bale handling systems to cope with the low labour availability on UK farms; gone are the days of competitive pitching of small bales onto trailers on most farms!

Bales are often stored in too great a stack. Strategically placed (i.e. convenient for points of use) stacks of, say, 1000 small bales are quite large enough. If properly covered with re-usable heavy-duty sheets, then all bales will be fit to use later and the fire insurance risk will be spread. The old thatched straw ricks are a rare sight now (except overseas) but a pleasant memory (Plate 13.10).

Burning

The advantages of burning:

- It reduces bulk of trash to ease subsequent cultivations by a quick, cheap means. It encourages surface drying and tilth formation.
- It removes a potential source of toxins and of some pathogens and weed seeds.
- It returns potash (K_2O), magnesium and other mineral salts to the soil.

The disadvantages of burning:

- It is a waste of straw, for which potential uses exist.
- It is a potential hazard to operators, landscape and wildlife, traffic on roads.
- It is a nuisance — to neighbouring communities as ash (washed clothes on lines, fresh paint on window frames, etc.) and to soils where excessive ash can adsorb and render ineffective some herbicides. This last point is particularly significant for soil-acting grassweed killers in autumn.

Methods of burning

It is necessary to follow the National Farmers' Union Code as given in Figures 13.10 and 13.11. The degree of success depends on ample dry material burning thoroughly and quickly (Table 13.5), and the ash must be incorporated by a statutory scratch of the field within 36 hours. Self-policing by farmer groups, such as the South Oxfordshire Responsible Burning Group, is to be welcomed. Members agree on fines for transgression of codes of practice.

TABLE 13.5   Temperatures at the soil surface beneath burning straw

Amount of spread straw tonnes/ha	Temperatures at soil surface Peak temp °C	Temperatures at soil surface Duration (secs) of temp above 200°C
2.1	143	0
4.2	225	10
6.3	270	35

Source: MAFF.

Incorporation

The advantages of incorporation are as follows:

- Long-term replenishment of soil organic matter has been expected, but medium-term evidence from ADAS experimental husbandry farms does not support this. Higher cereal yields have returned proportionately more root matter anyhow and reduced cultivation systems have conserved OM.
- It can be pursued immediately and planned as an integral part of harvesting the cereal crop. It is much less dependent on weather than baling and burning are.
- It can be expected to improve not only the consistence of some extreme-textured soils (clays and sands) over the long term, but also infiltration rates for rain each season and nutrient-supplying power. Some farmers are already confirming these effects.

The disadvantages of incorporation:

- Energy use and other costs increase. Energy use may be 45 per cent more than traditional cultivations and more than tenfold some direct-drilling situations such as Schleswig-Holstein work.
- Slow rotting of straw is a potential source of problems: organic acids and other toxins may build up, and it may develop into a reservoir for diseases (e.g. *Septoria* and net blotch) and pests, especially slugs (Figure 9.1 and Figure 13.12).
- Subsequent cultivations and the establishment of the next crop may be hindered.
- Available soil nitrates are immobilised.

The severity of these disadvantages of incorporation depends on how well or how badly the job is done. Incorporation remains a last resort for the majority of UK cereal growers, but in both France and West Germany much more is already incorporated, though generally into deeper loamier soils.

(text continued on page 190)

a) CHECKLIST

**Every time you burn
straw or stubble,
check before you burn:**

√ The weather forecast.

√ Your neighbours know of your plans.

√ You are fully insured.

√ The county fire control and district council have been informed, if required.

√ Your firebreaks are correct.

√ Five hundred litres of water and five fire beaters are on site.

√ Where help can be found in an emergency.

√ The block to be burned does not exceed 10 hectares.

√ Each block is at least 150 metres from any other being burned.

**Remember you *may not* start
a fire before sunrise or later
than one hour before sunset.**

b) NFU CODE

***Never* burn**

× When damage or annoyance could result.

× On weekends or bank holidays.

× In unsuitable weather.

× Without 2 people to supervise each fire.

× Without effective firebreaks: 15 metres for hedgerows, trees and telegraph poles, 25 metres for other features.

Always

√ Clear straw beneath powerlines and around above-ground power installations.

√ Burn against the wind if possible.

√ Inform your neighbours.

√ Inform the county fire control and district council if required.

√ Incorporate ashes within 36 hours.

√ Use the checklist when preparing to burn.

**Know the law —
including your local byelaws**

c) FIREBREAKS

FEATURE	STRAW CLEARANCE	ADDITIONAL ACTION
10 hectare block to be burned	Clear a 5 metre strip	Cultivate or plough entire strip
Hedgerows Trees Telegraph poles	Clear a 15 metre strip and move straw at least 25 metres from the feature	Cultivate entire strip or plough 5 metres of it
Residential buildings Thatched buildings Buildings, structures, fixed plant or machinery made of glass or combustible material Stacks of hay or straw Accummulations of combustible material Standing crops Woodland and nature reserves Scheduled monuments made of combustible material	Clear 25 metre strip	Cultivate entire strip or plough 5 metres of it
Other ancient monuments, e.g. standing stones	Clear 3 metre strip	
Non-combustible electricity, oil or gas installations	Clear 15 metre strip	

FIGURE 13.10 Guidelines for straw burning — subject to revision from time to time; also check local byelaws. (Source: After National Farmers' Union)

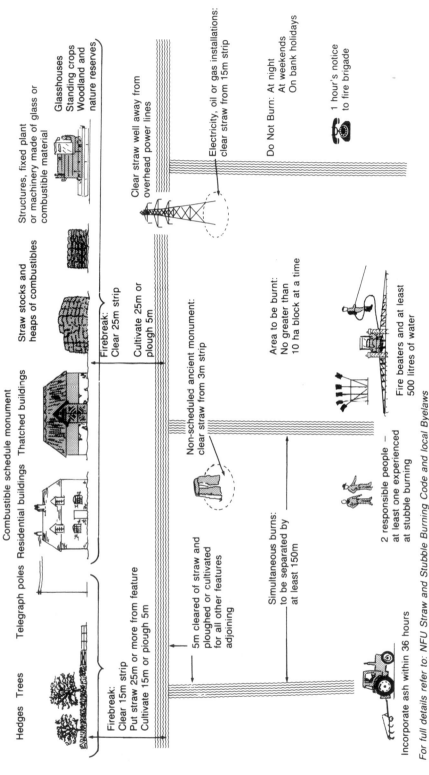

FIGURE 13.11 A pictorial guide to straw burning, based on model byelaws and the NFU code. (Source: ADAS, Phoenix newsletter)

189

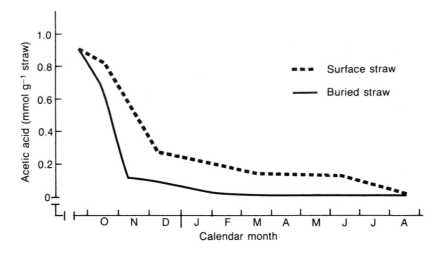

FIGURE 13.12   The decline in potential for acetic acid production from decomposing wheat straw under field conditions. (Harper & Lynch, 1981)

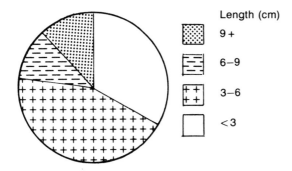

FIGURE 13.13   Combine-chopped straw: proportions of different length material recorded. (D.G. Christian, Letcombe Laboratory)

TABLE 13.6   Chop length and rate of decay of wheat straw

Chop length	Days to lose 50% weight @ 20°C
< 1 mm	14
0.5 cm	29
1.0 cm	30
2.0 cm	47
5.0 cm	54

Source: Letcombe Laboratory.

Methods of incorporation

Chopping is a necessary pre-requisite though undue fineness is undesirable (chaff, the fine glume residues, etc., are most potent sources of toxins). Figure 13.13 and Table 13.6 indicate chop lengths and decomposition patterns.

Mouldboard ploughing is currently the most widely used method. Various implements comprising disc and tyne combinations have been used with varying degrees of success. There seems no reason why the disc plough so commonly used in the tropics should not prove satisfactory and cheaper than conventional ploughing in some systems.

Products based on enzymes or on micro-organisms are available to accelerate straw and stubble digestion, but amply nurtured soils and cropping sequences should not need these to cope.

Straw needs 8–10 kg N/tonne incorporated to compensate for its low carbon-nitrogen ratio (around 45:1), which means that soil microbes will use up available soil N to produce humus at its standard C:N of 10:1, so that extra N improves decomposition rate (obviously, the most active phase is the first 1 to 2 months post harvest). Scandinavian countries and Germany all advocate this extra N at least in the first year of incorporation. Subsequently there may be a straw-decomposing flora encouraged and nutrient reserves should build up from previous incorporation. Undersowing with a green crop prior to incorporation also assists decomposition.

PLATE 13.11   Heavy-duty rotary cultivator (3m). (Courtesy of Falcon Ltd)

STUBBLE CULTIVATIONS

These are carried out directly after harvest on some fields, indeed as an integral part of harvesting, i.e. day one, remove grain; day two, remove straw; day three, cultivate stubble. The purposes are:

- to create a false seedbed to stimulate shed corn and weed seeds to germinate quickly so that these can be destroyed
- to incorporate residues to allow maximum time for decomposition before winter sets in (Plate 13.11)
- to allow proper infiltration of autumn rains

Priority should be given to fields due to be re-planted with early autumn-sown crops such as winter

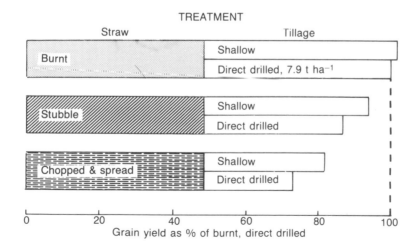

FIGURE 13.14   Effect of crop residues and method of tillage on yield of winter cereals on clay soils; mean results 1976–82. (Cannell, 1983)

oilseed rape and winter barley. Heavy land is also a priority because trash takes longer to decompose and, if autumn turns wet, fit cultivation conditions only return much later on a frost. Some land may well be left for later ploughing overwinter on frosts, especially calcareous loams. Slurry or FYM and/or lime may be given before any cultivation. Stubbles can be more reliably travelled upon later in a wet autumn than they can once stirred by cultivations, so some prefer to leave them untouched until ready to plough and prepare the next seedbed together. Stubble cultivation favours survival of certain weed seeds, notably wild oat, and it can also critically deplete moisture if overdone prior to autumn sowing. Untouched stubbles harbour weeds and diseases and slugs but they can also attract pheasants!

Treatment of previous crop residues can substantially influence performance of subsequent crops (Figure 13.14).

Chapter 14

CEREAL CROPPING SYSTEMS

No aspect of cereal cultivation demands thought and review on the part of the farmer more than the overall system adopted. Technical and economic advice on its component parts is readily available but the farmer must synthesise *all* considerations into an appropriate system. This chapter seeks to guide the reader through basic considerations with some measurable results. Long-term measurement is both expensive and painstaking to do; hence common-sense inference and farm experience are more abundant than trial results.

What is certain is the abiding truth of the words of Charles Dickens following his visit to the Royal Agricultural College in 1868, 'That part of the estate of a farmer or landowner which pays best for cultivation is the small estate within the ring fence of his skull; let him attend to his brains and it will be well with his grains.'

Although cereals are annual crops, proper husbandry must be dominated by longer-term considerations. The consequences of any particular cropping policy must be considered in relation to five factors:

1. Soil The conservation of soil against erosion and the sustenance of fertility in its widest sense are paramount.

2. Crops The effect of one crop upon another and the occurrence of beneficial or detrimental consequences must be considered.

3. Staff Both numbers and morale of staff depend upon the techniques and cropping sequence adopted.

4. Nation The national interest is affected in four main ways by cropping policy:

- *Environment* Cereals occupy a significant proportion of the land area. The appearance and quality of the landscape in terms of biological diversity depend upon the pattern and methods of cereal cultivation to a significant extent.
- *Energy* The adoption of an energy-efficient cropping policy is clearly a matter of national concern. The UK record is alarming by contrast with the so-called less developed nations (see Table 14.8).
- *Social* Rural depopulation is a matter of political concern, above all to farmers. There are strengths in a vibrant, agriculturally employed rural community.
- *Cost* The cost of taxpayer support is now very high in the EC.

5. Profits Without economic success, no cropping policy, however beneficial to soil, environment or society, can be pursued. Blind altruism without any profit has to die quickly.

Three levels limit the cereal system adopted:

- The Common Agricultural Policy of the EC with its centralised decisions tends to limit the choice of economically feasible cereal systems.
- The economic and political climate of member countries tends to limit severely the numbers of people employed in cereal cultivation and hence enforce mechanised, low-labour techniques.
- The circumstances of the individual farm determine what other enterprises are technically and economically feasible. For instance, a large

rental or overdraft charge usually enforces a higher output, higher pressure system.

Before Roman times a common rotation was as follows:

Year 1   Autumn-sown cereal (usually wheat)
Year 2   Spring sown cereal (often barley)
Year 3   Fallow

The Romans introduced a fallow/wheat/beans rotation as well.

In 1730 Viscount Townshend of Raynham, Norfolk introduced his famous four-course rotation (Figure 14.1) which remains in spirit if not in rigidity the basis of many rotations today. It was designed for light and medium soils. The system contained 50 per cent of the land in cereals, the two species being suitably separated by crops of other families.

Landlords have exercised significant control over the cropping sequence adopted by their tenants. Indeed one of the great strengths of the landlord-tenant system lies in the long-term concern for the soil and landscape borne by the landlord, whilst the tenant can give attention to the management of each year's crop. However, during the present century tenants have exercised greater freedom of cropping; it could be argued that shorter-term expediency has become a stronger influence in cropping policy decisions. Furthermore, landowners have been under pressure to specialise in cropping and shed labour however reluctantly.

The trend has been towards greater simplification of cropping sequence and the intensification of

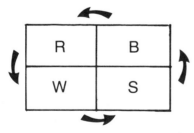

YEAR 1:   R = ROOTS, originally turnips
YEAR 2:   B = BARLEY, spring sown
YEAR 3:   S = SEEDS, often red clover for hay/grazing
YEAR 4:   W = WHEAT, autumn sown

FIGURE 14.1   Norfolk four-course rotation.

cereal cultivation. Once output per man has been maximised, a farm business then looks for more land to expand. Although average farm size has steadily risen during this century, this has not allowed the range of enterprises to remain diverse on each holding. Hence the size of each enterprise has become much larger.

The original concepts of rotation merit review here, even though the cropping sequences which constituted a full circle or rotation may not be so rigidly followed in the UK today. Early rotations attempted to be more rigid in respect of precise crops, their order and the number of years to complete full-circle.

The principles include:

1. *Substitution* of alternative crops for the fallow (uncropped) period, i.e. full use of land. There is still scope for introduction of this principle to some places in the world. On the other hand fallowing is being advocated by some as a means of reducing cereal area in the EC as well as reducing weed control costs and fertiliser needs (Table 14.1).

2. *Balance* between exhaustive and restorative crops. Exhaustive crops are those which deplete soil nutrients and tolerate declining humus quantity and quality without sudden failure. Cereals, other than paddy rice, come into this category. Restorative crops are those which enrich the soil or require ample replenishment of the soil if they are not to perform miserably. A period under a ley or other longer term crop allowing soil structure stabilisation or restoring organic matter is considered restorative. Root crops needing enriched soil conditions can be restorative. Whilst legumes can have a restorative role by adding nitrogen compounds to the soil, their continuous cultivation would be very unsuccessful in the UK and has proved very exhaustive where practised overseas.

3. *Breakage* of the life cycle or population build-up of soil-borne pests (e.g. cyst nematodes); diseases, especially soil- and trash-borne ones (e.g. eyespot); and weeds, especially those cereal relatives, the grass weeds, which cost heavily to control chemically.

4. *Cover.* The object is to maintain continuous soil cover, so minimising exposure of bare earth to weed colonisation and erosion liability in some areas. This last point is obviously of greater

TABLE 14.1 Effects of bare fallow on the grain yield (t/ha) of a following crop of winter wheat on Broadbalk field at Rothamsted

Period and cultivar	Fertiliser N applied to wheat (kg/ha)				
	None	48	96	144	192
1935−64					
cv Squarehead's Master					
4th wheat after fallow	1.43	1.78	2.26	2.58	nt*
1st wheat after fallow	2.35	2.71	2.94	3.07	nt
1956−67					
cv Squarehead's Master					
wheat grown continuously†	1.52	1.98	2.70	3.02	nt
1st wheat after fallow	2.40	2.64	2.75	3.04	nt
1970−78					
cv Cappelle Desprez					
Wheat grown continuously†	1.67	3.48	4.81	5.13	5.49
1st wheat after fallow†	3.56	4.60	5.10	4.96	4.79

* nt Not tested.
† With selective weedkillers.

Source: Data from A.E. Johnston, 1987 (Span 30,2).

importance in many tropical areas, though by no means irrelevant in the UK.

5. *Diversity.* Some of the advantages of producing a range of different crops can be attained by growing them in blocks within the same farm, not necessarily by rotating them all within each field. However, individual field rotation where possible secures all benefits simultaneously.

Diversity of cropping leads to a number of things:

- Less economic risk if the season or market is bad. It is unlikely that all will be equally affected. Undue dependence on one product can make one politically vulnerable too. The advantage of keeping to a definite rotation is that the farm may carry more or less the same proportion of each crop every year (depending on variability of field size or block size grown), and returns therefore tend to be steadier.
- Less biological risk. A diverse range of residues replenishing the soil is more likely to preserve the size and diversity of the soil population. Furthermore the likelihood of many weed, pest or disease infestations is reduced.
- Greater variety of diet and crop by-products available locally, thus reducing dependence and costs of transport in the overall economy.
- Greater variety of work, therefore improved job interest. The opportunity for specific tasks such

as subsoiling or liming may logically fit before the more responsive crops in the sequence.

- Better spread over the year of work and hence of labour demand.
- Less waste because different enterprises can interact.

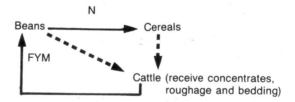

- Somewhat higher yields for lower inputs of energy-consuming biocides and fertilisers (Table 14.2 and Figures 14.2 and 14.3). Standards of crop quality can be higher, e.g. less grass weed contamination is likely in corn from mixed arable. However, labour requirements and certainly management skills are likely to be higher though better spread: this may mean less pressure for staff and increased challenge and interest for management.

The greatest aggregate of the above advantages of rotations could be expected to accrue from the widest contrast between the crops included, i.e. representation in the rotation of as many different

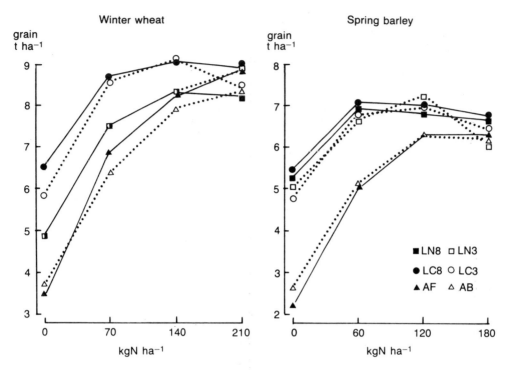

Key
LN & LC leys followed by wheat and barley
   LN8, LN3 8- and 3-year grass leys with fertiliser N
   LC8, LC3 8- and 3-year grass-clover leys without fertiliser N
AB AF all-arable rotations
   AB barley barley beans wheat barley
   AF fallow fallow beans wheat barley (fallowing was intended to diminish soil organic matter and lessen soil-borne pathogens)

FIGURE 14.2   Effect of leys on the yields of a first winter wheat followed by spring barley. (Courtesy of A.E. Johnston, Rothamsted)

plant families as sensible for the circumstances of the farm. In fact the major UK farm crops are derived from only five botanical families:

Gramineae	cereals and grasses
Leguminosae	pulses and legume fodder crops such as clover and lucerne
Cruciferae	green fodders, e.g. kale; fodder roots, e.g. turnips; vegetables, e.g. sprouts; oilseed rape
Solanaceae	potatoes
Chenopodiaceae	sugar beet, fodder beet, mangolds

Other families are represented by minor crops, e.g. *Linaceae*, linseed and flax; *Compositae*, sunflowers.

Factors Leading to Decline of Rigid Rotations

1. Fertilisers

Ever since the early work of Lawes and Gilbert at Rothamsted during the 1840s, the use of fertilisers has increased. It has escalated during the past fifty years and especially during the past twenty on cereals. When one can replenish nutrients in a particular ratio, the need to balance soil supplies using an assortment of crops and carefully returning all residues is lessened. Undoubtedly, fertilisers support the high average cereal yields in modern rotational systems but especially sustain continuous cereal systems.

2. Machines

Relatively larger areas of cereals or any single crop

TABLE 14.2 Preceding crop effect on first wheat: relative wheat yields at Bridgets EHF, Hants., 1965–75. (Mean yield = 4.55 t/ha = 8.5% above mean UK yield for the decade)

Continuous winter wheat	100
Alternated with barley	100
After continuous barley	103
2nd wheat after ley	109
After oats	112
After oats and beans	117
1st wheat after ley	117
After forage maize	122
After beans	125
After potatoes	126
After oilseed rape	137

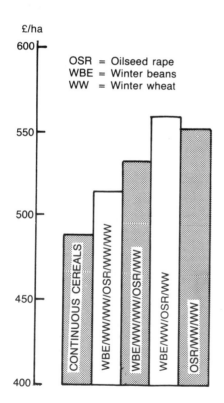

FIGURE 14.3 Relative average annual gross margins for different cropping sequences in the eastern counties on chalky boulder clays. (Based on data from Boxworth EHF, Cambridgeshire)

enterprise are needed to justify possession of specialised, increasingly large equipment. Crops of lesser importance have in some cases proved more technically difficult to mechanise than corn: in all cases the smaller market has been a lesser incentive for the development of specialist machines for other crops.

3. Management
Specialisation in fewer or even single crops allows the concentration of skills: one is no longer a jack-of-all-trades and master of none. The argument runs, perhaps somewhat insultingly to farmers' intelligence, that growing each crop is so 'technical' nowadays that one can scarcely cope with two! Furthermore, the simplification of cropping to just cereals affords greater leisure and, perhaps, easier management in some ways.

4. Economies of scale
It can be argued, up to a point, that the acquisition of contracts and the bargaining power in purchasing inputs and selling outputs are enhanced by running a larger cereal enterprise.

5. Plant breeding
By developing more resistant varieties, the plant breeder reduces farmers' dependence on diversity of cropping to lessen disease pressure. Furthermore, the disadvantage of one species cropping is less owing to the existence of a greater selection of cereal varieties with somewhat different required dates of sowing and harvesting.

6. Government policies
Governments can alter the balance of species cropped by price-fixing arrangements, subsidies and other production incentives. Above all, cereals are the mainstay of most agricultural economies: collapse in the cereal market would be disastrous for British agriculture as a whole. Therefore, market support for cereal production has had to be consistently encouraging, significantly reducing the economic risk of cereal monoculture.

7. Biocides
The development during the past fifty years of crop protection chemicals against weeds, pests and diseases of cereals has provided a powerful alternative to cultural means of control. In fact, rightly and economically used, biocides are not a substitute but rather a supplement to good general husbandry.

Biocides and fertilisers together have made continuous corn technically sustainable.

8. Tenancy agreements
Much greater freedom of cropping is now generally allowed by landlords.

9. Labour
The decline in the agricultural labour force has been rapid and continuous. Cereals by contrast with other crops have rather lower labour requirements per hectare (though these vary with the methods adopted, of course). General farm workers — GFWs — are becoming a rarer species! Specialists are being trained. Roles even within the cereal enterprise are often quite demarcated.

10. Targets
Economic performance and market objectives are more emphasised. Fixed resources can be focused on a single enterprise, whereas an assortment of different crops may compete leading to a clash of priorities. Furthermore, mixed cropping can lead to muddled management, and 'passenger' (uneconomic) enterprises may be tolerated because financial analysis is not incisive enough to detect them. Of course, a mix of crops can be properly costed but a specialist cereal farm is clearly more straightforward to evaluate.

PRACTICAL PLANNING OF CROPPING SEQUENCE

A good plan avoids an anarchy of cropping. However, it should not be so rigid as to exclude modification as circumstances and commonsense dictate. There are a multitude of possible recipes, possible permutations and combinations of cropping. Ideal material for a computer, some think! It is in fact done, and computers fed with partial budgets for changes of crop can give rapid calculation of the numerical impact of such changes. However, the computer pronouncement is only as relevant as the data it receives. The 'recipe' decided upon is only as good as the 'chef' who implements it!

When carrying out a review of cropping policy or planning it for a new farm, the following criteria need consideration:

1. Objectives
(both short and long term). What are the desired, expected outcomes of the policy? What does one wish to produce? What targets of soil or farm improvement/maintenance need to be set?

2. Environmental constraints
(the limitations of soil, site and climate). What alternatives to cereals are possible anyway?

3. Economic factors
(profitability, reliability, demand). How do the possible candidate crops compare in these respects? It may be better to have cropping of lower *potential* seasonal profitability if yield is more reliable from year to year and demand more consistent.

4. Technical options
(varieties, mechanisation, chemicals). What resources and methods are available for the contemplated cropping both on the farm and for purchase — and at what cost?

5. Human questions
(preference, supply of labour, skills). Allowance must be made for differences between farmers. Which of the candidate crops does the farmer/manager/staff really *want* to grow? It is certain that half-hearted commitment to a crop choice is no recipe for success. A 'new' crop can pose a challenge to management and stimulate staff. What casual or regular labour force is available? What skills exist or need to be acquired for the contemplated crops — and what local means exist for their acquisition? The neglect of human questions robs any business of enjoyment and long-term fruitfulness.

6. Husbandry factors
(rotation, interaction, by-products). How will the cropping sequence work — how will it fit the fields as they are, when will the bottlenecks of work occur? Will the crops complement each other either biologically or economically (in terms of shared use of fixed resources, for instance)?

7. History
(experience, records). What local experience exists about the proposed crops? He is a wise young farmer or manager indeed who, on taking a new farm, consults local farming and country people in the district. It is good practice to consult an obviously tidy and successful farmer when new to a district: his or her wisdom is likely to be worth a thousand printed documents. If farm records exist, they should be consulted for past performance of proposed crops.

There is a case, of course, for departures from local experience and practice to pioneer novel crops or techniques. If that sort of thing had not occurred, Britain would probably never have grown cereals, nor would we have potatoes, red clover and most of our crops!

8. Legal constraints (laws, contracts, quotas). The frequency of cropping or juxtaposition of some crops (e.g. cereals for seed) is restricted as to variety. It is always possible for the government or the EC to introduce quotas for categories of cereal as exist already for some other crops, e.g. sugar beet.

Examples of Cropping Sequences Featuring Cereals

There are many recipes but their effective management is paramount. Some possibilities are given below:

- Alternate husbandry or ley farming consisting of, for example, *three year ley/two winter wheats/ winter barley*. This rotation allows soil stabilisation and organic matter accumulation under the ley phase, particularly if it is grazed. Once livestock parasites and grassland weeds have accumulated, the ley is ploughed and fertility cashed in for one or two wheat crops. The ley must be released by the livestock enterprises of the farm in adequate time to allow preparation without high pest risk and for timely sowing of the next wheat crop. Grass weeds may build up, particularly meadow grasses. The cereal sequence is ended before arable problems can accumulate seriously and the winter barley allows early direct sowing of the ley to follow. The sequence could be extended by having a catch crop such as stubble turnips after the winter barley and following these with a spring barley crop undersown to the next ley. The system provides for a balanced, safe cropping policy. On deeper, lighter land a root crop or potatoes may also be interposed.
- *Two or three winter wheats/winter beans* — a traditional heavier land rotation, often now appearing as a longer run of wheat with winter oilseed rape as an alternative as well as beans.
- *Sugar beet/spring malting barley/peas/winter wheat* — a good, balanced lighter loam sequence emulating the famous Norfolk four-course rotation.
- *Winter wheat/potatoes/sugar beet* — a traditional Fenland sequence where the wheat is really the poor relation. Celery, onions and other vegetables may often be found now, with wheat as an infrequent break.
- *Potatoes/winter wheat/Brussels sprouts/spring wheat* — a balanced sequence found suitable on deeper sandy clay loam land.
- *Two or three winter wheats/winter barley/winter oilseed rape* — this is a typical sequence to allow a combinable alternative crop into an otherwise all-cereal sequence. Sometimes on lighter land only winter barley may appear, perhaps with spring barleys or triticale and the intermittent break crop of oilseed rape, linseed or pulses every fifth year.

A double break from cereals allows a cereal seed crop to follow afterwards if other conditions of field hygiene are met.

Continuous Cereals

Wetland or paddy rice has been successfully grown continuously for several thousand years. However, this has not been the case with dryland rice or with other cereals.

Arguments and counter-arguments about the continuous growing of dryland cereals include the following:

1. It is *biologically narrow*, leading to the accumulation of specific pests, diseases and grass weeds coupled with a relative microbial stagnation due to the lack of variety and quantity of raw organic matter arriving in the soil. Soil structural stability may thus be weakened. The amount of organic matter present can be conserved by reducing cultivations and therefore soil aeration so that decomposition is slower. However, long-term reduced cultivation for continuous winter cereals is dependent upon burning straw, at least more often when direct drilling is to be practised, thus depriving the land of a source of OM. Opportunity to add external sources of organic matter to the system is offered by the periodic insertion of a spring cereal into the sequence; this also relieves pressure of grass weed build-up and allows a cheaper control possibility. In addition, the interchange of different cereal species in the sequence can relieve specific disease and pest pressure relative to continually cultivating only one species of cereal. Proponents of continuous cereals would claim technical sustainability in the medium term (decades, not millenia!) using the range of available crop protection chemicals (biocides) and fertilisers now available in some parts of the world, and the take-all fungus declines after a few years (Figure 14.4) but recovers again once a break crop is introduced.

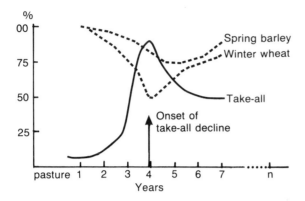

FIGURE 14.4 Yield and take-all in continuous cereals. (After Lester, 1969)

2. It is argued that it is *economically narrow*, with dependence on inputs from outside the farm and on direct sales of grain off the farm. These drawbacks can be offset if the cereals are grown continuously not over the whole farm but only within a block of land on the farm whilst other parts of it grow different crops — perhaps grass in a grazing block and a cutting block. Alternatively, even the all-cereal farm could diversify markets by feeding at least some of the grain via non-land-using enterprises such as pigs, poultry or intensive beef, at the same time adding substantially to the value of that grain. However, on individual fields cropped continuously there will tend to be an inevitable increase in variable costs for both sprays and fertilisers per tonne of yield, generally resulting in a less energy-efficient system (Table 14.3). Nevertheless, the proponent would argue the benefit of economies of scale in fixed costs per tonne by specialising in a larger area of cereals. It must be

remembered, though, that if the scale is too large — i.e. beyond the managerial capacity of the farm team — financial benefits may be lost.

3. Another argument is that it is *managerially narrow*, producing a simple cropping system which is less flexible, though it is easier to manage in several ways (specialised knowledge and skills are more attainable if only cereals are grown). It is easier to expand into alternative crops if there is already some prior experience of growing them in rotation on the farm. A diverse range of crops adds interest and challenge to work and offers the chance of a better spread of work or labour profile over the year. However, continuous cereals enable the farmer to reduce labour used per tonne produced, i.e. fewer men to control and to pay. On the other hand, this trend can be seen as detrimental to rural employment prospects.

Undoubtedly those farmers who have succeeded in continuous cereal production over several decades have done so by exploiting the simplicity, specialisation and scale advantages of the system and by sustaining crops by detailed attention to soil management and crop protection. The long-term sustainability of such systems is clearly open to question as is the desirability of such systems. Some soils and regions are particularly vulnerable to damage by continuous cereal cropping. A rotation of crops generally reduces such risks (Table 14.4). In any case, yields of first winter wheat crops after a break are generally 10−15 per cent higher than second and third wheats although this may often be used in practice to justify planting a lower-yielding but higher-quality variety as a first wheat. Some ADAS experience in trying to measure rotational effects on cereal performance was summarised by Bowerman and Jarvis (Table 14.5).

TABLE 14.3 Fertiliser practice on winter wheat and spring barley after different numbers of cereal crops

	Consecutive cereal crops	% area getting				Average actual kg/ha			% crop area
		N	P	K	FYM	N	P	K	
Winter wheat	1	98	74	68	9	154	25	47	34
	2	100	94	87	7	168	25	45	25
	3−4	100	96	85	15	175	26	44	19
	5+	100	98	81	11	185	25	47	22
Spring barley	1	98	88	88	21	83	17	34	27
	2	98	95	96	24	91	17	37	15
	3−4	99	97	97	27	97	18	37	33
	5+	98	96	93	23	107	20	39	25

Source: Church and Leech, 1983.

TABLE 14.4 Average annual maize yields, total water loss and total soil loss under various treatments on Rideau clay for 12 years at Ottawa

Crop and treatment	Total soil loss t/ha	Total water loss t/ha	Average annual relative yield
Rotation – manured – grown on level land	0	0	100
Rotation – manured – grown on contour on 10% slope	103	1186	79
Rotation – manured – grown up and down 10% slope	138	1372	63
Continuous – manured – grown up and down 10% slope	486	5475	36
Continuous – no manure – grown up and down 10% slope	653	6616	25

Source: After OMAF, Canada.

TABLE 14.5 Cropping policy and yields of wheat (t/ha at 15% moisture content) at two levels of nitrogen fertiliser

	113 kg/ha N	151 kg/ha N	Mean
First wheats (1973–75) after:			
Continuous wheat	5.13	5.08	5.10
Beans*	5.72	5.69	5.70
Oats	5.24	5.57	5.41
Spring barley	5.19	5.49	5.34
Spring wheat	5.33	5.46	5.40
Two years of spring barley	5.56	5.66	5.61
Two years of spring wheat	5.68	5.73	5.71
Second wheats (1974–76) after:			
Continuous wheat	4.50	4.75	4.63
Beans	4.77	4.92	4.84
Oats	4.67	4.94	4.81
Spring barley	4.43	4.81	4.62
Spring wheat	4.64	4.82	4.73
Two years of spring barley	4.63	4.96	4.80
Two years of spring wheat	4.86	5.07	4.97
Third wheats (1975–77) after:			
Continuous wheat	4.19	4.27	4.23
Beans	4.46	4.51	4.49
Oats	4.32	4.51	4.42
Spring barley	4.34	4.54	4.44
Spring wheat	4.44	4.42	4.43
Two years of spring barley	4.45	4.35	4.40
Two years of spring wheat	4.46	4.63	4.55

* For the first wheat crops after beans, the rates of nitrogen were reduced to 83 kg/ha N and 121 kg/ha N.
Source: Data of Bowerman and Jarvis, 1982.

BREAK CROPS

A break crop is any change of crop from the one customarily planted on a particular field. It may even be so described if a spring cereal is planted where the winter version of the same species is usually grown. Some use the term to mean only those crops which can be combine-harvested but are non-cereal.

The following is a checklist of the purposes of break crops:

1. Weed control A major cost, hindrance and yield detractor of intensive cereals can be weeds, especially grass weeds. A break crop may be sought to allow both a cultural and a chemical opportunity to hit the offending weeds.

2. Pest control A change of crop to starve out a soil pest may be the cheapest means of arresting it.

3. Disease control Soil- and trash-borne diseases can be dealt with by interposition of a non-host crop under which the infective cereal trash can thoroughly decompose.

4. Restorative effect Is a change of diet due for the soil organisms − or an improvement in soil conditions?

5. Economic diversity If the market looks less promising for some of the cereals and/or has been boosted for another crop, then it may be time to include it. Such a reason accounts in large part for the dramatic expansion of oilseed rape in Britain starting in the 1970s and the recent stimulus of interest, albeit lesser, in dry peas and field beans.

6. Human value Is a break crop needed to interest the staff?

The following characteristics are those sought for a suitable break crop from cereals:

A. *Growth habits* preferably a different botanical family with a different growth pattern
B. *Value* profitable in its own right or at least offering reliable and acceptable returns
C. *Fixed costs* produced with equipment similar to that needed for cereals (e.g. the combine)
D. *Compatibility* one which provides a good lead-in to the following cereal, which is usually winter wheat

Possible break crops, in the widest sense of the term, are considered broadly in terms of these characteristics A to D:

PRINCIPAL BREAK CROP CANDIDATES

These are all combinable.

Oilseed Rape

This crop, formerly called coleseed, has become a profitable arable crop in its own right owing to a continued EC deficit in vegetable oils, though the future is now less favourable economically, and from 1991 only double-zero varieties are supported (i.e. oil low in both erucic acid and glucosinolate).

The price per tonne has been over two-and-a-half times that for wheat, but fear of oilseed surpluses is reducing EC price support. The average gross margin is likely to equal that for winter wheat but shows a much wider range.

The British climate favours production of good-quality oil and there are prospective varieties with 'safe' protein cake by-product for human diet. So far rape meal after oil extraction from single zero varieties also needs to be used in relatively low proportions in animal feed compounds. The crop is considerably more successful in the eastern half of Britain than in the west.

The dramatic expansion of winter barley cultivation has provided the ideal lead-in since its July harvest allows time to establish winter oilseed rape during the optimum period (15−31 August). Whilst the interest of Hampshire growers in the 1960s was focused on spring varieties, present cultivation is predominantly of winter oilseed rape; spring varieties allow for autumn working of the ground and can have a role. Yields of the following wheat are generally

Crop	A	B	C	D	Remarks
Potatoes	√	√	×	?*	OK if land contracted out for early-lifted crop to another
Sugar beet	√	√	×	?	farmer with appropriate kit of his own
Vining peas	√	√	×	√	OK if consortium of growers or if land contracted out
Ley	×	√	×	?	OK if herbage seed *or* contracted out for hay or grass-drying and provided it is ploughed out early enough
Another cereal	×	√	√	√	
Field beans	√	√	√	√	Recent EC protein scheme encourages these
Dried peas	√	√	√	√	
Oilseed rape	√	√	√	√	EC oilseed deficit has ensured good prices
Linseed	√	√	√	√ (?)	
Seed crops	√	√ (?)	√	√ (?)	Specialist crops need contracts, e.g. sugar beet for seed, brassicas for seed
Catch crops	√	√	?	?	OK for shooting and game cover; can be folded quite cheaply with temporary fencing; usually enforce spring corn to follow

* ? means questionable

improved unless the previous wheat was continuous and take-all returns.

Oilseed rape requires good soil physical conditions to do well and its inclusion may encourage improved standards of soil structure from which cereals will benefit too. The crop provides a good opportunity to selectively kill grass weeds and cereal volunteer plants.

But it has a number of practical snags:

- Pigeon attack can be severe.
- Pests and diseases increased as the hectarage of the crop expanded some hundredfold between 1970 and 1987.
- There is a probable harvest clash with winter barley and harvesting can be difficult.
- The seed is small and easily sheds from pods or leaks from trailers (rapidly becoming a weed on the farm perhaps).
- Seed has to be dried to 8.5 per cent moisture content and leaks around the drier easily too!
- The haulm is of little value to the soil, often difficult to burn and encourages slugs, frequently necessitating slug pellets for the following winter cereal crop.

In spite of its snags, oilseed rape is currently the most popular of the combinable break crops. Whilst it is grown in two successive years in North Germany, the UK recommendation is not more than once in four years on account of diseases and pests. (See Ward et al., 1985.)

Linseed

This crop is the oilseed version* of flax (which is also grown as a minor specialist crop). Unlike oilseed rape, which has had a high price determined by the EC, linseed growing is encouraged by a hectarage subsidy; this has tended to be less attractive than the oilseed rape support. Nevertheless, the EC is currently substantially undersupplied. Yields of linseed are much lower than oilseed rape, and gross margins are perhaps equivalent to spring barley; it is cheap on fertiliser.

The crop is spring planted and is a poor competitor

---

* Other oilseed crops are still being researched in Britain but so far are scarcely feasible commercially: e.g. oilseed sunflowers (suitable for southern Europe) and oilseed lupins (Continental, Russian and Australian varieties). Evening primrose and borage offer small scope for pharmaceutical/health-food markets, being rich in gamma-linolenic acid (GLA) which is polyunsaturated.

for weeds but fairly hardy in terms of pests (it is wireworm resistant!) and diseases (Botrytis can be damaging and Sclerotinia may affect it). Harvest is late (September, clashing with autumn drilling; stubble is tough and of no use; seed must be dried to 9 per cent moisture content.

Field Beans

Field beans (Vicia faba) for animal feeding have traditionally been a heavy land alternative to wheat. The beans contain some 25–28 per cent crude protein. They are invariably of benefit to the following wheat, not chiefly because of residual nitrogen but rather the well-structured rootbed they require.

To save on imported soya beans, the EC supports field bean growing. An activating price is determined by the EC, and growers receive (via the merchants who purchase registered crops for stockfeeding) 45 per cent of the difference between this price and the actual market price if it is lower. Contracts are also available for tic beans (spherical field beans) for pigeon-fanciers and export.

Yields are notoriously variable, though promising varieties include Troy spring beans. A problem is the dependence on bee pollination for up to one-third of the seed set (in arable areas, beans need at least a hive per hectare).

Chocolate spot (Botrytis fabae) can be very damaging to all plant parts, and pigeons can be troublesome in a predominantly cereal district.

Winter varieties often achieve higher yields at slightly lower crude protein content than most spring beans. However, spring beans allow longer for preplanting soil treatments and suffer less from chocolate spot disease and waterlogging. Nevertheless, spring beans attract more aphids (blackfly) and can be late to harvest, though of shorter straw.

Bulky trash strains the combine (and driver!), and drying has to be done slowly and thoroughly at low temperatures (around 43°C) to attain 15 per cent mc or below. The haulm is of little value except as freedraining bedding for older beef cattle; it attracts slugs, so that following cereals often need slug pellets.

Overall the potential for beans looks more attractive on paper than for some time, though their actual record hitherto is best described as unreliable. Average gross margins are likely to lie between winter wheat and spring barley, if not below the latter, but may now equal oilseed rape. (See Hebblethwaite, 1983.)

Dried Peas

These are otherwise known as combining or protein peas. The majority of varieties grown are marrowfat peas; there is some market for large blues and particularly for small blues. They attract EC support as protein import savers by the same scheme as mentioned for field beans.

Virtually all are spring sown; winter varieties developed so far have not proved commercially successful. They will not tolerate wet land, but they need well-structured, moist soils. Pigeons are attracted to young crops. Like beans, they benefit the following cereal.

Harvest is early, and harvest weather is rather more critical than for corn. They can be difficult to pick up with the combine, but newer varieties have less leaf. They need slow drying as for beans to a final moisture content of 14.5 per cent for storage.

Peas should not be grown closer than one year in four or five on account of pea cyst nematode and soil-borne diseases; pea moth can seriously infest crops — follow ADAS warnings based on Rothamsted trap samplings. The crop can host *Sclerotinia* as can oilseed rape and spring field beans.

New varieties plus market demand and support make this a currently more attractive option than previously. Average gross margins should equal or exceed winter barley. Quality dried peas sold for human consumption attract variable premiums. (See PGRO Pea Handbook.)

Herbage Seeds

Once established, grass sown for seed should give two or more harvests. (Italian ryegrass, however, usually gives only one seed crop.) Amenity species can also be considered for lawnseed or turf. Ryegrasses, especially perennial, have been the most in demand, although they have been in surplus recently. Prices and contracts are somewhat variable and since yields are small profitability varies considerably. Beware!

A significant attraction to farms carrying a grazing livestock enterprise, especially sheep, is the provision of periods of clean grazing between seed crops.

Harvesting can be tricky; timing is critical. It may often clash with cereal harvest or, if fortunate, it can be slotted in between winter barley and winter wheat/spring barley harvests.

The crop does not allow cleaning the land of grass weeds which are likely to be a major reason for a break from intensive cereals: indeed, herbage seed must *not* be attempted where grass weeds are at all a problem — contamination is disastrous. A period under such a ley will undoubtedly help to stabilise a poorly structured soil.

Red clover for seed, offering bonus hay for sale (horses perhaps!), plus soil nitrogen and structural improvement, has attractions. But take care over desiccation and combining and avoid late hay cuts, which will mean that the subsequent seed harvest is late.

Other Cereals

The admixture of cereal species in a rotation has been discussed earlier in this chapter under 'Examples of Cropping Sequences'. Key points on the less usual cereals — oats, rye, triticale and durum wheat — are summarised in the notes on individual cereals later.

Provided that grass weeds are not out of control, a cereal change may be the most sensible and least disruptive first option for the all-cereal farm. However, some pests and diseases are common to all cereals.

Specialist Seed Crops

Sugar beet for seed on contract may be tried away from beet sugar-producing districts, but the crop is in the ground a long time — a full twelve months or over. Profits can be good but as with all crops with strictly limited demand for seed of each variety, growers may be instructed to destroy good crops. If that occurs, adequate compensation does not make up for the months of careful husbandry which see no useful result, and the consequent effect on farm morale is negative.

Brassica seeds offer another possibility. These often occupy the ground longer than sugar beet seed, e.g. kale may be there for eighteen months. Yields are low and each kilogram is of very high value, hence risks are enormous. These crops are best left to seed-crop enthusiasts!

CATCH CROPS AND THE CEREAL ENTERPRISE

Catch cropping means quickly taking a short-term crop between two main crops. In the present context it usually means after a relatively early-harvested cereal and before a subsequent spring corn crop. Catch cropping is most appropriate for lighter land in the southern half of England. Some catch crops may be undersown, e.g. trefoil in spring barley for subsequent green manuring (rarely practised nowadays).

Possible candidates would include:

Month sown	Crop
July/August	Forage rape (possibly late kale also) for autumn use
July–September	Stubble turnips (grazing varieties also) for winter/ early spring use. Aerial sown *just* before harvest or into stubbles; also direct drilled (Plate 14.1)
August/September	Forage rye and Italian rye-grass for early spring/spring use
July/August	White mustard for early autumn use for game cover, green manure, fodder

Some of the above may be usefully mixed, e.g. rape/turnips; rye/ryegrass.

PLATE 14.1 *Stubble turnips — four weeks after direct drilling into winter barley stubble in August.* (Courtesy of Arable Farming)

The most relevant and strategic use of catch crops is obviously for livestock (especially sheep), and they are normally folded on the field necessitating fencing (usually temporary and electrified on predominantly cereal farms where permanent boundaries if present do not justify stock-proofing costs). Cereals themselves may be beneficially grazed if their growth is excessive or sheep fodder is short (Table 14.6). In many parts of the world such as N.S.W. Australia and southern Africa, cereal stubbles are important for grazing by sheep (which select shed ears and grains) and cattle (which are less selective).

The case *in favour of* catch crops in the cereal rotation can be summarised thus:

- Make *fuller use* of land.
- Fill *fodder gaps* for livestock enterprises.
- Allow *stock* into the arable rotation without reducing the cereal hectarage.
- Maintain *soil cover*, thus helping to smother weed flushes and keeping soluble nutrients cycling through catch crop plants within the upper horizon of soil.
- Maintain *soil organic matter* status and relieve the monotony of cereal root and stubble diet for soil organisms. They may be specifically grown as green manure: this effect can be a stimulus to the organic activity in sandy soils, though a single such crop will not produce a lasting effect.
- May replace a *failed crop*. A wise farmer can always think of something at virtually any time of year to occupy redundant land beneficially. Catch crops are often still cheap to grow (broadcast on, cheap seed, perhaps some nitrogen fertiliser to boost yield, e.g. 75–100 kg of N per hectare for stubble turnips).

The case *against* catch crops in the cereal rotation can be summarised thus:

- There may be competition for *resources* needed by the main crop, especially labour and management time. Time of sowing the next cereal may thus be delayed or insufficient time may be available for proper land preparation.
- If *weeds* are troublesome in the cereal rotation, a catch crop simply occupies the time when cleaning operations could be carried out. Catch crops are not usually competitors with the really troublesome weeds.
- The *performance* of the catch crop may be disappointing through insufficient attention, yet it will

TABLE 14.6   **Sheep grazing of cereals in New Zealand**

Cultivar	W = Winter S = Spring	Ear type	Apex height (cm)	Apex length (mm)	Pregraze yield	Residual yield (kg DM/ha)	Apparent intake
Barley							
Illia	W	6-row	0.9	2.2	1640	650	990
Priver	W	2-row	0.9	2.0	2350	960	1390
Gwylan	S	2-row	2.3	3.1	1860	1310	550
Hassan	S	2-row	6.6	4.5	2080	1910	170
Kakapo	S	6-row	1.0	2.5	1500	770	730
Koru	S	2-row	12.2	4.6	2870	1610	1260
Magnum	S	2-row	5.2	3.3	2370	1860	510
Mata	S	2-row	11.2	4.0	2910	2170	740
Triumph	S	2-row	4.4	2.9	2090	1780	310
Zephyr	S	2-row	11.2	4.2	3020	2570	450
Oats	—	—	0.7	1.0	1860	650	1210
Rye	—	—	5.7	5.7	2600	1480	1120
S E mean			0.7	0.1	185	138	233
CV%			25.2	7.1	16.3	18.6	59.3

Notes: Sown 7 April 1983, Lincoln College, Canterbury, NZ and grazed 8–11 August 1983 with 600 ewes/ha. The first 10 cultivars listed are barleys. Cultivars showing pronounced reproductive development were disliked by the sheep; cv Illia gave only 3.5 per cent lower yield of grain after grazing (11.5 t/ha grain) than ungrazed plots.

Source: From W.R. Scott, 1984.

still have complicated the overall management of cropping.

In Surrey 150 years ago the following sequence using catch crops allowed the production of 15 crops in 12 years, apparently with great practical success:

Year 1   Wheat after sainfoin
Year 2   Tares (vetches) cut or fed off very early, and immediately ploughed up for turnips
Year 3   Barley lightly ploughed in (undersown) with ryegrass and trefoil harrowed in
Year 4   Early grazing by ewes and lambs; afterwards dunged
Year 5   Wheat
Year 6   Peas or early beans
Year 7   Forage rye followed by turnips
Year 8   Barley undersown with clover
Year 9   Red clover for hay and grazing
Year 10   Red clover for grazing and later for a seed crop
Year 11   Wheat
Year 12   Forage rape followed by oats or barley with sainfoin

I am not suggesting we all emulate this degree of complexity but offer it as an interesting challenge!

ORGANIC HUSBANDRY FOR CEREALS

The United States Department of Agriculture (USDA) defines organic farming as 'a production system which avoids or largely excludes the use of synthetically compounded fertilisers, pesticides, growth regulators and livestock feed additives. To the maximum extent feasible, organic farming systems rely on crop rotations, crop residues, animal manures, green manures, off-farm organic wastes, mechanical cultivation, mineral-bearing rocks, and aspects of biological pest control to maintain soil productivity and tilth, to supply plant nutrients and to control insects, weeds, and other pests.'

In the United States such methods have been stimulated by concern over pollution by agrochemicals as well as by analyses of the declining energy efficiency (gross output of energy against support energy used) of conventional agriculture, in spite of dramatic yield increases. The case of maize in the United States between 1950 and 1970 illustrates this: whilst labour use was halved, nitrogen fertiliser use increased over sevenfold, insecticide use tenfold, herbicide use twentyfold, crop-drying energy use fourfold and electricity use sixfold whilst transport energy use more than doubled. Overall energy efficiency declined by 11 per cent. Pimental *et al.*

(1983) reckoned organic wheat in the US to yield 4 per cent below conventionally grown crops with 35—47 per cent better energy efficiency but 26—49 per cent lower labour efficiency (i.e. more man-hours needed per tonne).

In Missouri in 1974, Commoner, when comparing sixteen organic with sixteen conventional farms growing maize, wheat, oats and soya beans, found that organic farms used only one-third of the energy input and also had a different cost structure (Table 14.7).

In Britain, the Soil Association was founded in 1946 as a charity 'to promote a fuller understanding of the vital relationship between soil, plant, animal and man' since 'life on earth depends on the soil. To respect and nurture the soil is essential if the quality of life and life itself are to be maintained'. In Europe, organic farming has long been promoted in Switzerland and Germany with Professor Hardy Vogtmann in the chair of alternative agriculture at Kassel University. Movements also exist in France, Holland, Scandinavia and elsewhere. In 1977, the International Federation of Organic Agriculture Movements (IFOAM) held its first international symposium and in the same year, the International Institute of Biological Husbandry (IIBH) was inaugurated in London. Around the same time, the Soil Association instigated discussions to agree standards required for organically grown produce to claim the premium available on the market with a symbol to indicate its acceptability.

It must be emphasised that organic produce must also satisfy normal market requirements on cereal quality, such as Hagberg value and protein content standards for milling wheats.

Procedures

Organic Farmers and Growers Limited (OFG) has been set up to market produce. Their 'pure organic grade' cereals must fulfil the following requirements (inspected by IIBH) to qualify:

- come from a field which has not received agro-chemical-grown crops for at least two previous seasons
- have no soluble fertilisers given except up to 250 kg/ha of Chilean nitrate of soda (which gives 37 kg N and 65 kg Na/ha). This product is justified as being a totally natural salt deposit
- have no seed dressing or synthetic agrochemicals applied to it

On the positive side, such a crop may receive:

- additional N from livestock manures and slurries, hoof-and-horn, dried blood, previous legume crop, composts, sewage sludge and various bacterial/organic preparations (some overpriced and overrated!)
- additional phosphate as micro-ground rock phosphate (Gafsa), basic slag and bone meal
- potash from slurry, wood-ash and Highland potash (Adularia shale)
- lime as ground chalk or limestone, including magnesian limestone (dolomite)
- seaweed preparations as foliar feeds or otherwise. These provide trace elements and can also give a protective effect against pests and diseases
- derris or pyrethrum as insecticides
- sulphur or copper as simple contact fungicides. Wettable sulphur powder is moderately effective against cereal mildew

Establishment can be very difficult in wet, slug-infested seedbeds: 25—30 per cent only is not uncommon. Neither is the need to re-drill.

Effective weed control is perhaps the most difficult aspect for the purist organic farmer in a low-labour economy. Control is attempted by crop rotation, cultivations including shallow (10—15 cm) ploughing and the use of fine-tined weeders, bastard fallows, smothering with undersown green manures such as trefoil, and roguing. Overzealous cultivations before sowing can critically deplete moisture; post-emergence harrowing can damage an advanced crop, whilst the use of minimal cultivations can encourage persistent grass weeds. The use of a suitable weeder can be quite effective (Plate 14.2).

TABLE 14.7 Costs as percentage value of products for organic and conventional cropping

	Fertilisers	Field operations	Manuring	Biocides	Total
Organic	4	7	1	0.2	19
Conventional	10	6	1	3.0	27

Source: Data of B. Commoner, St Louis, Missouri, USA.

PLATE 14.2   Tearaway weeder, available from 1.8 to 6.3m wide. It has four rows of tines and these are spaced 30mm apart to effect full ground cover. The tines have three adjustments — hard pressure, light pressure and a 'lifted' position for row crop work, so that combing can be tailored to individual crop and ground conditions. Each tine unit is spring loaded, and this gives it a vibratory effect and stops material blocking. Tractors in the 50—90hp range are quite adequate to operate the weeder, and suggested land speed is between 5 and 7mph depending on the operation needed. (Courtesy of J.A. Harvey Ltd, Bassingham)

Both grass weeds and disease pressure are eased by delayed sowing of winter cereals — say late October instead of late September — but this is likely to incur a marked yield penalty, especially on droughtier and difficult soils. Sowing in earlier October at conventional densities to get a good competitive crop established, followed by grazing with sheep later on to remove diseased foliage as well as weeds, can be a workable compromise. The need for rotations and the avoidance of high doses of soluble N sources reduce the incidence of grass weeds markedly by contrast with intensive winter cereal cropping. It is also noticeable that the less lush foliage of organic cereals may be host to a wide spectrum of diseases but at tolerable levels. Organically grown cereals at Rothamsted have been found to have fewer aphids and more predators — notably rove beetles, springtails and mites, which comminute the added manures, sewage sludge being especially favourable for this. An example rotation

might be a three year ley, winter wheat, spring oats, spring beans undersown with the ley again.

Prognosis

OFG budgets suggest:

- yields at two-thirds of conventional crops
    Winter cereals 5—6.25 t/ha
    Spring cereals 4.35—5 t/ha
- variable costs at two-thirds of conventional crops
- prices at 40 per cent premium or so over conventional crops (for milling wheats for instance); see Table 15.9

In a 1980 study of five organic cereal farms in England, Hardy (unpublished postgraduate thesis, Royal Agricultural College) found wheat yields down 26 per cent below conventional crops but with an average 5 per cent improvement in gross margins owing to much reduced (minus 73 per cent) variable costs (though he did not allow for extra manure application and cultural weed control costs) and an average premium price of plus 23.5 per cent over conventional grain. In their 1981 survey, Vine and Bateman reckoned organic cereals to yield only 10 per cent below conventional in many cases but up to 40 per cent below when only given low inputs.

Purist organic producers fervently wish to establish in perpetuity an alternative system of cereal growing devoid of agrochemical and soluble fertiliser dependence but based on long-term feeding of the soil, whilst organic consumers are prepared to pay premiums for the absence of synthetic materials in their food. Therefore, standards have to be set and adhered to by both parties. The British government has introduced a UK register of organic food standards (UKROFS).

If the management of organic crops is not radically different from that of conventional cereals, then nutrient supplies to them will be inferior and weeds especially could soon become very competitive. Many British farmers are interested in organic techniques in the face of EC cereal surpluses, as evidenced by the standing-room-only sell-out of an NAC Conference in 1986. It was then reckoned that the organic market was 70 per cent undersupplied for cereals and that demand for whole-cereal and bran products was growing at some 10 per cent per annum.

However, many farmers would be quite happy to compromise and use selected agrochemicals as necessary to maintain control over such problems as grass weeds. Also, there is a grade of produce to be marketed during the two years' transition to fully organic standard for a field as specified by the OFG.

Any talk of compromise is frowned upon by purists but the Guild of Conservation Food Producers has produced contracts for such 'compromise crops', which are to be produced under conditions trending towards those favoured by the pure organic lobby and resulting in cleaner food at some 20 per cent premium for breakfast oats at 70 per cent less chemical input cost. Additional fertilisers such as nitro-chalk, calcium ammonium nitrate and calcium nitrate are allowed. Pyrethroid insecticides may be used against aphids, and some herbicides are permitted (MCPA, MCPB, CMPP, asulam for docks and bracken, and glyphosate, i.e. Roundup, pre-drilling, i.e. after the harvest of the previous crop).

Opponents of organic farming claim that productivity will be inadequate, but cereals are in surplus and there are enough nutrients available if properly recycled (over 2 million tonnes NPK equivalent per year of human and animal excreta alone in the UK).

Management is said to be too complex, and weeds, pests and diseases will eventually take over, many fear — this is a real possibility unless attention to appropriate husbandry is fully given. Many feel the premiums would soon disappear if too many producers adopted organic techniques to meet the current market demand. Against this point is the apparently irrevocable increase in health and diet-consciousness in America and Europe, coupled with growing alarm over pollution from chemical residues. The EC limit for nitrate nitrogen in public water supplies is now 50 mg/l, for instance. As long ago as 1948, Sir John Russell reported a 25 per cent reduction in nitrate nitrogen in drainage water from Broadbalk field at Rothamsted when N was given as FYM at 35 t/ha rather than the equivalent 96 kg/ha nitrate-N fertiliser.

There is a genuine appeal for the vast majority of farmers in a system of farming which stresses the positive interdependence of organisms, including man; which reduces pollution, including fouling from massive doses of slurry or other organic matter dumped on nearby land; and which seeks to promote positive crop health through a vigorous organic cycle rather than piecemeal cures of crop ailments.

It is no longer adequate nor is it acceptable to adopt a 'muck and mystery' or a mystical approach to organic farming. What is needed is controlled management of the organic cycle to maximise photosynthesis, promote decomposition and minimise waste. To achieve this requires:

- skill to manage several integrated enterprises
- patience to wait for cumulative rewards
- willingness to research the appropriate mix of complementary cropping for a particular farm
- a positive approach to crop health
- courage, determination and faith to adopt new techniques

Successful tropical cereal systems offer lessons for the temperate zone in the above respects. Data reported by Spedding and Walsingham (1975) support this point (Table 14.8). However, success

TABLE 14.8 Energy ratios for cereal production (i.e. gross energy in product over support energy input)

Temperate zone		Tropics	
Maize	2.8	Maize (Southern Africa)	40
Wheat	2.2	Rice (Fiji)	20
Oats	2.0	Dryland rice (Sarawak)	34
Barley	1.8		

Source: After Spedding and Walsingham, 1975.

TABLE 14.9 Cereal facts checklist

Soil preference and tolerance

Place in cropping sequence

Main uses/ markets:	*Standards required*
Varieties:	Main groups
Sowing:	Date/rate/depth/method(s)
Manuring:	Basal NPK : other elements?
Population targets:	Seeds/m^2 : % establishment and plants/m^2 : Tillers/m^2 Heads/m^2 : grains/head
Treatments:	Weeds, pests, diseases Growth regulators N top dressing (doses/timing) After cultivations (roll/harrow)
Harvesting:	Readiness: method − snags Stubble cultivations Straw? Burn/bale/chop
Drying:	Options
Storage:	Conditions
Marketing:	Opportunities
Economics:	Yield/price (premium?) ∴ *Gross output* minus *variable costs* equals *gross margin* (deduct *fixed costs*/ha equals '*profit*')

depends on labour intensity and effective nutrient recycling as revealed in Professor F.H. King's 1911 classic about Far East farming ('Farmers of forty centuries'). See Barry Wookey's 1987 book on his experience with organic cereal production in Wiltshire.

CHOOSING THE CEREAL TO GROW

It must be remembered that the relative desirability of growing a particular type of cereal depends on varieties available, markets and individual farm circumstances − all of which are constantly changing. The situation must therefore be regularly reviewed. The cereal facts checklist in Table 14.9 may be used and earlier chapters of this book consulted along with other sources.

Wheat

Common wheat (*Triticum aestivum*) is the chief crop of the UK cereal area. Winter varieties predominate, having increased their share of total wheat hectares from under 90 per cent in the 1960s to over 95 per cent in the 1980s. However, all the winter varieties commonly grown have a very low vernalisation requirement and so can be sown up to February or even March in some cases, though with yield penalties. Spring varieties would be the preferred choice to be sown when soil conditions allow from January/February onwards. In the south, autumn sowing of spring wheat is also practised. Spring wheats have higher quality than winter wheats and the UK is over-supplied with feed wheats but still short of home-grown bread-making wheat. Winter wheat can be typically expected to yield 25 per cent more than spring wheat under comparable soil conditions and standards of husbandry but at some 25 per cent greater variable costs. However, spring wheats typically attract some 10−12 per cent premium on their price over feed wheats and some 3−4 per cent over average milling winter wheat prices (see Wibberley, 1984).

Wheat is the normal first crop after a restorative break crop. This would usually be winter wheat and the intrinsically lower-yielding milling varieties may well be sown as first wheats after the break, high-yielding feed wheats as second wheats and, if a third wheat is being grown, a more disease-resistant and thrifty variety − e.g. Avalon or Mercia, followed by Longbow, and Galahad as a third wheat.

A typical winter wheat crop would be sown using dual-purpose dressed seed in early October at some 400 seeds/m^2 to achieve about 300 plants/m^2. The field would be monitored for slugs to determine whether or not to treat.

Grass weeds, such as blackgrass, would be controlled where necessary just post-emergence of the crop in autumn. Up to 50 kg N/ha may be applied as an early spring dressing at the end of February to March. In late March/early April, a combined treatment may be given for stem-based diseases (e.g. eyespot) if necessary, plus chlormequat to prevent lodging later, plus broad-leaved herbicide. Fungicide may be delayed where possible to GS 37 to aim for only a two-spray total programme. The main nitrogen top dressing may be given during mid-April to early May by which time a foliar disease treatment is often needed. In any season and in any crops with high disease pressure revealed by crop monitoring, a further fungicide may be needed within around five weeks of the first spring spray, making a programme of three foliar sprays against diseases by harvest. Once the ears emerge, there may be a bonus dose of up to 50 kg N/ha given during June to milling varieties, an ear-protectant fungicide would normally be worthwhile and the crop would be monitored during June and into July for the need to control aphids on the ear, which may arise dramatically in some seasons. Specific fields may need spring grass weed control also, e.g. for wild oats, or a later spray of glyphosate just pre-harvest for couch. Harvest would normally occur during August to early September depending on the district of the UK and on the particular season. Winter wheat harvest would normally coincide, more or less, with spring barley, durum wheat, triticale and winter oats. Winter barley would be up to a month earlier and early March-sown spring wheat, up to a month later.

Spring wheats rather than spring barleys logically follow late-harvested vegetables or root crops on heavier land, or can be sown to cheapen weed and disease control in a run of winter wheat crops. Quality is generally high and autumn sowing satisfactory for many varieties.

Durum wheat (*Triticum durum*) is a distinct species really adapted to Mediterranean climates and requiring the better cereal soils, but French varieties are grown in the UK for semolina production to supply the expanding pasta food trade. Major world producers include the USA, Canada, the USSR, Central and South America. In Europe, production is mainly in Spain and Italy, followed by Greece and France. Durum has been sown in the autumn but overwinter survival has sometimes been poor and spring sowing is an option since no vernalisation is required. The crop appears thin-leaved and yields are generally expected to be only two-thirds of those for common wheat and per hectare costs of production are similar if not up to 15 per cent more.

Sowing should take place during October to early November using 500 seeds/m^2 to achieve 400 plants/m^2 established. Good clay loams are favoured, especially in eastern England, and sowing durum wheat as a first cereal after a break crop is desirable. There are herbicide restrictions compared with winter wheats. The crop requires a high level of nitrogen fertiliser per tonne of expected yield (up to 40 kg N), some of this given late. It requires growth regulator to combat lodging and may suffer from pests such as slugs, wheat bulb fly and *Opomyza* since it has rather low tillering ability. Durum is eyespot-susceptible and should be protected during rapid spring growth on high-risk sites.

Attaining quality is all important for this lower-yielding, higher value species of wheat, and most is grown on contract in the UK with specified targets (Table 14.10) and husbandry advice given. It can be

TABLE 14.10 Typical contract requirements for durum quality

Moisture	< 15%
Hagberg	> 220
Specific weight	> 78 kg/hl
Protein	> 12.5%
Vitreous grains*	> 70%
Broken grains	< 4%
Heat-damaged grains	< 0.55%
Common wheat	< 4%
Other cereals	< 3%
Foreign material	< 3%
Shrivelled grain	< 6%

* These must be hard, amber and translucent (not *mitadine*, i.e. opaque and soft)

expected to attain some 5 per cent better specific weight than common wheat under similar conditions.

Barley (*Hordeum sativum*)

Barley crops are reckoned thriftier than wheats, though they are less tolerant of heavier, structurally inferior land. Winter barley only equalled spring barley in the UK at harvest 1986, spring barley having dominated over 95 per cent of the whole barley hectares fifteen years earlier. Spring barley is a very flexible crop, allowing a wide range of sowing dates and thus fitting into a wide range of cropping sequences. However, there are rising yield penalties for crops sown after mid-March and up to early May. Conversely, if frost tilths allow good drilling conditions, sowing as early as January can be beneficial in southern Britain, particularly if followed by a dry spring, which would have hindered crop establishment from March sowings, or a wet one, which would have delayed drilling until late. Many spring barleys also offer the attractive malting premium. Costs of growing spring barley may be 25−30 per cent below winter barley but yields are usually some 15−20 per cent lower and are also much more variable.

Winter barley crops (mainly two-row, as are spring barleys) have become steadily more popular since the adoption of systemic fungicides to control mildew and other diseases from 1970 onwards. Higher, more consistent yields than spring barley arise from the ability to sow from mid to late September/early October at around 350 seeds/m^2 to establish up to 300 plants/m^2 (spring barleys sown in autumn can work but only in milder districts and seasons and with inferior results). The earlier development of roots enables winter barley to better withstand spring/early summer droughts. Its early harvest during July — a month before spring barleys and winter wheats — provides early cash-flow if the crop is immediately sold; early straw removal (clean straw is suitable for feeding or for sale); early land preparation for an early-sown following crop such as winter oilseed rape, grass ley or more winter barley; early catch crops to follow such as stubble turnips or forage rape; time for ample sub-soiling or weed control before the next autumn season.

Winter barley produces such a dense leaf canopy in autumn that it can smother out weeds with the help of autumn grass weed control on infested fields and autumn broadleaved weed control alone (where wheat needs spring control of broadleaved weeds). Aphids capable of carrying BYDV may be more likely to need control on winter barleys than on winter wheats. Disease damage may also arise earlier and broad-spectrum seed-dressing use has been commoner than on wheats. Most treatments to the crop will occur just prior to those on winter wheats. The crop is more likely to suffer from late lodging and ear loss than wheat and should be protected accordingly, e.g. with ethephon.

The total fertiliser input to winter barley is likely to be less than for winter wheat since the yield potential is lower, both of the crop and of its customary lighter land sites. It is easy to spend as much per hectare on protecting winter barley as on wheat and the economics of so doing are seldom justified, except for malting or seed crops set to yield well.

Newer, bold-grained varieties of the six-row winter barleys such as Gerbel and Plaisant have increased its popularity but it still accounts for only some 5 per cent of barley. Six-row varieties yield somewhat better than two-row varieties — 5 to 10 per cent — but have traditionally suffered an alarmingly high proportion of small grains leading to low specific weights and high incidence of volunteers left on the field to provide a 'green bridge' for diseases onto the next crop.

The chief advantage of winter barley in a cropping sequence arises from the earliness of its harvest at a time when staff and equipment are not tied up with other crops except perhaps winter oilseed rape.

Oats (*Avena sativa*)

Oats once occupied half the UK cereal area — they now account for under 3 per cent. They have not been eligible for intervention buying in the EC so prices have fluctuated more than for the other cereals in direct relationship between supply and demand. There can be variable, sometimes attractive, premiums for feed oats, which account for 60 per cent of oat demand. Milling oats typically attract a 10–15 per cent premium over feed oats off the combine, and contracts are available. The UK has not been meeting home demand for oats, leaving Germany as the leading producer in the EC, whilst Russia produces over one-third of the world oat crop.

Oats continue to be taller than wheats and barleys with an inferior harvest index. Unfortunately, dwarfness in the oat crop is associated with compact heads and lower yields. However, tallness often leads to serious lodging in diseased crops, especially in the east of Britain. This leads to reduced yields, difficult combining, increased weediness and extra drying costs.

Winter varieties are preferred over spring oats in the Midlands and southern Britain, owing to their superior kernel content, resistance to stem nematode, lower susceptibility to frit fly, better drought tolerance and earlier maturity. However, oats are not winter hardy so spring varieties are more common further north. Overall, sowings are often 50:50 between spring and winter crops now in the UK.

Any oat crop is a camouflaging hazard on a cereal farm which is attempting to control wild oats or where considerable sums have been spent for years to reduce the weed following the UK National Wild Oat Year in 1974.

As a minor combinable alternative crop to wheat and barley, it is more straightforward to handle than some of the exotic species being tried by UK farmers. In long runs of either barley or wheat, it can yield well as a break and may form part of a successful rotation of combinable crops, e.g. winter oilseed rape/winter wheat/winter oats/winter barley. It is less at risk from eyespot or take-all but cyst nematode susceptibility could be a problem, especially with spring oats.

Winter oats should be sown from early to mid-October at 450 seeds/m^2 to give some 350 plants/m^2 and 750 panicles/m^2. There are restrictions on the residual herbicides which can be used on oats. They typically respond to about 18 kg N/t of expected yield. They usually receive chlormequat at GS 32, coupled with a fungicide such as tridemorph for mildew. A further fungicide may not be justified for average crops unless crown-rust develops seriously on susceptible varieties, though some growers give sulphur post-flowering to improve acceptability for milling as well as keeping late mildew at bay.

Spring oats should be sown as early as possible,

preferably during February (March in the north) and at 550–600 seeds/m^2 to give about 500 plants/m^2 established. Where frit fly is a risk on later-sown crops or crops sown too soon after ploughed-out grass (within less than a month to six weeks), then crops should be carefully monitored and given sprays of chlorpyriphos or triazophos. Such treatment may protect up to one-third of the yield in frit fly situations. Oats are also very susceptible to BYDV, turning purple-pink. The spring naked oat (e.g. cv Rhiannon, developed by the Welsh Plant Breeding Station) has the top overall concentration of nutrients of any cereal with 8.5 per cent oil, ME (metabolisable energy) of 15, CP of 14 per cent (4 per cent of this being lysine), but it is dusty to handle and low-yielding, though attracting price premiums up to 40 per cent. Kynon is a winter naked oat.

All oat crops are liable to shedding when ripe and tend to be slow to harvest.

Rye (*Secale cereale*)

Only autumn-sown varieties such as Animo and Dominant are used in the UK for grain production. Forage varieties sown for early-bite spring grazing are not suitable for grain production since they give low seed yields. However, early-sown grain varieties can be winter grazed and still taken for grain later.

Rye's reputation for moderate resistance to such common problems as drought, acidity and rabbit-grazing does not mean a lack of response to better conditions. Crispbread contracts are available, as are limited opportunities for thatching straw from binder-cut crops.

Rye should be sown from mid to late September at up to 600 seeds/m^2 on better soils since it tillers poorly and some 700 ears/m^2 are ultimately sought. However, on sands at Gleadthorpe EHF in Nottinghamshire, much lower seed rates improved lodging resistance without incurring yield penalties. It typically responds to some 20 kg N/t of expected yield. There are herbicide restrictions. All three plant growth regulators can be used and a two-split spring dosage is usually wise on this tall crop. The crop is susceptible to slugs, wheat bulb fly, take-all, eyespot and ergot.

Harvest should take place at 20 per cent grain moisture content, i.e. before sprouting in the ear occurs. Rye is quite prone to this problem which leads to reduced Hagbergs that may mean loss of milling quality.

Like oats, rye should be considered by the wheat and barley grower seeking an alternative combinable crop before turning to some of the more exotic minor crops.

Triticale

Tetraploid (4x) durum wheat is the usual mother of this hybrid, whilst diploid (2x) rye is the father. Artificial doubling of the chromosome number of the resulting sterile hybrid using colchicine gives hexaploid (6x) triticale.

The crop has been recently developed at the University of Manitoba, Canada, and at the International Maize and Wheat Improvement Centre (CIMMYT) in Mexico. In Ethiopia, it has shown suitability for poor, sandy and acidic soils whilst its superior resistance to leaf stripe and stem rust has given it double the yield expectation of established wheat varieties in Kenya. The use of triticale flour as a substitute for wheat in doughnuts was undetectable in a consumer survey by Egerton College in Kenya.

Polish varieties, Lasko and Salvo, have spearheaded its adoption in the UK as an autumn-sown crop. Many diverse varieties exist. It should be sown from late September to October at some 300–350 seeds/m^2 to give 250 plants/m^2. There is no vernalisation requirement. There are herbicide restrictions. Like rye, it is tall (often 1.25m or so) and weak-strawed, but all three plant growth regulators can be used (split dosed at GS 30 and GS 32 for chlormequat, with a later ethephon if necessary). N response is not unlike feed winter wheat, i.e. up to 25 kg N/t expected yield where lodging is controlled and in a predominantly cereal rotation. However, it is often sown on poorer land where it can only respond to lower N rates. It resists mildew and gets less take-all but can succumb to eyespot and ergot. It should need only a single fungicide treatment.

It combines wheat yield potential, including better tillering, with rye hardiness for land of 7.5 t/ha wheat potential and below. The crop grows profusely once it starts in spring, and the bulk of straw slows harvesting. Its ears, which generally look more promising than they prove to be, often bear 36–40 grains per ear in 22–24 spikelets.

Contracts are available, including those for special human consumption markets. Yields are somewhat variable, ranging from 3–9 t/ha but often 4.5 to 7.5 t/ha. There is no official intervention. Its appearance is like crumpled to normal wheat grains, depending on variety and sample. It is often around 72 kg/hl specific weight. Triticale has potential as a substitute for both feed and 'quality' wheats and barleys since it can be milled for bread and cake flour as well as malted, so it is a crop to watch with interest.

Chapter 15

CEREAL ECONOMICS

The term 'economics' is derived from the Greek 'oikos' meaning home and 'nomos' meaning law. As far as cereals are concerned, the 'home' may be one field, one farming business or the national crop of a particular country and the 'law' may be applied in the short term to one season's production or as far as the long term. Both short-term results and long-term trends are significant since the account books must balance so that outputs exceed or at least equal inputs. This applies whether the farmer seeks profits or merely to survive in business.

The question arises as to what units should be used to measure inputs and outputs. In a cash economy, money clearly has to be one of these units but it is only a relative term subject to such variables as inflation, differing exchange rates and alteration between comparative costs of inputs and outputs. Therefore, in addition to money, a more absolute unit of measurement is needed to keep a watching brief, especially on the longer term sustainability of cereal-producing systems. Such a unit is energy. It is possible to calculate the total energy inputs needed of solar radiation plus the energy involvement of manufacture, transport and application of all materials to the field and of all equipment used to grow, market and process a cereal to the threshold of consumption. These can be expressed per hectare (or per acre!) against the energy content of the usable yield of the crop. All cereal producing systems anywhere in the world can be validly compared using the criterion of energy input/energy output per hectare or per tonne produced. This is obviously not something the average individual producer can be expected to calculate but the strategist and thinking farmer will want to reflect on it. It

raises the frequent dilemma of the possible convergence of short-term interests of the individual producer and longer-term interests of the nation. There is an inescapable inter-relationship between:

- economy
- energy-efficiency
- employment
- environment/ecology
- ethics

The most energy-efficient cereal-growing systems in the world also happen to have the highest direct-labour involvement. These are well-managed farms found notably in parts of the Far East, including China. The FAO are promoting such systems elsewhere (Figure 15.1). I am not, of course, claiming that all labour-intensive farms are well-managed! However, the higher yield, low-labour systems of North American and West European cereal production (emulated in parts of Africa, Latin America, India and Oceania) have succeeded in physical and financial terms at the *expense* of energy-efficiency (Figure 15.2) and opportunities for direct work in agriculture, for rural residency with a real job and subsequently for the size of the rural versus urban voice in matters of public policy. The decline in direct employment per hectare or per tonne of cereal produced in the world has been paralleled by a growing urbanisation crisis coupled with serious environmental questions. Fewer farmers left in the countryside shoulder ever-bigger responsibilities for environmental management. Whilst it must be said that the majority of UK farmers do an excellent job in this respect, there is a growing communication gap between those who

Haiti has adequate water and good growing conditions for rice, but agricultural land is at a premium. FAO is helping introduce a continuous growing system — called rice gardens — which gives exceptionally high yields per unit area.

Farmers each divide their plots into 14 equal sections. Instead of planting the entire plot at once, they plant the sections at one-week intervals. Since the rice takes approximately 90 days to mature, by the time the last section is planted the first is ready to harvest and then plant again. As a result, input and labour requirements are distributed throughout the year, eliminating the traditional peak demands at planting and harvest. Using the rice garden system, a family of four workers can easily manage a one-hectare plot. In addition, the farmer obtains a steady, weekly income.

The rice garden system also limits crop damage caused by poor weather or pest infestation, because only a small portion of the crop matures at any given time. Annual yields are averaging over 22 tonnes per hectare, more than three times those of conventional rice farming.

FAO is improving the rice gardens even further by turning them into integrated farming systems. After each section is harvested, it is left bare for a week before being replanted. During that time, ducks are released on the section to feed on the rice that remains after harvesting and also on animal pests. In addition, they provide a source of natural fertiliser. At the end of the week, the ducks are simply shifted to the next section.

Finally, the rice straw is being used as a base for mushroom farming, giving the farmer yet another potential source of cash income.

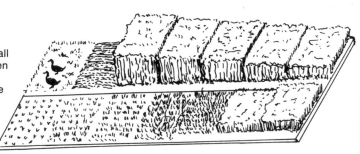

FIGURE 15.1   Rice gardens in Haiti. (FAO, 1985)

live and work in the countryside and the increasingly vociferous ecological lobby — some of it well informed and some seemingly devoid of rationality.

The ethical argument must begin with reference to God, who creates and sustains the framework within which man is responsible to manage cereal production, i.e. for its economy. Then, I contend that the most ethical system in practice (man being so liable to fail in his side of the contract!) is that in which individual and national interests most nearly coincide. In the UK, the landlord/tenant system of land tenure has traditionally balanced the long-term with the short-term considerations in land use for cereals. The family farm is recognisable as the ideal unit in which individual and national interests meet in many parts of the world. A relatively high, though not excessive, population directly employed in cereal growing increases the

people directly available to maintain the environment as a useful and aesthetic asset for their own daily life and for future generations.

If one agrees in any measure at all with the above ideals and the objectives of agriculture stated at the beginning of this book, how then can we more closely attain such a utopia? Firstly, I suggest, by a conscious decision that we want to trend in that direction. Secondly, by seeking more integrated farming systems (see Chapter 14) and, thirdly, by seeking and welcoming any willingness to pay higher prices for cereals relative to non-food goods. In practice, this needs to be backed up by some production control system once national surplus of the cereal has been reached. Price control is a crude implement with which to seek to influence production but it is an index of value relative to other goods. It can be argued that when a society has placed too low a value on primary production of

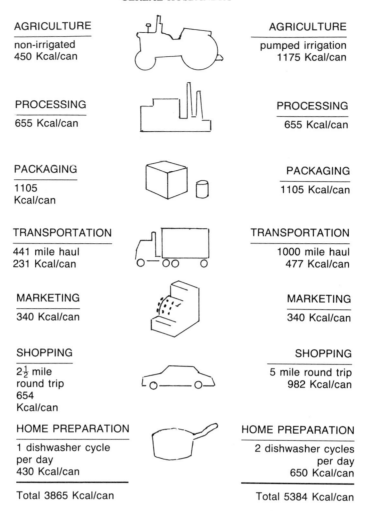

AGRICULTURE

non-irrigated
450 Kcal/can

AGRICULTURE

pumped irrigation
1175 Kcal/can

PROCESSING

655 Kcal/can

PROCESSING

655 Kcal/can

PACKAGING

1105
Kcal/can

PACKAGING

1105 Kcal/can

TRANSPORTATION

441 mile haul
231 Kcal/can

TRANSPORTATION

1000 mile haul
477 Kcal/can

MARKETING

340 Kcal/can

MARKETING

340 Kcal/can

SHOPPING

$2\frac{1}{2}$ mile
round trip
654
Kcal/can

SHOPPING

5 mile round trip
982 Kcal/can

HOME PREPARATION

1 dishwasher cycle
per day
430 Kcal/can

HOME PREPARATION

2 dishwasher cycles
per day
650 Kcal/can

Total 3865 Kcal/can

Total 5384 Kcal/can

FIGURE 15.2   Summary of energy inputs (Kcal) to a one pound can of whole kernel corn. The left column shows energy inputs based on rather conservative assumptions. The right column shows how these energy inputs can vary depending on the assumptions made. The digestible food energy of the corn is 269 Kcal. (Source: Data of Watt and Merrill, 1973, quoted by Brown and Batty, Transactions of the ASAE, Vol. 19, No. 4, 1976)

food or fuel crops, then cheap, often short-cut methods of production have resulted in some well-documented, serious, long-term consequences involving soil erosion as a prelude to the decline of that civilisation.

FACTORS AFFECTING CEREAL PRODUCTION LEVELS

Production (tonnes) = *area* (hectares) × *yield* (tonnes per hectare)

In a cash economy, financial return is the key factor in determining the amount that will actually be produced. If a totally free market is allowed to operate, demand will interact with supply to determine price, and production will continue until marginal cost equals marginal revenue (Figure 15.3), limited only physically by the land-holding's capacity to produce.

The level of maximum financial efficiency for any one holding will be the point where marginal

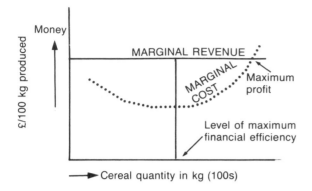

FIGURE 15.3   Financial limits to cereal production.

cost/tonne is lowest. Up to this point there will be *economies of scale* by increasing production.

However, cereal production is subject to the vagaries of climate so it is not possible to predict supply precisely for any one season or region. Nevertheless, despite great local fluctuations, world wheat production is reckoned to show an average annual coefficient of variation of only between 5 and 10 per cent. In spite of this no government wishes to risk the spectre of cereal shortages in view of their dominant role in diets and trade. Therefore, varying degrees of interference are implemented to regulate the operation of free market forces. Furthermore, the scale of the EC market involved in grain is now so huge that its behaviour more closely resembles an unwieldy super-tanker than a manoeuvrable rowing-boat.

There will be diminishing returns to further inputs and maximum financial output may occur before maximum physical output is reached.

THE COMMON AGRICULTURAL POLICY (CAP)

The CAP of the European Community has been largely based on cereals and so the cereal market has been highly organised and regulated. The objectives of the CAP as stated by the treaty of Rome in 1957 included:

* increasing agricultural productivity
* ensuring a fair standard of living for the agricultural community
* stabilising markets
* guaranteeing regular supplies
* ensuring reasonable prices in supplies to consumers

These objectives have been pursued by means of four CAP principles:

1. common organisation of the market with common prices for main products
2. free trade between member countries
3. a single trading system with non-member countries
4. joint financing of market support costs, subsidised exports to non-members and agricultural modernisation of members

The basic philosophy of the CAP for cereals was that the producer could earn his returns from the market itself rather than being directly subsidised by the taxpayer as formerly with the UK cereal deficiency payments given when average market prices for grain fell below an annually predetermined price-review level.

The system has accordingly involved the creation of a protected market with:

* import regulations for grain from non-members of the EC which have had to pay a levy into the common agricultural fund for any grain sold within the EC
* export subsidies from the common agricultural fund to EC producers who sell their cereals outside the EC at the lower grain prices assumed to be operating on world markets
* intervention buying into medium-term storage when the average market price for EC-produced grain has fallen below a predetermined intervention price. Intervention buying only applies to certain types of grain in quantities and at quality standards subject to revision by the EC (Appendix 3)

The protected market has been designed to seek an inter-relationship of cereal prices in favour of EC producers as shown in Figure 15.4.

The policy had by 1986 so encouraged EC cereal growers that surpluses of grain were widely agreed to have reached embarrassing proportions (Figure 15.5). However, these represented under 5 per cent of the approximately 400 million tonnes of grain then stockpiled in the world — nearly three-quarters of it in the United States. In the early 1970s the world had only about one week's grain reserve in stock.

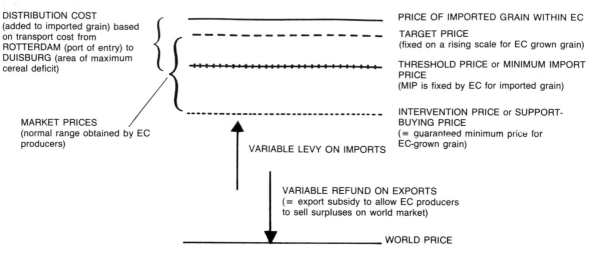

DISTRIBUTION COST
(added to imported grain) based
on transport cost from
ROTTERDAM (port of entry) to
DUISBURG (area of maximum
cereal deficit)

PRICE OF IMPORTED GRAIN WITHIN EC

TARGET PRICE
(fixed on a rising scale for EC grown grain)

THRESHOLD PRICE or MINIMUM IMPORT
PRICE
(MIP is fixed by EC for imported grain)

INTERVENTION PRICE or SUPPORT-
BUYING PRICE
(≡ guaranteed minimum price for
EC-grown grain)

MARKET PRICES
(normal range obtained by EC
producers)

VARIABLE LEVY ON IMPORTS

VARIABLE REFUND ON EXPORTS
(≡ export subsidy to allow EC producers
to sell surpluses on world market)

WORLD PRICE

FIGURE 15.4   EC cereal price scale.

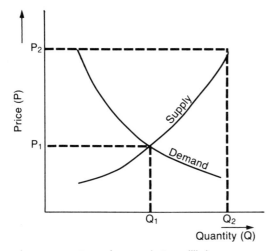

Notes: (a) P_1Q_1 = free market equilibrium
       (b) P_2Q_2 = disequilibrium created by CAP
           intervention system
       (c) Future supply/demand is calculated by
           economists using regression analysis

FIGURE 15.5   Supply and demand for cereals.

Means of Promoting Cereal Production

Production policies such as those outlined above for the CAP have proved effective, especially favouring larger producers. These have also applied for combinable non-cereal alternative crops like oilseed rape (which has price support) and peas and beans (price support to merchants using these as substitutes for imported soyabean protein in compound feedingstuffs). A protected (indirectly subsidised) market for cereals existed in a buoyant state and was largely unquestioned until the early 1980s.

The *area* grown to cereals on any one farm has been unrestricted.

Yields have increased steadily, trebling in the UK between 1945 and 1985 owing to two principal developments: the use of new varieties, especially winter-sown, and the adoption of novel inputs and use of generally higher input levels.

Means of Restricting Cereal Production

In the case of cereal oversupply, production policy can seek to deal directly with it, area grown can be limited, yield per hectare can be reduced or, in commonsense strategy, a combination of some of these approaches can be adopted voluntarily but with substantial statutory help!

Production policies
If price alone were allowed to slump suddenly from an artificially protected level to world grain price, then many smaller producers and heavily borrowed farmers would be bankrupted, which would be socially damaging to the rural economy. On the other hand, the number of producers could be reduced by a system of 'golden handshakes' coupled with a long-deferred, if any, return to cereal production.

Specific export subsidies can be given or market access can be permanently denied to surplus grain lots by denaturing (deformation of quality).

Stabilisers have been introduced by the EC whereby price is automatically reduced by an increased co-responsibility levy (equivalent to a tax on overproduction) if production exceeds a predetermined level (currently about 160 mt/year for the EC 12).

Production quotas could be introduced whereby each cereal grower would receive price support only on a fixed tonnage of cereal output — such a system can obviously be designed to favour smaller as against larger cereal producers. Such a policy is unfavourable to UK producers because on average UK cereal farmers produce over 100 acres (40 ha) each in contrast with only around 20 acres (8 ha) in Germany and below this average still for the whole EC.

Production policies could also be designed to favour *alternative crops* including linseed and other oilseeds to occupy cereal land. Finally, widening the differential revenues attainable between premium *quality* and average grain can be used to restrict the proportion eligible for top price achievement.

Area
The area to be grown to cereals can be restricted by means of *quotas* based on a percentage reduction on the mean area grown during the previous, say, five years. This policy can be coupled with an active *set-aside* policy whereby alternative uses of land are required and supported by the nation (see 'A Digression on Set-aside').

For instance, cereals and trees have been integrated in the past on a limited scale, e.g. wheat and matchstick poplars, and timber occupies under 10 per cent of UK land against 25 per cent in the twelve EC countries and one-third of the land in the world as a whole. The UK imported some 90 per cent of her timber requirements in 1985/6. Whether by more agroforestry or more straight farm woodlands, there is a need for the timber supply to be increased and cereal supply to be decreased in the UK and in the EC as a whole.

Yields
Spring cereals incur similar or lower costs per tonne to produce than their winter-sown counterparts but on average produce some 15−25 per cent lower yields. Top-quality breadmaking wheat is more readily obtainable from spring than from winter varieties and good malting samples are derived more from spring than from winter barleys. Voluntary planting of more spring corn could reduce an individual farm's costs by cheapening grass weed control amongst other things as well as reducing the bulk to dry, store and sell. It has been suggested that a statutory requirement not to sow before specified spring dates could be checked by annual hectare returns and policed by satellite surveillance using equipment already in place.

Yields are lowered by *reducing inputs* but this is unlikely to happen voluntarily unless either input costs rise dramatically relative to output value or there is a premium market for non-use of certain inputs as with organically grown cereals. Statutory reduction of inputs might be achieved by:

- quota imposition on key inputs such as nitrogen fertilisers (administratively cumbersome)
- legislation to ban or severely restrict the use of certain inputs as perhaps presaged by the new control of pesticides regulations introduced in Britain in 1986
- taxation of key inputs — again perhaps nitrogen fertilisers — to make them disproportionately and prohibitively costly to use at yield-maximising rates. However in 1988 even with a 25 per cent drop in cereal prices and a 50 per cent increase in nitrogen fertiliser cost by taxation, the economic optimum rate for winter wheat would only be down some 10 per cent or around 15−20 kg N/ha (scarcely within the accuracy of many fertiliser-spreaders!), so any such taxation policy would need to be at a much higher rate

Finally, farmers can operate with a lower yield per hectare of cereals by *adding value* to their product. This can be done by growing:

- premium quality to supply traditional markets (10−15 per cent extra price for milling over feed wheats; 15−20 per cent extra for malting over feed barleys; some 20−25 per cent extra for seed crops)
- special quality to supply particular markets (currently an extra 25−30 per cent or more is available on ex-farm price per tonne for organically grown cereals; special markets exist for such commodities as oats for racehorses and durum wheat for the expanding pasta-food market)

A Digression on Set-aside

In 1987 the EC required member governments to propose some sort of scheme to reduce cereal surpluses, and the UK government put forward an extensification scheme for cereals in December 1987.

SUMMARY OF THE EXTENSIFICATION SCHEME

Proposals

1. Largely by area set-aside reducing the cereal area by at least 20% as at entry.
2. Divert to fallow (possibly rotational), afforestation, non-agricultural uses or a mixture of these.
3. Voluntary participation; 5-year schemes plus; payments (25% from EC sources)

Aims

1. Free some land from intensive agriculture to put into trees, conservation, amenity and new farm enterprises.
2. Change the pattern of land use to better accommodate wildlife, landscape and environmental interests generally.
3. Adopt rules that are simple and cost-effective.

Payments

1. Annual in arrears, probably with no regional differentiation.
2. *Either* single flat rate of £150–200/ha *or* farmers could be invited to tender rates they would accept to take specified hectarages out and EC would accept from lowest upwards to get its required quota.

Discussion of options

1. Fallowing (a) Code of practice for management (by NCC and Countryside Commission perhaps?)
                 (b) No fertilisers; no agricultural production
                 (c) Whole or part fields allowed, e.g. 6m borders, permanent or rotational.
2. Forestry – link to existing Farm Woodland Scheme or devise another one?
3. Non-agricultural uses – *not* development but tourism, wildlife, sporting, horse-based enterprises (not meat), non-agricultural crops such as coppiced willow.
4. Alternatives to removing land from cereal cultivation were mentioned, including the possibility of incentive payments to produce organically which would lower production.

SOME VIEWS ON THE SCHEME

1. Set-aside should be a temporary measure since the public would not accept indefinite funding of farmers to do nothing. Indeed positive alternative land uses and creation of rural employment should be favoured.
2. The scheme should be voluntary (though the NFU has advocated compulsion).
3. £200/ha is totally inadequate to attract enough people to the scheme. £300/ha might attract some marginal land from cereals but a higher figure would be needed to attract better land.
4. A sealed tender system would be preferable. The initial figure tendered should be adjusted annually for inflation, i.e. index-linked to the retail price index.
5. The choice of fallow system should give as much freedom to the landowner as possible. Opportunity to increase sporting value is important.
6. Guidance as to how set-aside land should be managed should be provided by MAFF through ADAS. Interference by outside pressure groups or other government departments would not be acceptable to farmers.
7. The woodland scheme should go hand in hand with the set-aside scheme to further encourage the planting of trees on arable land. Indeed, more generous support for mixed woodland/agroforestry/biomass might go a long way towards achieving the EC aims.
8. Alternatives to set-aside
    (a) The introduction of support for organic farming is an indirect way of limiting overall nitrogen use and would attract some farmers, but beware of obliterating retail organic premiums.
    (b) Consider promoting spring cereals with 'sow after' dates.
    (c) Some conservation activities will still require additional incentive schemes.
9. One of the principal aims of any policy should be to arrest decline in the rural working population and indeed to reverse it.
10. Strategically sensible reserves must be kept – remember Joseph in Egypt and 'Chernobyl' in Russia.

POSTSCRIPT

A voluntary UK scheme was introduced for which farmers had to register by the end of October 1988 whether or not they took up the option later. See *Setaside*, the MAFF rule book on the subject (20 pp, 1988).

● supplies sold with limited additional processing such as grinding or rolling and small-scale packaging. Such additions can readily treble or more the ex-farm price by contrast with twenty-tonne loads of grain

Of all the voluntary measures, every cereal farmer should welcome the growing market for organically grown cereals and for whole-wheat bread as a means of limiting cereal supplies. Organic methods rely on nutrients from sources other than soluble mineral fertilisers and thus necessitate renewed integration of livestock into cereal-growing areas. Moreover, dependence on rotational/cultural practices to control weeds rather than by herbicides further enforces a dilution of the cereal hectarage on the organic producer. Add to this the expectation of yields at two-thirds to three-quarters of those achievable with more chemical inputs and there is a formula for reducing cereal surpluses. At the same time the consumer is spending an increasing proportion of disposable income on basic food, so indicating a reversal of the trend for society to place too low a value on primary production goods. The farmer meanwhile is able to indulge in more environmentally friendly production systems (which must incidentally provide more farm work if husbandry standards are not to deteriorate against the easier nutrient replenishment and field hygiene achievable with more chemical inputs).

CEREAL FIELD ECONOMICS

The prudent farmer will want to cost not only each cereal species on the farm but each variety and probably also each field of that variety.

Total costs can be conveniently split into the now commonly used *variable* and *fixed* costs associated with cereal growing. The *variable costs* are those which can be specifically allocated to the cereal enterprise and which vary directly with the number of hectares grown, i.e. seed, fertilisers, sprays. The deduction of these variable costs from the gross output from sales of all produce per hectare (grain, plus straw if this is also sold) gives the *gross margin* (£/hectare) for each cereal, variety or field (depending on how carefully each field is separately costed). The gross margin gives a rough guide to the relative performance of different crops (Table 15.1). Gross margin thus depends on yield of grain,

TABLE 15.1 Leading UK arable crops — typical gross margins (£/ha) for good crops

	1977–8	1981–2	1985–6	1988
Winter wheat	400	600	700	500
Spring barley	330	400	420	350
Oilseed rape	340	550	800	480
Maincrop potatoes	1200	1200	1200	1300
Sugar beet	700	800	850	800

selling price of grain and variable costs of production.

These three in turn depend upon season, markets and cereal management.

Yield

If considerable effort is to be devoted to the growing and financial monitoring of the cereal crop, it is logical that yield per field should be measurable fairly accurately. Yet it seems that many otherwise carefully costed farms continue to rely on 'guestimation' for yield or do not even bother to assess treatment, variety or field performance separately. It is true that grain quantities can be calculated from volumes stored, corrected even by sample hectolitre weighing in some cases. However, direct weighing of grain is possible on combine weigh-meters (Plate 15.1), trailers, weighbridges or associated with the drying and storage equipment at the barn.

Volumetric calculations are likely to prove ±10 per cent accurate in many cases, though some would claim greater consistency of precision according to the effort made. Weighbridges are perhaps the most accurate at ±2 per cent or less. Yields must be corrected to a standard moisture content (usually 15 per cent, perhaps 14 per cent now); for instance, 7.5 t/ha at 18 per cent mc corrects to about 7.2 t/ha at 15 per cent moisture, whilst 7.5 t/ha at 12 per cent mc corrects to about 7.8 t/ha. For volumetric calculations, it is necessary to correct for bushel weight or hectolitre weight since cereals vary typically as follows:

Cereal	*Hectolitre weight range (kg/hl)*
Oats	< 50–63
Six-row barley	< 58–69
Two-row barley	< 63–73
Wheat	< 70–84

PLATE 15.1   On-combine weigher.

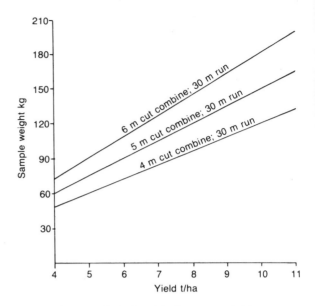

FIGURE 15.6   Relationship between yield per hectare and sample weight for trial or tramline treatment assessments.

Sample measurements can also be taken from field trials or tramline-treatment trials using the normal farm combine but discharging measured runs into separate containers, even within the normal trailer loads, for later assessment back at the farmyard. Weights to be expected from typical sample cuts are indicated in Figure 15.6. Samples are usually replicated three to six times for accuracy.

Since it costs typically around the same amount to produce a low and a high yield, gross margin can be shown to vary sensitively with yield and price (Table 15.2). Break-even yields need to be carefully calculated with each season's cost structure, covering *total* costs of cereal production. They are likely to be around 6 t/ha for spring cereals and over 6.75 for winter cereals in the UK now when selling at feed prices.

Yield responses of all factors need to be known (Table 15.3).

Price

The selling price clearly exerts a powerful influence on cereal economics. It is especially and increasingly influenced by the quality of the sample for

TABLE 15.2   Sensitivity of spring cereal gross margin to yield and price, assuming variable costs of £160/ha

Price (£/t)	Yield (t/ha)									
	3.0	3.5	4.0	4.5	5.0	5.5	6.0	6.5	7.0	7.5
90	110	115	200	245	290	335	380	425	470	515
100	140	190	240	290	340	390	440	490	540	590
110	170	225	280	335	390	445	500	555	610	665
120	200	260	320	380	440	500	560	620	680	740
130	230	295	360	425	490	550	620	685	750	815
140	260	330	400	470	540	600	680	750	820	890

TABLE 15.3 Rothamsted's multifactorial trials on winter wheat. Yield response (t/ha) to the main treatments 1979–81

	Rothamsted (clay loam)	Woburn (sandy loam)
Overall mean	9.2	8.2
Benefit of		
Early sowing	+0.45	+1.02
Extra nitrogen	−0.16	+0.18
Divided nitrogen*	+0.13	+0.36
Early times nitrogen	−0.31	+0.16
Irrigation	−0.25	+0.96
Autumn pesticide	+0.15	+0.72
Summer aphicide	+0.46 ⎱	not tested
Fungicide	+1.14 ⎰	applied basally
Best 8 plot mean	10.5	9.2
Given by	Early sowing	Early sowing
	Later timed N	Irrigation
	Aphicide	Aphicide
	Fungicide	Fungicide

* Split N dressings

Source: AFRC.

particular market requirements. This is characteristic of an oversupplied market which can afford to be selective and to specify high-quality standards. Wheat prices in average market terms and real terms have varied as shown in Table 15.4, which also details costs and margins (see also Figure 15.7). Attention to detailed husbandry of each field and treatment is paramount in securing premium-quality prices (Table 15.5).

Variable Production Costs

Costs need to be set against the yield response expected from each and therefore costed in grain equivalents. It is important to work out costs per tonne to produce each cereal on the farm (e.g. Table 15.6). Careful crop monitoring and attention to husbandry detail can avoid unnecessary insurance treatments:

- Seed accounts for <20->30 per cent of all variable costs.
- Fertilisers account for <45->50 per cent of all variable costs. Considerable economies can be made by judicious purchasing of fertilisers. For instance, the cost per kilogram of nitrogen can vary by a factor of as much as two depending on the type and source of material used.

- Sprays account for <25->35 per cent of all variable costs. Clearly organic producers make considerable savings here but these are probably somewhat offset by extra labour costs elsewhere for them. Some spray costs, e.g. for wild oat control, are not legitimately chargeable only to that crop in that field in that season because they pose a long-term threat to the entire cereal enterprise and so should be costed accordingly.

Gross margins per se also fall down in that contract machinery and casual labour are chargeable per hectare whilst regular farm labour is not easy to allocate per hectare, except on all-cereal farms, and so is not included in gross margin calculations. Gross margins should therefore be used and quoted with due qualification. As a quick, comparative index between crops and enterprises they are useful, provided that their limitations are recognised. Gross margin/tonne of yield can be used as an approximate 'efficiency' index.

Fixed Costs

Fixed costs normally amount to around double the variable costs for winter cereals on an average-sized mainly cereal-growing farm in the UK and almost treble those for spring cereals. The chief elements are machinery (including fixed equipment in granaries, etc.), labour and rent or rent equivalent. In addition, there is a share of general farm overhead costs. A disturbing trend in the UK between 1977 and 1987 has been the rate of increase in real terms in bank-borrowing by farmers to quite alarming levels in some cases. It is useful to calculate output per £100 of labour and per £100 of machinery used and compare this with published farm data.

Fixed costs too can be helpfully expressed per tonne produced and savings here on cultivations must be pondered. It is wise to be aware in costing that not all valuable things or activities on a farm can be reliably costed but this does not exclude them from a true assessment of worth, often over the long-term and sometimes aesthetic as well as utilitarian.

Profitability

Cereal profitability has varied enormously and in recent years a top performance enterprise might have expected total profits to approach total costs, i.e. gross output to be nearly double total costs or more.

Table 15.4 Winter wheat: output, variable costs and gross margin in current and real terms, 1970/71—1985/86

Year	Yield t/ha	Price Current	Price Real	Gross output Current	Gross output Real	Seed cost Current	Seed cost Real	Fertiliser cost Current	Fertiliser cost Real	Spray cost Current	Spray cost Real
		£ per tonne								£ per hectare	
1970/71	4.11	30.9	163.1	127.0	670.5	8.2	43.3	10.4	54.9	3.5	18.5
71/72	4.66	31.2	149.2	145.5	695.8	8.6	41.1	11.4	54.5	4.4	21.0
72/73	4.50	35.4	161.2	159.4	725.9	8.9	40.5	13.8	62.8	6.2	28.2
73/74	4.31	59.2	193.2	255.0	832.2	9.9	32.3	16.1	52.5	9.1	29.7
74/75	5.26	58.8	154.6	309.4	813.4	14.8	38.9	18.3	48.1	13.1	34.4
75/76	4.46	66.0	159.9	294.3	712.9	18.0	43.6	26.4	64.0	18.3	44.3
76/77	3.92	83.1	162.4	325.7	636.5	19.2	37.5	27.2	53.2	23.8	46.5
77/78	5.22	77.8	131.7	404.6	684.7	25.3	42.8	32.6	55.2	30.8	52.1
78/79	5.65	92.4	152.9	522.4	864.2	26.0	43.0	40.2	66.5	44.0	72.8
79/80	5.46	98.8	146.6	539.7	800.7	29.9	44.4	49.7	73.7	55.7	82.6
80/81	6.42	106.0	140.5	680.3	901.4	29.8	39.5	57.9	76.7	64.9	86.0
81/82	6.29	112.4	135.4	706.7	851.2	33.5	40.4	69.9	84.2	73.7	88.8
82/83	6.45	119.6	134.5	771.3	867.5	37.5	42.2	78.9	88.7	75.0	84.4
83/84	6.87	126.8	133.3	871.1	916.1	38.7	40.7	83.5	87.8	79.5	83.6
84/85	8.36	113.5	114.9	948.9	960.5	40.3	40.8	90.6	91.7	81.5	82.5
85/86	6.53	112.5	112.5	734.2	734.2	40.8	40.8	99.8	99.8	82.4	82.4
86/87	7.20	110.0	106.8	792.0	768.9	41.3	40.1	100.8	97.9	83.1	80.7

Year	Misc. cost Current	Misc. cost Real	Total VCs Current	Total VCs Real	Gross margin Current	Gross margin Real
1970/71	0.2	1.1	22.3	117.7	104.7	552.8
71/72	0.5	2.4	24.9	119.1	120.6	576.8
72/73	0.7	3.2	29.6	134.8	129.8	591.1
73/74	0.7	2.3	35.8	116.8	219.2	715.4
74/75	1.0	2.6	47.2	124.1	262.2	689.3
75/76	1.3	3.1	64.0	155.0	230.3	557.9
76/77	2.3	4.5	72.5	141.7	253.2	494.8
77/78	2.3	3.9	91.0	154.0	313.6	530.7
78/79	4.7	7.8	114.9	190.1	407.5	674.1
79/80	6.7	9.9	142.0	210.7	397.7	590.1
80/81	6.1	8.1	158.8	210.4	521.5	691.0
81/82	6.4	7.7	183.5	221.0	523.2	630.2
82/83	7.7	8.7	199.2	224.0	572.2	643.6
83/84	8.0	8.4	209.7	220.5	661.4	695.6
84/85	10.5	10.6	222.9	225.6	726.0	734.9
85/86	5.8	5.8	228.8	228.8	505.4	505.4
86/87	5.9	5.7	231.1	224.4	560.9	544.6

Current = when all items are valued at 1970—85 market prices
Real = when all items are valued at 1985—86 input prices

Source: M.C. Murphy, 1986.

However, although the trend was for both yields and prices to go up until the early 1980s, gross incomes remained relatively static (Figure 15.8). Oversupply portends reductions in price and restrictions on production, perhaps directly or indirectly on cereal yields per hectare themselves.

Costings date rapidly so the reader is referred to the many regular market commentaries and management annuals, including the Scottish Agricultural Colleges' *Farm Management Handbook*, the *Farm Management Pocketbook* by Professor John Nix at Wye College, University of London, and

TABLE 15.5 Annual average market prices 1886–1986: corn returns for England and Wales

Year	Wheat	Barley	Oats	Year	Wheat	Barley	Oats
1886	7.14	7.30	6.73	1936	7.05	8.12	6.23
1887	7.46	6.97	5.74	1937	9.19	10.74	8.45
1888	7.30	7.71	5.91	1938	6.64	10.01	7.46
1889	6.81	7.14	6.23	1939	4.92	8.69	6.81
1890	7.30	7.87	6.56	1940	9.84	17.88	13.12
1891	8.53	7.79	7.05	1941	14.43	23.62	14.43
1892	6.97	7.22	6.97	1942	15.67	44.95	14.68
1893	6.07	7.05	6.64	1943	15.99	30.92	15.42
1894	5.25	6.73	6.07	1944	14.68	26.00	15.99
1895	5.33	6.07	5.09	1945	14.19	24.03	16.16
1896	5.99	6.32	5.25	1946	14.60	23.87	15.99
1897	6.89	6.48	5.99	1947	16.49	23.62	17.96
1898	7.79	7.46	6.48	1948	20.67	26.41	20.50
1899	5.91	7.05	5.99	1949	22.88	25.43	20.67
1900	6.15	6.89	6.23	1950	25.43	27.48	21.24
1901	6.15	6.97	6.48	1951	29.20	38.22	25.75
1902	6.48	7.05	7.14	1952	29.12	32.07	26.33
1903	6.15	6.23	6.07	1953	30.67	29.61	24.20
1904	6.48	6.15	5.74	1955	27.80	25.34	22.23
1905	6.81	6.73	6.15	1955	22.55	25.59	25.84
1906	6.48	6.64	6.48	1956	25.10	25.26	24.28
1907	7.05	6.89	6.64	1957	21.24	22.80	22.47
1908	7.38	7.14	5.91	1958	21.41	22.55	23.46
1909	8.45	7.38	6.64	1959	20.67	22.23	22.23
1910	7.30	6.40	6.15	1960	21.00	20.91	22.14
1911	7.30	7.55	6.64	1961	20.26	19.52	19.11
1912	7.96	8.45	7.63	1962	21.49	22.64	22.64
1913	7.30	7.55	7.63	1963	20.59	20.34	20.67
1914	8.04	7.46	7.38	1964	21.57	20.83	21.08
1915	12.14	10.25	10.66	1965	22.14	22.47	22.31
1916	13.45	14.76	11.81	1966	21.98	21.73	22.31
1917	17.39	17.80	17.63	1967	21.24	20.46	19.26
1918	16.73	16.24	17.47	1968	22.62	21.55	19.80
1919	16.73	20.83	18.45	1969	23.24	21.44	19.50
1920	18.54	24.61	20.09	1970	27.10	28.41	24.68
1921	16.40	14.35	12.06	1971	23.78	24.30	20.17
1922	10.99	10.99	10.25	1972	35.99	31.58	26.72
1923	9.68	9.27	9.43	1973	58.39	52.21	47.17
1924	11.32	12.88	9.60	1974	57.66	58.16	54.75
1925	11.97	11.56	9.60	1975	65.64	65.14	60.45
1926	12.22	10.17	8.86	1976	84.95	83.19	76.14
1927	11.32	11.56	8.94	1977	80.30	78.85	68.56
1928	9.84	10.83	10.25	1978	91.98	85.17	76.55
1929	9.68	9.76	8.69	1979	98.13	93.42	95.96
1930	7.87	7.79	6.07	1980	104.98	95.50	97.00
1931	5.66	7.79	6.15	1981	112.38	106.32	99.79
1932	5.82	7.46	6.89	1982	119.08	114.56	104.60
1933	5.25	7.79	5.50	1983	124.33	121.97	111.94
1934	4.76	8.53	6.15	1984	111.26	110.05	114.57
1935	5.09	7.79	6.56	1985	112.20	106.02	92.95
				1986	112.70	111.30	114.10

Prior to 1967 prices relate to calendar years. From 1967 to 1985 prices are for August/July marketing years. For 1986 prices are for July/June marketing year, and are gross of co-responsibility levy.

NB: In real terms this represents declines of only 50–60% of value a century ago

Source: H-GCA.

TABLE 15.6 Cost of winter wheat production per tonne 1971–90 (at constant 1985 prices)

Year	Variable costs £	Fixed costs £	Total costs £	Price per tonne £	Net margin £
1971	27.3	71.6	98.9	159.3	60.4
1978	36.3	71.8	108.1	165.0	56.9
1979	41.2	78.2	119.4	156.3	36.9
1980	34.9	68.2	103.1	149.7	46.6
1981	37.2	70.3	107.6	143.4	35.9
1982	36.8	66.2	103.0	142.5	39.5
1983	34.1	61.2	95.3	141.2	45.9
1984	28.3	52.4	80.7	120.4	39.7
1985	35.1	60.1	95.2	112.5	17.3
1986	32.0	55.6	87.6	106.8	19.2
1990*	26.6	48.5	75.1	84.0	8.9

* Obviously projected

Source: M.C. Murphy, 1986.

regional agricultural economics bulletins such as those published by certain universities, including Exeter, Reading and Cambridge.

Table 15.7 attempts to present relative costings. Table 15.8 shows recent performance of cereal crops on ten English farms, whilst 15.9 indicates the performance of organically grown cereals.

Figure 15.9 shows the actual relationship of variable costs to gross margin for recorded barley crops, whilst Figure 15.10 presents a means of

displaying comparative crop budgets against the average fixed costs of a farm. It is obviously preferable to calculate net margin by attempting to allocate fixed costs to each crop (Table 15.10).

CEREAL MARKETING

The farmer must know about market requirements and marketing in order to understand and pursue those qualities needed in cereal crops. Qualities sought for barley and wheat are summarised in Appendix 3. They are subject to alteration as market circumstances may dictate. The wider context must be studied as well as the home market.

Europe

France has produced over one-third of EC 12 grain but exports over one-half of total grain exported by the EC. West Germany has been the second major producer just ahead of the UK which has produced one-seventh of the EC 12 total. When Spain and Portugal joined the EC in 1986 they added a further 35 per cent to EC land area, 40 per cent to EC farm population and a further amount of grain equivalent to some 95 per cent of UK production. The potential to improve production in the Iberian peninsula is considerable. Wheat yields still average about one ton per acre (2.5 t/ha), yet with irrigation the climate is favourable if husbandry is improved. Spain and Portugal now import far less US grain but depend

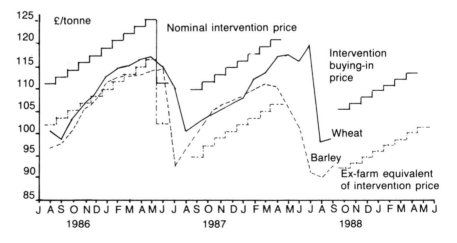

FIGURE 15.7 UK ex-farm prices of wheat and barley and support price levels for 1985/86 and 1987/88 showing typical seasonal trends. (Source: H-GCA)

TABLE 15.7   Expected UK cereal performance: relative yields and prices (feed wheat = 100)

Crop	Use	Yield High	Yield Average	Yield Low	Price
Winter wheat	Feed	100	75	50	100
	Milling	95	72	48	115−120
Spring wheat	Milling	75	58	40	118−122
Durum	Pasta	70	50	35	165−175
Winter barley	Feed	90	70	48	95−100
	Malting	80	62	45	115−120
Spring barley	Feed	75	60	40	95−100
	Malting	72	60	40	115−122
Winter oats ⎫		80	65	45	
⎬	Feed (milling)				95−102 (105−115)
Spring oats ⎭		70	55	40	
Rye	Milling	75	60	40	118−125
Triticale	Feed (milling)	80	65	45	98−100 (120)

Notes: (1) This table is for general guidance only and does not indicate the extremes of which any species may be capable; figures are *relative*.

(2) Feed wheat yield of 100 might reasonably be taken as equivalent to 10 t/ha.

(3) Price depends on market, duration of storage and does not include any added value through processing or packaging.

TABLE 15.8   Cereal performance on ten English farms with grade 3 land

Crop	Average yield (range) (t/ha) 1977−86	1986 yield (range) (t/ha)	1986 variable costs (range) (£/ha)
Winter wheat	7.00 (6.07−8.25)	6.69 (5.50−7.50)	211 (200−300)
Winter barley	6.35 (6.01−6.70)	6.62 (6.10−7.21)	202 (182−223)
Spring barley	5.26 (4.90−5.60)	5.75 (5.10−6.20)	147 (137−161)

more on the EC instead. The EC 12 can be expected to produce 175 M tonnes of grain (over 10 per cent of world total) without allowing for potential improvements in Spain and Portugal, but structural changes in the CAP may limit this figure. Indeed, the *stabiliser* figure agreed to limit EC overproduction is 160 Mt per year. If this amount is exceeded, co-responsibility levy is increased (it amounts to a tax of about 7 per cent of the grain price on sales at the time of writing). Yield increases in the EC 12 have been at around 3 per cent/year though cereal hectarage has remained at about 35 M hectares.

Exports

In order to protect North American interests, as well as those of Argentina and Australia, a 1982 understanding was reached with the EC to limit EC exports to 13 per cent of the world market. The UK

is nevertheless expanding its share of cereal exports and continuing to export variable amounts of cereal products such as flour, malt and whisky. The Middle East and North Africa continue as attractive markets, as do East European countries. Britain still imports high-quality wheat from Canada and elsewhere to improve breadmaking grists. An indication of the export position is given in Figure 15.11; see also Tables 15.11 and 15.12.

Intervention

Intervention buying accounted for 30 per cent of the UK wheat sold ex-farm in 1984/5 and 17 per cent of the barley. A co-responsibility levy of £3.37/tonne (equivalent to about 3 per cent of feed grain value then) was introduced for all grain sold into intervention or exported. This was not enough of a price penalty to deter overproduction. Accordingly

TABLE 15.9 Performance data of commercial crops grown organically and conventionally

	Winter wheat Conventional	Winter wheat Organic	Spring oats Organic	Winter barley Conventional
1985				
Variable costs/ha	£273	£54	£93	£222
Yield t/ha	6.55	5.18	5.25	7.18
Price £/tonne	£115	£144	£140	£107
Output/ha	£753	£746	£735	£769
Gross margin/ha	£480	£693	£642	£547
1986				
Variable costs/ha	£206	£49	£68	£223
Yield t/ha	6.42	5.97	4.07	6.10
Price /tonne	£112	£170	£150	£115
Output/ha	£719	£1014	£611	£702
Gross margin/ha	£513	£965	£543	£479
1987				
Variable costs/ha	£202	£52	£49	£196
Yield t/ha	6.52	5.60	3.50	7.21
Price/tonne	£115	£165	£156	£105
Output/ha	£749	£924	£595	£757
Gross margin/ha	£547	£872	£547	£561

Notes: (i) Organic crops were reckoned to incur an extra £10/ha cultivations cost.
　　　 (ii) It is not possible to produce organic cereals continuously.
Source: Courtesy of D.J. Ursell.

TABLE 15.10 UK cereals survey harvest year 1985–86: return for single farm shown along with regional averages

	Winter wheat Single farm	Winter wheat Regional average	Winter barley Single farm	Winter barley Regional average
Area of crop in hectares	176.1		19.1	
Yield in tonnes per hectare	6.60	6.17	7.18	6.03
Value of output	£	£	£	£
Grain	733.0	684.6	790.0	
Straw	0.0	10.7	5.4	
Total	733.0	695.3	795.4	688
Variable costs				
Seed	53.7	42.7	49.8	
Fertiliser	107.5	110.8	94.4	
Sprays	92.2	73.0	77.7	
Contract	0.0	16.0	0.0	
Miscellaneous	0.6	3.1	0.0	
Total	254.0	245.6	221.9	222.1
Gross margin	479.0	449.7	573.5	465.9
Fixed costs				
Labour　　 — direct	23.9	26.8	27.8	
— overhead	3.6	1.7	4.2	
Machinery — direct	113.4	134.0	130.7	
— tractor overhead	0.0	0.1	0.0	
Rent and rates	115.0	102.0	115.0	
Other overheads	34.5	35.0	34.5	
Total	290.4	299.6	312.2	304

(continued)

| | Winter wheat | | Winter barley | |
	Single farm	Regional average	Single farm	Regional average
Net margin	188.6	150.1	261.3	161.9
Production cost per tonne	82.5	88.4	74.4	87.3
Fertiliser kg per hectare				
Nitrogen	211.2	210.6	185.4	161.0
Phosphate	52.4	64.2	52.2	61.0
Potash	52.4	71.7	52.2	70.0
Spray chemical cost £ per ha, by category				
for broad-leaved weeds	19.0	12.5	7.6	10.0
annual grass/wild oats	8.4	13.3	0.0	7.8
mixture BL weed/grass	16.8	11.8	16.8	14.2
couch/perennials	10.3	1.1	3.7	.3
with fungicides	21.9	29.4	36.0	23.9
growth regulators	8.7	1.7	5.3	1.2
insecticides/nematicides	3.3	2.0	3.4	2.6
other chemicals	3.9	1.2	5.0	0.8
Man hours per hectare	7.6	8.5	8.8	8.8
Tractor hours per hectare	3.9	6.0	5.0	6.2

Source: Department of Agricultural Economics, Reading University.

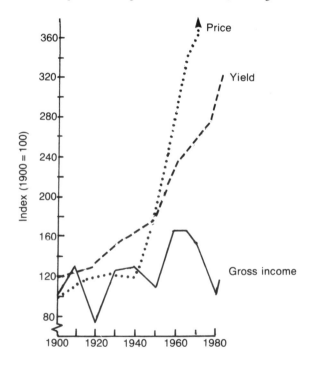

FIGURE 15.8   Yield (in kilograms per hectare), price (per kilogram) and gross income of one hectare of wheat in the period 1900–83 in the Netherlands. The value of the year 1900 is set at 100 for all three and the gross income is adjusted for the rate of inflation. Although the average yields and the prices are increased considerably, the gross income of one hectare of wheat shows only a small increase, which means that the buying capacity of the farmer has hardly changed. (Source: P. Buringh, Trans. of the Royal Society, 1983)

for 1986/7, intervention quality standards were tightened (Appendix 3) and sales into intervention can only take place between September and April and payment is deferred. Together with handling costs, these measures are combining to decrease the attractiveness of the intervention stores as the market which they were never intended to be.

Grain Deficit Countries

It is no fair solution to EC surplus grain production to dump our unwanted grain in countries of high deficit since this tends to depress local initiatives to develop farming output which must be the long-term solution to food shortages there. Furthermore, nearby less-rich countries may be able to sell some of their grain to needy neighbours but they cannot afford to just donate it, so that an EC dumping policy could depress whole regional economies.

Provision of EC or US grain for cases of famine as short-term relief aid is quite another matter but great care has to be taken to judge when oversupply of relief aid is beginning to delay indigenous development efforts.

Marketing Functions

The functions associated with the marketing process include:

- assembly
- grading
- pricing
- storage
- processing, blending or assortment

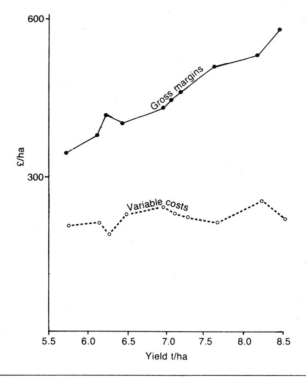

Crop structure and performance: Cotswold winter barley crops (1978–80)					
	Seed rate (kg/ha)	Seeds/m²	Plants/m²	Ears/m²	Yield (t/ha)	Gross margin (£/ha)
Higher-yielding group	183	367	326	906	7.59	486
Lower-yielding group	165	333	289	812	6.42	374
% extra for higher-yielding group	11	10	13	12	18	30

Apart from achieving higher ear populations, the better yielding crops were associated with:
- better control of grass weeds in the previous crop
- a wider spread (plus one month) between the earliest and latest fungicide treatments (*not* more chemicals used)
- more sensible timing of nitrogenous fertilisers in relation to seasonal weather and crop activity

FIGURE 15.9   Winter barley: impact of yield and variable cost levels on gross margins for ten English farms.

- risk-bearing
- providing market intelligence information
- financing

Market Information

In the UK, the Home-Grown Cereals Authority (H-GCA) was set up under the Cereals Marketing Act (1965) to provide market intelligence and technical information, so promoting orderly and relevant marketing. The H-GCA produces weekly, monthly and annual documents and classifies wheat varieties for breadmaking suitability as:

Class I   preferred
Class II   acceptable
Class III   unsuitable (this material is classified for suitability in biscuitmaking or as standard feed wheat, SFW)

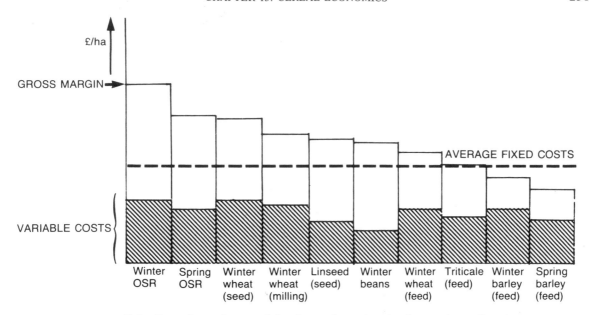

Note: Farm circumstances obviously vary from place to place and over time, but the annual preparation of such a chart gives an instant overview of relative expectations. High fixed-cost livestock enterprises ought not to be included in the average!

FIGURE 15.10   Cereals and alternative combinable crops: actual budget for an English farm, 1987 (area over 500 ha). (Courtesy of A. John)

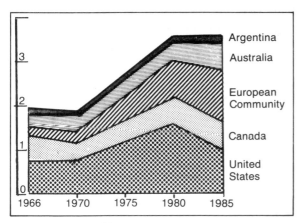

Note: EC TRADE (Mt, 1986)

	Imports	Exports
Wheat	18.82	25.18
Maize	13.06	10.04
Barley	8.04	13.93

FIGURE 15.11   Wheat exports: the changing proportions. The amount of wheat exported by the five major exporters, in billions of bushels.
(Source: New York Times, 4 August, 1986, quoted by Butler and Saylor, 1986)

Other sources of information available to farmers on market requirements include NABIM (National Association of British and Irish Millers Ltd), FAB (Flour Advisory Bureau), FMBRA (Flour-Milling and Baking Research Association), IOB (Institute of Brewing), UKASTA (UK Agricultural Supply Trade Association) and FFB (Food From Britain), all of whose regular and periodic publications should be consulted.

Marketing Methods

There are several options for marketing. Some farmers enjoy marketing their own grain without feeling that this distracts them from the field end of the job. Others detest the marketing aspect or prefer to take advice from and devolve responsibility to a grain broker, a marketing agency (such as the farmer-owned Grainlink Limited) or an agricultural merchant. Considerable confidence and loyalty can be developed over the years in relations with such people, though smaller firms are rapidly becoming absorbed by larger ones which may offer less personal service. If these agencies fail to satisfy the

to consumption. The FAO has estimated that enough grain is lost per annum in Africa alone due to rats, weevils and mites to feed 50 to 60 million people (the equivalent of the UK population). One rat will eat 10 to 12 kg of grain per year and spoil and contaminate three times as much.

Vigilant husbandry is needed, from land preparation before sowing right until grain is safely despatched. If this is given there is no reason why the world's population cannot be fed by hard work and fair sharing of its fruits.

Appendix 1

Selected References and Further Reading

These have been arranged by chapter for convenience, though some obviously overlap more than one chapter.

Chapter 1

BROWN, L.R. and ECKHOLM, E.P. (1975) *By Bread Alone* (Pergamon, 272pp)

FAO (1986) *World Food Report 1985* (UN, Rome)

GALLAGHER, E.J. (1984) *Cereal Production*, Proc. Second International Summer School in Agriculture, Dublin (Butterworths, 354pp)

GOMPERTZ, M. (1927) *Corn from Egypt* (Howe, London, 88pp)

KENT, N.L. (1983) *Technology of Cereals* (Pergamon, 221pp)

LOCKHART, J.A.R. and WISEMAN, A.J.L. (1988) *Introduction to Crop Husbandry* (6th ed, Pergamon, 319pp)

MCKEY, J. (1981) *Cereal Production*, Proc. Conf. 'Cereals, a renewable resource — theory and practice', in Copenhagen (eds Pomerantz and Munck, pp 5–24)

SANDERS, H.G. (1945) *An Outline of British Crop Husbandry* (Cambridge University Press, 348pp)

STOSKOPF, N. (1986) *Cereal Grain Crops* (Reston, Virginia, USA, 516pp)

Chapter 2

ALLEN, H. (1981) *Direct Drilling and Reduced Cultivations* (Farming Press, 219pp)

BOWLER, D.G. (1980) *The Drainage of Wet Soils* (Hodder and Stoughton, Auckland, NZ, 259pp)

DAVIES, D.B., EAGLE, D.J. and FINNEY, B. (1982) *Soil Management* (4th ed, Farming Press, 287pp)

GREEN, R.D. ed (1984) *Soils and Their Use* (Soil Survey of England and Wales, bulletins 10–15). A series of six regional books integrating the impact of soils and climate

LYNCH, J.M. (1983) *Soil Biotechnology: Microbiological Factors in Crop Productivity* (Blackwell, 191pp)

MORGAN, R.P.C. (1986) *Soil Erosion and Conservation* (Longman, 298pp)

SOANE, B.D. ed (1982) *Compaction by Agricultural Vehicles: A Review* (Scottish Institute of Agricultural Engineering)

STANHILL, G. (1976) 'Trends and deviations in the yield of the English wheat crop during the last 750 years', *Agroecosystems* 3, 1–10

STRUTT, N. (1970) *Modern Farming and the Soil* (MAFF report, HMSO, 119pp)

WILD, A. (1988) *Russell's Soil Conditions and Plant Growth* (Longman, 1008pp)

Chapter 3

AUSTIN, R.B. et al. (1980) 'Genetic improvements in winter wheat yields since 1900 and associated physiological changes', *J. Agric. Sci.* 94, 675–689

BARLING, D.M. (1982) *Development of Winter Wheat Crops* (NAC Cereal Unit, 31pp)

BENZIAN, B. and LANE, P. (1979) 'Some relationships between grain yield and grain protein of wheat experiments in South-east England and comparisons with such relationships elsewhere', *J. Sci. Food Agric.* 30, 59–70

BINGHAM, J. (1969) 'The physiological determinants of grain yield in cereals', *Agric. Progress* 44, 30–42

BISCOE, P.V. and GALLAGHER, J.N. (1978) 'A physiological analysis of cereal yield', *Agric. Progress* 53, 34–70

BRIGGS, D.E. (1978) *Barley* (Chapman and Hall, 612pp)

ELLIS, F.B. and BARNES, B.T. (1980) 'Growth and development of root systems of winter cereals grown after different tillage methods including direct drilling', *Plant and Soil* 55, 283–295

EVANS, L.T. (1975) *Crop Physiology* (chapters on maize, rice and wheat) (Cambridge University Press, 374pp)

FBC (1986) *Guide to Grain Quality*, 94pp

GEDYE, D.J. et al. (1982) *A Farmer's Guide to Wheat Quality* (NAC Cereal Unit, 50pp)

JENKINSON, R.H. and WIBBERLEY, E.J. (1986) 'Investigations on the nitrogen requirements of winter barley', *J. RASE* 147, 76–89

JUNIPER, B.E. (1979) 'The structure and chemistry of straw', *Agric. Progress* 54, 18–27

KETTLEWELL, P. (1987) *Cereal Quality — Aspects of Appl. Biol. 15* (AAB, 580pp)

KIRBY, E.J.M. and APPLEYARD, M. (1983) *Cereal Development Guide* (2nd ed, NAC Arable Unit, 95pp)

LARGE, E.C. (1954) 'Growth stages in cereals. Illustrations of Feekes scale', *Pl. Path.* **3**, 128−129

LUPTON, F.G. (1987) *Wheat Breeding* (Chapman and Hall, 566pp)

PERCIVAL, J. (1921, reprinted) *The Wheat Plant* (Duckworth)

SRIVASTAVA, J.P. ed (1988) *Drought Tolerance in Winter Cereals* (John Wiley, 384pp)

THORNE, G.N. (1965) 'Photosynthesis of ears and flag leaves of wheat and barley', *Ann. Bot.* **29**, 317−329

TOTTMAN, D.R. (1987) 'The decimal code for the growth stages of cereals, with illustrations', *Ann. Appl. Biol.* **110**, 441−454

ZADOKS, J.C., CHANG, T.T. and KONZAK, C.F. (1974) 'A decimal code for the growth stages of cereals', *Weed Research* **14**, 415−421

Chapter 4

FENWICK-KELLY, A. (1988) *Seed Production of Agricultural Crops* (Longman, 250pp)

HAYWARD, P.R. et al. (1978) *Developments in the Business and Practice of Cereal Seed Trading and Technology* (RHM/ The Gavin Press, 149pp)

JORNA, M.L. and FLOOTMAKER, L.A.J. (1988) *Cereal Breeding Related to Integrated Cereal Production* (Pudoc Wageningen, 244pp)

NIAB (annual) *Recommended Varieties of Cereals*, Leaflet No. 8

NIAB *Detailed Descriptions of Cereal Varieties*

NIAB *Handbook of Seed Identification*

RIGGS, T.J. et al. (1981) 'Comparison of spring barley varieties grown in England and Wales between 1880 and 1980', *J. Agric. Sci.* **97**, 599−610

SILVEY, V. (1978) 'The contribution of new varieties to increasing cereal yield in England and Wales', *J. Nat. Inst. Agric. Bot.* **14**, 367−384

Chapter 5

ADAS (1982) *Sowing Cereals* (Booklet 2073, 17pp)

BOYD, D.A. (1952) 'The effects of seedrate on yield of cereals', *Emp. J. Expl. Agric.* **20**, 115−122

CANNELL, R.Q. et al. (1978) 'The suitability of soils for sequential direct drilling of combine harvested crops in Britain: a provisional classification', *Outlook on Agric.* **9**, 306 −316

CANNELL, R.Q. (1983) 'Crop establishment in relation to soil conditions and cultivations', pp 33−48 in *Yield of Cereals* (RASE)

GRAHAM, J.P. and ELLIS, F.B. (1980) 'The merits of precision drilling and broadcasting for the establishment of cereal crops in Britain', *ADAS Quarterly Review* **38**, 160−169

HOLLIDAY, R. (1960) 'Plant population and crop yield of cereals', *Field Crop Abstracts* **16**, 71−81

HUDSON, H.G. (1941) 'Population studies with wheat II. Propinquity', *J. Agric. Sci.* **31**, 116−144

KIRBY, E.J.M. (1969) 'The effect of sowing date and plant density on barley', *Ann. Appl. Biol.* **63**, 513−521

Chapter 6

ADAS (1984) *The Nitrogen Requirements of Cereals* (Ref. book 385, 298pp)

ADAS (1988) *Fertiliser Recommendations* (Ref. book 209)

ARCHER, J. (1985) *Crop Nutrition and Fertiliser Use* (Farming Press, 278pp)

CHURCH, B.M. and LEECH, P.K. (1983) *Fertiliser Use on Farm Crops in England and Wales, 1982* (MAFF, London)

COOKE, G.W. (1967) *The Control of Soil Fertility* (Crosby Lockwood, 526pp)

ELSMERE, J.I. ed. (1988) *Survey of fertiliser practice: fertiliser use in England and Wales, 1987* (via Rothamsted Stats. Dept., 34 pp plus appendices)

Chapter 7

HAWKINS, A.F. and JEFFCOAT, B. (1982) *Opportunities for Manipulation of Cereal Productivity* (British Plant Growth Regulator Group, Monograph No. 7, 209pp)

HUMPHRIES, E.C. (1968) 'CCC and Cereals', *Field Crop Abstracts* **21**, 91−99

McLAREN, J. (1982) *Chemical Manipulation of Crop Growth and Development*, Proc. 33rd Easter School, University of Nottingham (Butterworths)

Chapter 8

ATTWOOD, P. ed (1986) *Crop Protection Handbook — Cereals* (BCPC)

GAIR, R., JENKINS, J.E.E. and LESTER, E. (1987) *Cereal Pests and Diseases* (4th ed, Farming Press, 268pp)

IVENS, G.W. ed (1989) *The UK Pesticide Guide* (CAB/BCPC, 434pp)

MAFF (1988) *Pesticides* (Ref. book 500, HMSO, 399pp)

Chapter 9

EMPSON, D.W. (1982) *Cereal Pests* (MAFF Bulletin 186, HMSO)

JONES, F.G.W. and JONES, M.G. (1984) *Pests of Field Crops* (3rd ed, Arnold)

Chapter 10

ADAS/MAFF *Managed Disease Control* (leaflets 831, 843, 844)

HESSAYON, D.G. (1982) *The Cereal Disease Expert* (Pan Britannica)

JONES, D.G. and CLIFFORD, B.C. (1984) *Cereal Diseases, Their Pathology and Control* (2nd ed, BASF)

LOCKE, T. (1983) 'Resistance to fungicides in farm crops', *Agric. Progress* **58**, 1–10

PRIESTLEY, R.H. and BAYLES, R.A. (1982) *Identification and Control of Cereal Diseases* (NIAB, 41pp)

SCHERING (1987) *Cereal Diseases Guide* (87pp)

WIESE, M.V. (1977) *A Compendium of Wheat Diseases* (Amer. Soc. Phytopathology, Minnesota)

Chapter 11

BEHRENDT, S. and HANF, M. (1979) *Grass Weeds in World Agriculture* (BASF, 160pp)

CHANCELLOR, R.J. (1982) 'Dormancy in weed seeds', *Outlook on Agriculture* **11** (2), 87–93

CHANCELLOR, R.J. and FROUD-WILLIAMS, R.J. (1984) 'A second survey of cereal weeds in central southern England', *Weed Research* **24**, 29–36

CUSSANS, G.W. (1981) 'Weed control in cereals – a long term view', Proc. Grass Weeds in Cereals in the UK conf., pp 355–361

ELLIOTT, J. (1980) 'The economic significance of weeds in the harvesting of grain', Proc. British Crop Protection conf. (Weeds, pp 787–797)

GWYNNE, D.C. and MURRAY, R.B. (1985) *Weed Biology and Control* (Batsford, 258pp)

HANF, M. (1983) *The Arable Weeds of Europe* (BASF, 494pp)

MOSS, S.R. (1980) 'The agro-ecology and control of blackgrass in modern cereal growing systems', *ADAS Quarterly Review* **34**, 170–191

ROBERTS, H.A. (1982) *Weed Control Handbook* (7th ed, Blackwell)

SALISBURY, Sir E. (1964) *Weeds and Aliens* (2nd ed, Collins, 384pp)

SCHERING (1986) *Weed Guide* (91pp)

Chapter 12

BARLING, D.M. (1981) 'Winter barley development work on the Cotswolds', *JRASE* **141**, 156–164

ENGLEDOW, F.L. (1926) 'Investigations on yield in cereals', *J. Agric. Sci.* **16**, 166

FARMAN, C. ed (1987) *Cereal Crop Manual* (NAC Arable Unit)

SAULL, M.A. ed (1985) A Directory of Arable Farmer Groups (NAC Arable Unit, 86pp)

WIBBERLEY, E.J. (1984) *Spring Wheat* (NAC Arable Unit, 65pp)

WIBBERLEY, E.J. (1988) 'Developments in arable management through farmer groups', *JRASE* **149**, 133–147

Chapter 13

BUTTERWORTH, B. (1985) *The Straw Manual* (Spon, 212pp)

CATLING, H. ed (annual) *The Green Book of Farm Machinery* (Guardian, 700pp)

HARPER, S.H.T. and LYNCH, J.M. (1981) 'The kinetics of straw decomposition in relation to its potential to produce phytotoxin acetic acid', *J. Soil Sci.* **32**, 627–637

MCLEAN, K.A. (1980) *Drying and Storing Combinable Crops* (Farming Press, 281pp)

NASH, M.J. (1985) *Crop Conservation and Storage* (2nd ed, Pergamon, 286pp)

REXEN, F. and MUNCK, L. (1984) *Cereal Crops for Industrial Use in Europe* (EEC Commission Report, EUR 9617 EN, 242pp)

STANIFORTH, A.R. (1980) *Cereal Straw* (Oxford University Press)

STANIFORTH, A.R. (1982) *Straw for Fuel, Feed and Fertiliser* (Farming Press, 154pp)

WHITE, D.J. (1984) *Straw Disposal and Utilization: a Review of Knowledge* (MAFF, 94pp)

Chapter 14

BOWERMAN, P. and JARVIS, R.H. (1982) 'Breaking a run of continuous winter wheat crops', *Expl. Husb.* **38**, 20–24

DIXON, P.L. and HOLMES, J.C. (1987) *Organic Farming in Scotland* (Edinburgh School of Agriculture)

DUCKHAM, A.N. and MASEFIELD, G.B. (1970) *Farming Systems of the World* (Chatto and Windus, 542pp)

FWAG (Farming and Wildlife Advisory Groups). Advisers operate throughout the UK and will discuss compatibility of cereal growing techniques and wildlife habitats/landscape value.

JOHNSTON, A.E. (1986) 'Soil organic matter: effects on soils and crops', *Soils Use and Management* **2** (3), 97–104

KING, F.H. (1911) *Farmers of Forty Centuries* (Rodale Press, 441pp, reprint)

LEACH, G. (1976) *Energy and Food Production* (IPC, Guildford, 137pp)

ORGANIC FARMERS AND GROWERS (1987) *The Organic Handbook*

PYE-SMITH, C. and NORTH, R. (1984) *Working the Land* (Temple-Smith, 138pp)

SLEE, W. (1987) *Alternative Farm Enterprises* (Farming Press, 208pp)

SPEDDING, C.R.W. (1979) *An Introduction to Agricultural Systems* (Appl. Sci., 169pp)

SPEDDING, C.R.W. and WALSINGHAM, J.M. (1975) 'The production and use of energy in agriculture', *J. Agric. Econ.* **27**, 19–30

VINE, A. and BATEMAN, D. (1981) *Organic Farming Systems in England and Wales* (UCW Aberystwyth report)

WARD, J.T., BASFORD, W.D., HAWKINS, J.H. and HOLLIDAY, J. (1985) *Oilseed Rape* (Farming Press, 312pp)

WIBBERLEY, T. (1918) *Farming on Factory Lines: Continuous Cropping for the Large Farmer* (Pearson, 267pp) — on mixed farming and cropping

WOOKEY, C.B. (1987) *Rushall: the Story of an Organic Farm* (Basil Blackwell, 209pp)

Chapter 15

GILES, A.K. and STANSFIELD, J.M. (1980) *The Farmer as Manager* (Allen and Unwin, 199pp)

MURPHY, M.C. (1986) Royal Agricultural College conference paper, Dec. 1986

NIX, J.S. ed (annual) *The Farm Management Pocketbook* (Wye College, University of London)

STRATTON, J.M. (1978) *Agricultural Records* (Baker, 259pp)

TAYLOR, M.R. and TURNER, J.C. (1989) *Applied Farm Management* (Collins, 304pp)

WILLIAMS, N.T. and HYDE, C.A. (1987) *Fixed Costs: the Key to Profit?* (Wye College, University of London, 24pp)

Farm Business Data (annual) Reading University

The Scottish Agricultural Colleges (annual) *Farm Management Handbook*

Agricultural departments of commercial banks publish annual brochures of costings.

Some Useful Addresses

ADAS Great Westminster House
Horseferry Road
London SW1P 2AE
Tel: (01) 233 3000

AGRA Europe
25 Frant Road
Tunbridge Wells, Kent
Tel: (0892) 33813

ARC Ltd (Arable Research Centres)
Royal Agricultural College
Cirencester, Glos GL7 6JS
Tel: (0285) 652184

Arable Farming Magazine, Farming Press Ltd
Wharfedale Road
Ipswich IP1 4LG
Tel: (0473) 43011

British Crop Protection Council
20 Bridport Road
Thornton Heath
Surrey CR4 7QS
Tel: (01) 683 0211

British Oat and Barley Association
6 Catherine Street
London WC1B 5JJ
Tel: (01) 836 2460

Department of Agriculture, Ireland
Agriculture House
Kildare Street
Dublin 2
Tel: (0001) 789011

Department of Agriculture and Fisheries for Scotland
Chesser House, 500 Gorgie Road
Edinburgh EH11 3AW
Tel: (031) 443 4020

Department of Agriculture, Northern Ireland
Dundonald House, Upper Newtonards Road
Belfast BT4 3SB
Tel: (0232) 661033

EC Information Office
Abbey Buildings, 8 Storeys Gate
London SW1
Tel: (01) 222 8122

FAO (Food and Agriculture Organisation of the United Nations)
Via delle Terme di Caracalla
00100 Rome, Italy

Flour Advisory Bureau
21 Arlington Street
London SW1A 1RN
Headquarters of NABIM (National Association of British & Irish Millers)
Tel: (01) 493 2521

Food from Britain
301–344 Market Towers, New Covent Garden Market
London SW8 5NQ
Tel: (01) 720 2144

Farming and Wildlife Advisory Group
The Lodge
Sandy, Beds SG19 2DL
Tel: (0767) 80551

Farming and Wildlife Trust Ltd
Brentwood, Wragby Road
Sudbrooke, Lincoln LN2 2QU
Tel: (0522) 750006

GAFTA (Grain and Feed Trade Association Ltd)
24–28 St Mary Axe
London EC3A 8EP
Tel: (01) 283 5146

H-GCA (Home-Grown Cereals Authority)
Hamlyn House, Highgate Hill
London N19 5PR
Tel: (01) 263 3391

Institute of Brewing (IOB)
33 Clarges Street
London W1Y 8EE
Tel: (01) 499 8144

Institute of British Bakers
50 Sandygate Road
Sheffield S10 5RY
Tel: (0742) 686323

International Wheat Council
Haymarket House, 28 Haymarket
London SW1Y 4SS
Tel: (01) 930 4128

Irish Cooperative Organisation Society
The Plunkett House, 84 Merrion Square
Dublin 2
Tel: (0001) 764783

Long Ashton Research Station (IACR)
Long Ashton
Bristol BS18 9AF

MAFF (Ministry of Agriculture, Fisheries and Food)
Whitehall Place
London SW1A 2HH
Tel: (01) 270 3000

National Agricultural Centre Arable Unit
Stoneleigh, Kenilworth CV8 2LZ
Tel: (0203) 555100

NIAB (National Institute of Agricultural Botany)
Huntingdon Road
Cambridge CB3 0LE
Tel: (0223) 276381

NSDO (National Seed Development Organisation)
Newton Hall, Newton
Cambridge
Tel: (0223) 871167

Organic Advisory Service
Elm Farm Research Centre, Hamstead Marshall
Newbury, Berkshire RG15 0HR
Tel: (0488) 59298

Rothamsted Experimental Station (IACR)
Harpenden, Herts AL5 2JQ
Tel: (05827) 63133

Scottish Crops Research Institute (SCRI)
Invergowrie, Dundee DD2 5DA
Tel: (0382) 562731

UKASTA (United Kingdom Agricultural Supply Trade Association)
3 Whitehall Court
London SW1A 2EQ
Tel: (01) 930 3611

World Bank (IBRD)
1818 H Street NW, Washington, DC 20433, USA
66 avenue d'Iena, F-75116, Paris, France

Appendix 3

Typical Market Standards

WHEAT

	Specific weight min kg/hl	Moisture content max %	Impurities/ admixture max %	Protein %	Hagberg falling number min	Varieties by H-GCA class
Breadmaking	76	16	2	11+ preferred	250+	1 and 2
Cake flour	74	16	2	max 10.5	200+	some 2 and 3
Other milling (including biscuit)	74	16	2	9.2	180+	1, 2 and some 3 for biscuit flour
Standard feed wheat	68	16	12*	—	—	—
Feed (DNQ)	70	16	2	—	—	—
GAFTA futures	72.5	16	—*	—	—	—
Export: Milling	76–77	16	2	11–11.5	220–250+	1 and 2
Feed	72–73	16	2	—	—	—
Intervention: Bread	72–76†	15‡	10*	10.5+	220	Machine test/Zeleny 20
Feed	72	15‡	12*	—	—	—

BARLEY

	Specific weight min kg/hl	Moisture content max %	Impurities/ admixture max %	Nitrogen %	Germination min %	Variety
Feed barley	60–62	16	2	—	—	Any
GAFTA futures	62.5	16	—*	—	—	Any
Feed exports	62–63	16	2	—	—	Any
Intervention	64	15‡	12*	—	—	Any
Malting	92–95% of grains pass over 2.2 mm sieve	16¶	2	1.75% max	95+	Purity of variety is essential: varieties Pipkin, M. Otter, Triumph, Atem, Blenheim, Kym, Natasha recommended by Inst of Brewing for Eng & Wales
Malting exports	90% pass over 2.25 mm Max 4% pass 2.2 mm	16	2	1.84% max (Protein 11.5% max)	95+	As above plus certain other 2-row varieties

* Individual limits for different categories of impurities.
† Penalties if < 76.
‡ Sometimes 14 has been specified.

¶ Traded at 16% but higher levels are accepted with a price adjustment.
NB: Standards are subject to alteration as market circumstances may dictate.
Source: Based on MAFF/ADAS charts.

ADAS Guidelines for Cereal Seed Production

FIELD STANDARDS

	Level	Category Basic	C1	C2
1. Varietal purity	minimum	99.9%	99.7%	99.0%
	HVS (higher voluntary standard)	99.95%	99.9%	99.7%
2. Species purity	minimum	No standards		
	HVS	99.99%	99.99%	99.99%
3. Wild oats plants per ha wheat	minimum	7	625	625
barley		7	125	125
oats		nil	nil	nil
wheat	HVS	7	7	7
barley		7	7	7
oats		nil	nil	nil
4. Lodging	minimum	No standard but see 6 below		
Maximum permitted lodging	HVS	One-third laid		

5. Isolation   The minimum isolation requirements are a physical barrier (e.g. a continuous hedge, ditch, fence or road), two metres of a non-cereal crop or two metres of clean fallow ground between the crop and another cereal.

6. Crop condition   Good enough to inspect properly.
7. Loose smut   Minimum standard is not > 0.5% infection (HVS = 0.2 and 0.1 for basic seed).
8. Ensure proper labelling as per current legislation.

STANDARDS FOR THE SEED

		Minimum	HVS
Sample size to be examined	all categories	500 g	1 kg
Wild oats	all categories	nil*	nil
Other cultivated cereal species, max. no. of seeds	Basic	1†	0
	C1	7	1
	C2	7	3
Wild radish and corn cockle, max. no. of seeds	Basic	1	0
	C1	3	1
	C2	3	1
Total all species other than cultivated cereals, max. no. of seeds	Basic	3	1
	C1	7	1
	C2	7	2
Total all other species, max. no. of seeds	Basic	4	1
	C1	10	2
	C2	10	4

		Minimum	HVS
Ergot, max. no. of seeds	Basic	1	0
	C1	3	1
	C2	3	1
Analytical purity, %	Basic	99	99
	C1	98	99
	C2	98	99
Germination, %	Basic	85‡	85‡
	C1	85	85
	C2	85	85

* The presence of one seed of wild oat in a 500 g sample shall not be considered an impurity if a second sample of 500 g is free.
† Where the maximum content of seed of other cereal species is fixed at one seed, a second seed shall not be regarded as an impurity if a second 500 g is free.
‡ Except that if the Basic seed does not meet the required minimum germination, it may still be marketed under special provisions.

C1—Certified Seed, 1st generation
C2—Certified Seed, 2nd generation

UK Cereal Seed Certifications[1]

		WHEAT VARIETIES					*tonnes*
	Class[2]	1984/85	% Total wheat	1985/86	% Total wheat	1986/87[3]	% Total wheat
WINTER WHEAT							
Aquila	III	6,633	2	2,872	1	960	*
Armada	II	5,236	2	1,832	*	566	*
Avalon	I	72,383	24	61,146	20	65,880	24
Brigand	III	6,136	2	2,847	1	1,673	*
Brimstone	II	—	*	16,631	6	13,217	5
Brock	III	—	*	8,249	3	29,164	10
Copain	U/C	179	*	214	*	—	*
Fenman	III	23,664	8	19,697	7	5,823	2
Flanders	II	4,269	1	1,272	*	348	*
Galahad	III	39,327	13	65,329	22	67.625	24
Hammer	U/C	513	*	191	*	45	*
Hobbit	U/C	479	*	115	*	6	*
Kador	II	494	*	465	*	381	*
Longbow	III	46,229	15	39,585	13	17,385	6
Maris Huntsman	U/C	3,043	1	929	*	269	*
Mercia	I	—	*	—	*	10,725	4
Mission	II	18,408	6	22,131	7	7,726	3
Moulin	I	—	*	4,281	1	7,499	3
Norman	III	45,749	15	33,951	11	16,857	6
Rapier	III	17,823	6	9,180	3	2,364	*
Renard	U/C	193	*	922	*	322	*
Slejpner	III	—	*	542	*	18,879	7
Stetson	U/C	4,749	2	836	*	102	*
Virtue	III	1,773	1	902	*	384	*
Vuka	U/C	502	*	225	*	199	*
Other winter		1,137	*	1,569	*	2,623	*
TOTAL WINTER		298,922	99	295,913	99	271,022	97
SPRING WHEAT							
Alexandria	I	—	*	—	*	939	*
Axona	II	—	*	783	*	3,065	1
Broom	II	889	*	962	*	453	*
Jerico	II	58	*	555	*	866	*
Minaret	II	2,251	1	571	*	574	*
Musket	U/C	761	*	96	*	—	*
Tonic	I	—	*	222	*	1,359	*
Wembley	II	—	*	269	*	2,066	*
Other spring		482	*	63	*	45	*
TOTAL SPRING		4,338	1	3,521	1	9,367	3
TOTAL WHEAT		303,260	100	299,525	100	280,526	100

(continued)

BARLEY VARIETIES

tonnes

	1984/85	% Total barley	1985/86	% Total barley	1986/87[3]	% Total barley
WINTER BARLEY						
Concert	—	*	1,224	*	3,276	1
Gerbel	2,454	1	1,971	*	1,345	*
Halcyon	7,184	3	6,340	2	6,241	2
Igri	88,064	32	76,722	30	65,800	26
Kaskade	1,774	1	1,929	*	1,138	*
Magie	—	*	8	*	2,512	1
Marinka	—	*	—	*	1,829	*
Maris Otter	11,874	4	8,602	3	5,323	2
Monix	950	*	863	*	471	*
Panda	19,235	7	26,271	10	29,885	12
Plaisant	574	*	1,683	*	5,751	2
Sonja	6,639	2	3,980	2	2,052	*
Tipper	8,799	3	2,232	*	543	*
Other winter	2,260	2	2,506	*	6,522	3
TOTAL WINTER	149,233	55	134,331	52	132,681	52
SPRING BARLEY						
Apex	2,853	1	4,965	2	608	*
Atem	25,145	9	27,419	11	26,665	11
Celt	1,656	1	1,093	*	180	*
Corgi	783	*	1,315	*	1,491	*
Digger	—	*	—	*	1,723	*
Doublet	—	*	1,655	*	7,160	3
Egmont	1,218	*	331	*	109	*
Flare	636	*	44	*	—	*
Golden Promise	15,378	12	11,379	4	8,405	3
Goldmarker	3,725	3	777	*	414	*
Golf	6,058	5	5,845	2	9,883	4
Klaxon	3,409	1	12,447	5	15,007	6
Kym	7,554	3	3,481	1	1,821	*
Mazurka	216	*	58	*	—	*
Natasha	1,099	*	11,716	5	11,144	4
Patty	5,192	2	1,393	*	575	*
Piccolo	585	*	376	*	133	*
Regatta	—	*	149	*	2,358	*
Tasman	511	*	55	*	—	*
Triumph	45,164	16	36,091	14	30,204	12
Tweed	986	*	1,460	*	279	*
Other Spring	1,193	1	2,542	*	4,172	2
TOTAL SPRING	124,630	46	124,591	48	122,331	48
TOTAL BARLEY	273,863	100	258,922	100	255,019	100

(1) Excluding basic and pre-basic seeds which are used for seed multiplication.
(2) As classified in the H-GCA 'Marketing Guide for Milling Wheat'.
(3) For example seed harvested 1986 to be planted for 1987 crop.
*   Less than 1%.
U/C Unclassified.

Source: MAFF and H-GCA

UK ESTIMATED AREAS OF SPRING AND WINTER CEREALS

'000 hectares

	WHEAT			BARLEY			OATS		
	Area	*%W*	*%S*	*Area*	*%W*	*%S*	*Area*	*%W*	*%S*
1968	978	86	14	2,401	5	95	380	15	85
1969	833	77	23	2,413	5	95	381	16	84
1970	1,010	86	14	2,243	7	93	375	21	79
1971	1,097	91	9	2,288	8	92	362	27	73
1972	1,127	92	8	2,288	9	91	315	32	68
1973	1,146	92	8	2,267	10	90	281	32	68
1974	1,233	92	8	2,214	10	90	253	29	71
1975	1,035	88	12	2,345	10	90	232	29	71
1976	1,231	96	4	2,182	14	86	235	43	57
1977	1,076	94	6	2,400	14	86	195	40	60
1978	1,257	95	5	2,348	18	82	180	39	61
1979	1,371	96	4	2,343	25	75	136	40	60
1980	1,441	97	3	2,330	31	69	148	34	66
1981	1,491	97	3	2,329	36	64	144	46	54
1982	1,664	97	3	2,221	41	59	130	48	52
1983	1,695	98	2	2,143	43	57	108	50	50
1984	1,939	99	1	1,978	53	47	106	50	50
1985	1,902	98	2	1,966	52	48	134	52	48
1986	1,997	99	1	1,916	50	50	97	60	40
1987	1,992	97	3	1,836	53	47	100	42	58

Based on: 1968 to 1978 — Proportions of winter and spring varieties recorded in the Seed Sales Survey; 1979 to 1987 — proportions of winter and spring varieties recorded in the Seed Certifications Scheme, except for 1985, 1986 and 1987 when details for barley were obtained from June Census results.

Source: MAFF and H-GCA.

Appendix 6 Herbicides used for Weed Control in Cereals

GENERAL GUIDE

Crops: w = wheat, b = barley, o = oats, r = rye, d = durum, t = triticale

Trade names — see product leaflets for rates, timing, and safe use.

Active ingredients — all are post-emergence sprays except isoxaben (Flexidor), which is a soil-acting residual and must be applied before the weeds emerge.

Crops	Herbicides	Trade names	black bindweed	corn buttercup	charlock	chickweed	cleavers	cranesbills	dead-nettle, red	fat-hen	forget-me-not	fumitory	groundsel	hemp nettle	knotgrass	marigold, corn	mayweeds	nettle, small	pansy, field	parsley piert	penny-cress	poppy, common	radish, wild	redshank	shepherd's purse	sow-thistle	speedwell, common	speedwell, ivy-leaved	spurrey, corn	Venus's looking glass
w b o r	MCPA	various	r	r	S	s	R	R	r	R	S	s	—	—	s	R	R	R	s	s	R	S	S	s	S	S	s	r	s	—
w b o	mecoprop (CMPP)	various	r	s	S	S	S	r	S	S	R	r	r	s	r	R	R	s	R	R	S	S	S	r	S	s	r	r	r	—
w b o	dichlorprop (2,4-DP)	various	S	S	S	S	S	s	R	S	R	s	s	r	r	R	S	S	R	R	S	S	S	s	S	r	r	r	s	—
w b o	ioxynil	Mate	S	—	s	s	r	—	S	S	S	S	S	s	r	r	s	s	S	r	—	S	S	s	S	—	S	S	—	s
w b o	ioxynil+bromoxynil	various	S	—	S	s	—	—	S	S	S	S	S	S	S	s	S	S	R	—	S	S	S	s	S	S	S	S	—	S
w b o	ioxynil+bromoxynil+mecoprop	e.g. Swipe	S	—	S	S	s	s	S	S	S	S	S	S	S	s	S	S	S	—	S	S	S	S	S	S	S	S	r	—
w b o	ioxynil+bromoxynil+benazolin	Asset	S	r	S	s	s	—	—	S	S	S	s	r	S	s	S	s	r	r	—	S	S	S	S	s	S	S	s	S
w b o	ioxynil+mecoprop	Mylone	S	S	S	S	S	—	S	S	S	S	S	S	S	S	S	S	s	s	S	r	S	S	S	—	S	S	S	S
w b o	bentazone+dichlorprop	Basagran DP	S	S	S	S	S	S	r	S	S	S	S	s	S	S	S	S	R	r	—	S	S	s	S	—	S	S	S	—
w b o	bentazone+dichlorprop+MCPA	Triagran	S	S	S	S	S	S	—	S	S	S	S	S	s	S	S	S	r	—	S	r	S	s	S	—	S	r	S	s
w b o spring	clopyralid+ioxynil+mecoprop+bromoxynil	Crusader	s	s	S	S	S	—	s	S	S	s	S	S	S	s	s	S	r	r	S	—	S	s	S	s	S	s	r	s
w b o	clopyralid+MCPA+dichlorprop	Lontrel Plus	S	S	S	S	S	s	—	S	S	S	S	S	S	S	S	S	—	s	S	—	S	S	S	s	S	S	S	—
w b o	clopyralid+cyanazine	Coupler	S	—	S	s	S	—	s	s	S	S	S	S	S	s	S	S	—	—	s	s	S	S	S	s	S	S	s	—
w b	clopyralid+ioxynil+bromoxynil+fluroxypyr	Crusader S	S	—	S	S	S	—	S	S	S	s	S	S	s	S	S	s	S	r	—	s	S	S	S	s	S	S	s	s
w b o t winter	fluroxypyr	Starana 2	s	—	R	S	S	—	s	R	R	s	r	S	s	R	R	r	—	—	R	R	R	r	R	—	r	r	—	—
w b	fluroxypyr+ioxynil	Stexal	S	—	S	S	S	s	—	—	S	S	S	S	S	—	—	S	s	—	—	—	s	r	—	—	s	s	S	—
w b	fluroxypyr+ioxynil+bromoxynil	Advance	S	—	S	S	S	—	s	S	S	s	S	S	S	S	S	S	r	s	s	s	S	s	s	s	S	S	s	S
w b o t	fluroxypyr+ioxynil+clopyralid	Hotspur	S	—	s	S	S	—	S	S	S	S	S	s	S	S	S	S	r	r	S	s	S	r	S	s	S	S	—	s
w b	chlorsulfuron+ioxynil+bromoxynil	Glean TP	S	—	S	s	s	—	S	S	S	—	S	S	s	s	s	s	s	s	S	S	S	r	S	s	S	—	S	S
w b o d t	metsulfuron-methyl	Ally	s	S	S	S	R	S	s	S	S	S	S	s	s	S	s	S	S	S	S	R	S	S	S	s	S	R	R	s
w b o	mecoprop+bifenox	Ceridor	s	S	S	S	s	—	S	S	S	s	s	S	s	r	r	s	S	—	S	—	S	r	s	r	S	R	S	—
w b o	mecoprop+cyanazine	Cleaval	S	S	S	S	S	—	S	S	S	S	S	S	S	S	S	S	S	s	s	S	S	S	S	s	S	S	s	S
w b o	MCPA+cyanazine	Envoy	S	S	S	S	S	r	S	S	S	S	S	S	S	R	s	S	r	r	R	S	S	s	S	r	S	S	—	S
w b o r t winter	isoxaben (pre-emergence of weeds)	Flexidor	—	S	S	S	s	—	S	S	S	S	S	S	S	R	S	S	S	S	S	S	S	R	S	s	S	S	S	—

246

AUTUMN-SOWN CEREALS

Symbols at foot: S = susceptible; s = moderately susceptible; R = resistant; r = moderately resistant

Crops	silky bent	canary grass	seedling ryegrasses	rough-stalked meadow	annual meadow	barren brome	blackgrass	wild oats	Active ingredients (1 pre-emergence, 2 post-emergence, * control at high rate)	Trade names — see product literature for rates, timing and safe use	code	black bindweed	charlock	chickweed	cleavers	cranesbills	dead-nettles	fat-hen	forget-me-not	field pansy	fumitory	knotgrass	mayweeds	parsley piert	poppy	redshank	shepherd's purse	speedwells
w b d t r	—	—	—	S	S	s	s	S	tri-allate liquid / granules	Avadex BW	1	—	R	R	R	R	R	R	R	R	R	R	R	R	R	R	R	R
w b d t r	—	S	S	S	S	s	S	S	diclofop-methyl	Hoegrass	2	R	R	R	R	R	R	R	R	R	R	R	R	R	R	R	R	R
w b r t (most vars)	S	R	s	S	S	—	S	s	isoproturon (IPU)	Arelon, Hytane, Tolkan	1 2	R	S	S	S	R	R	R	R	R	R	R	R	s	S	R	R	R
w b t d (most vars)	s	R	s	S	S	—	S	s	chlortoluron	Dicurane	1 2	—	S	S	S	R	R	S	S	R	R	R	S	s	S	R	S	R
w b	s	R	s	S	S	—	S	s	chlortoluron + bifenox	Dicurane Duo	1 2	S	S	S	s	s	S	S	S	S	S	s	S	S	S	S	S	S
w b	—	—	s	S	S	r	s	r	linuron + trifluralin	Chandor, Linnet, Trifluron, Warrior	1	S	—	S	s	—	s	S	S	s	r	s	S	S	S	S	S	S
w b	S	—	s	S	S	s	S	s	trifluralin + terbutryne	Lextra, Terbalin, Laurel	1	S	—	S	s	—	S	—	S	S	S	s	S	S	S	S	S	S
w b	—	—	s	S	S	—	s	r	linuron + trifluralin + trietazine	Pre-Empt	1	S	—	S	s	—	S	—	S	S	S	S	S	S	S	S	S	s
w only	—	—	s	S	s	—	R	R	linuron + bifenox	Alibi	1 2	s	s	S	S	r	S	s	S	s	S	s	S	S	S	S	S	S
w b o? d t r	s	s	s	s	s	—	R	R	terbutryne	Prebane	* 1	S	S	S	R	s	S	S	S	S	S	S	S	S	S	S	S	r
w b d t r	s	s	S	S	s	—	S	s	pendimethalin	Stomp	* 1	S	s	S	s	S	S	—	—	S	S	s	S	S	S	S	—	s
w b t r	S	s	R	S	S	—	r	R	methabenzthiazuron	Tribunil	1 2	S	—	S	R	S	S	S	S	s	S	s	S	r	S	r	s	S
w o?	S	S	S	S	s	R	S	R	methabenzthiazuron + chlorsulfuron	Glean C	* 1 2	—	S	s	s	S	S	S	S	S	r	r	S	s	S	S	S	S
w b o d t r	—	—	r	r	r	—	s	—	chlorsulfuron + metsulfuron-methyl	Finesse	w o 1 2 / b d t r 2	—	S	S	S	S	S	S	S	r	S	s	S	S	S	S	s	S
w b	—	—	S	S	S	—	S	r	IPU + trifluralin	Autumn Kite	1 2	—	S	S	r	—	S	S	S	S	S	S	S	S	S	r	—	S
w b	s	—	S	S	S	s	S	S	IPU + ioxynil + bromoxynil	Doublet, Astrol	2	S	S	S	s	—	s	s	S	R	S	S	S	S	S	S	S	S
w b	s	—	s	s	s	r	S	r	IPU + ioxynil + bromoxynil + mecoprop	Dictator, Terset	2	S	S	S	S	—	S	s	S	R	S	s	S	S	S	S	S	s
w b	S	—	S	S	S	s	S	s	IPU + ioxynil + mecoprop	Assassin, Post-Kite, Musketeer	2	S	S	S	S	—	S	S	S	S	S	S	S	S	S	S	S	S
w b	S	—	S	S	S	s	S	s	IPU + bentazone + dichlorprop	BAS 462	2	S	S	S	S	S	r	S	S	S	S	r	r	S	S	S	s	s
w b o	—	—	S	S	S	s	s	—	trifluralin + ioxynil + bromoxynil	Masterspray	* 2	S	S	S	S	—	S	s	S	S	S	s	S	—	S	S	r	S
w b	S	R	s	S	S	s	S	s	metoxuron	Dosaflo	2	s	S	S	s	—	s	s	s	R	s	—	s	s	s	s	S	R
w b	S	—	r	—	s	R	r	r	diflufenican	Javelin	1 2	s	S	S	s	—	s	s	S	s	r	S	s	s	s	r	S	S

Symbols: S = susceptible; s = moderately susceptible; R = resistant; r = moderately resistant
NB: Approval for products can be withdrawn and current clearance must be checked by users.

Source: Lockhart, J.A.R. and Wiseman, A.J.L. (1988), 6th edn

247

Conversion Factors: Metrication and Abbreviations

Length

cm (centimetres) to in	× 0.394
mm (millimetres) to in	× 0.0394
m (metres) to ft	× 3.29
m (metres) to yd	× 1.09
km (kilometres) to miles	× 0.621

Area

m² (sq metres) to sq ft	× 10.8
m² (sq metres) to sq yds	× 1.20
ha (hectares) to acres	× 2.47

Weight

g (grams) to oz	× 0.0353
g (grams) to lb	× 0.0022

kg (kilograms) to lb	× 2.20
kg (kilograms) to cwt	× 0.0197
t (tonnes) to tons	× 0.984

NB: Mt = million tonnes in the text

Temperature

(°C × 1.8) + 32 = °F

Some Double Conversions

lb/bushel × 1.25	=	kg/hectolitre
fertiliser units/acre × 1.25	=	kg/hectare
cwt/acre × 0.125	=	t/ha
lb/acre × 1.1	=	kg/ha
pints/acre × 1.4	=	litres/ha

Index

FARMING PRESS BOOKS

Below is a sample of the wide range of agricultural and veterinary books published by Farming Press. For more information or a free illustrated book list please contact:

Books Department, Farming Press Ltd, 4 Friars Courtyard, 30—32 Princes Street, Ipswich IP1 1RJ

Cereal Pests and Diseases
R. Gair, J. E. E. Jenkins and E. Lister
An outstanding guide to the recognition and control of cereal pests and diseases by two of Britain's foremost plant pathologists and a leading entomologist.

Farm Woodland Management
John Blyth, Julian Evans, William E. S. Mutch and Caroline Sidwell
A compendium for farmers in which all aspects of trees on the farm are considered: planting bare ground as well as dealing with existing woodland; financial aspects; environmental considerations.

Crop Nutrition and Fertiliser Use
John Archer
Gives details of uptake for each nutrient and then deals with the specific requirements of temperate crops from grassland and cereals to vegetables, fruit and nursery stock.

Soil Management
D. B. Davies, D. J. Eagle and B. Finney
Two soil scientists and a senior mechanisation officer with ADAS go into all aspects of the soil, plant nutrition, farm implements and their effects on the soil, crop performance, land drainage and cultivation system.

Potato Mechanisation and Storage
C. F. H. Bishop and W. F. Maunder
A comprehensive look at the latest techniques and equipment available for potato growers.

Oilseed Rape
J. T. Ward, W. D. Basford, J. H. Hawkins and J. M. Holliday
Contains up-to-date information on all aspects of oilseed rape growth, nutrition, pest control and marketing.

Intensive Sheep Management
Henry Fell
An instructive practical account of sheep farming based on the experience of a leading farmer and breeder.

Farm Machinery
Brian Bell
Gives a sound introduction to a wide range of tractors and farm equipment. Now revised, enlarged and incorporating over 150 photographs.

Outbursts
Oliver Walston
A provocative series of articles, mostly about arable farming.

Straw for Fuel, Feed and Fertiliser
A. R. Staniforth
Provides an excellent practical summary of the profitable disposal and utilisation of straw for arable and livestock farmers.

Pearls in the Landscape
Chris Probert
The conservation and management of ponds for farm and garden.

Farming Press also publish three monthly magazines: *Arable Farming, Pig Farming* and *Dairy Farmer*. For a specimen copy of any of these magazines please contact Farming Press at the address above.

FARMING PRESS